Contemporary Topics in

POLYMER SCIENCE

Volume 7

Advances in
New Materials

CONTEMPORARY TOPICS IN POLYMER SCIENCE

Sponsored by the American Chemical Society

A Continuation Order Plan is available for this series. A continuation order will bring delivery of each new volume immediately upon publication. Volumes are billed only upon actual shipment. For further information please contact the publisher.

Contemporary Topics in
POLYMER SCIENCE

Volume 7

Advances in
New Materials

Edited by

J. C. Salamone

University of Massachusetts
Lowell, Massachusetts

and

Judy S. Riffle

Virginia Polytechnic Institute and State University
Blacksburg, Virginia

SPRINGER SCIENCE+BUSINESS MEDIA, LLC

Library of Congress Cataloging-in-Publication Data

Advances in new materials / edited by J.C. Salamone and Judy S.
  Riffle.
      p.   cm. -- (Contemporary topics in polymer science ; v. 7)
    "Proceedings of an international symposium on advances in new
  materials held at the Fifteenth Biennial Meeting of the Division of
  Polymer Science of the American Chemical Society, held November
  17-21, 1990, in Fort Lauderdale, Florida"--Galley.
    Includes bibliographical references and index.
    ISBN 978-1-4613-6530-3   ISBN 978-1-4615-3456-3 (eBook)
    DOI 10.1007/978-1-4615-3456-3
    1. Polymers--Congresses.   I. Salamone, Joseph C., 1939-
  II. Riffle, Judy S. III. American Chemical Society. Division of
  Polymer Science. IV. Series.
  QD380.C63 vol. 7
  668.9--dc20   92-24362
                                                               CIP

Proceedings of an International Symposium on Advances in New Materials,
held at the Fifteenth Biennial Meeting of the Division of Polymer Science
of the American Chemical Society, held November 17–21, 1990,
in Fort Lauderdale, Florida

ISBN 978-1-4613-6530-3

© 1992 Springer Science+Business Media New York
Originally published by Plenum Press, New York in 1992

PREFACE

The Division of Polymer Chemistry, Inc. of the American Chemical Society held its 15th Biennial Polymer Symposium on the topic, "Advances in New Materials," November 17-21, 1990, at the Pier 66 Resort and Marina in Ft. Lauderdale, Florida. A three and one-half day program was presented by recognized leaders in major areas of new polymeric materials.

The topics of the Biennial Symposium included new high performance polymers, polymers for electronic applications, electrically conducting polymers, nonlinear optics, new polymer systems, and polymers derived from biological media. These are the subject areas of this volume of "Contemporary Topics in Polymer Science". The intent of the Symposium was to focus on recent advances in polymeric materials. The technical sessions were complemented by an initial poster session which augmented the various technical sessions.

A particular highlight of the meeting was the presentation to Professor Michael Szwarc of the 1990 Division of Polymer Chemistry Award by Dr. J. L. Benham, Chairman of the Polymer Division. During his Award address, Professor Szwarc described how he had become a polymer chemist and later developed "living polymers." Without a doubt, Professor Szwarc has made a profound contribution to the polymer field, which has yielded many new forms of living polymerization.

J. C. Salamone of the University of Massachusetts, Lowell, was Chairman of the symposium and he has been particularly grateful to his Co-Chairmen, K. J. Wynne of the Office of Naval Research and J. L. Benham of 3M Corporation, for their untiring efforts in developing the Biennial program.

The Organizing Committee was particularly grateful to the following corporations for their generous support of this Symposium: Allied Signal, American Cyanamid, Chevron Research and Technology, Ciba-Geigy, Dow Corning, Eastman Kodak, E. I. duPont de Nemours, Ethyl Corporation, General Electric, Hoechst Celanese, Inolex Chemical, Mobay, Philips Petroleum, Procter and Gamble, and Texaco.

J. C. Salamone
J. S. Riffle

PROFESSOR MICHAEL SZWARC

Winner of the 1990
Division of Polymer Chemistry Award
American Chemical Society

CONTENTS

NONLINEAR OPTICS

POLYMERS FROM BIOLOGICAL SOURCES

NEW POLYMER SYNTHESIS

PERSPECTIVES ON THE CONTRIBUTIONS OF MICHAEL SZWARC TO POLYMER SCIENCE

* J. C. Salamone and **J. S. Riffle

* University of Lowell
  Department of Chemistry
  Lowell, MA   02884

**Virginia Polytechnic Institute and State University
  Department of Chemistry
  Blacksburg, VA   24061-0212

Professor Michael Szwarc has published over four hundred papers, numerous reviews and three books. He has received a number of major awards, including the ACS Award in Polymer Chemistry (1969), the International Award in Plastics Science and Engineering (1972), the Gold Medal of the Benjamin Franklin Society (1978), and, at this Biennial Symposium, the 1990 Division of Polymer Chemistry Award given by the ACS Polymer Division. More recently, Professor Szwarc has been awarded the prestigious Kyoto Prize for 1991.

The research activities of Professor Michael Szwarc started in 1945, after the second World War, when he joined the research group of Professor Polanyi in Manchester, England. There he developed a pyrolytic technique (toluene carrier method) that allowed him to determine bond dissociation energies of over fifty polyatomic molecules. The work done until 1950 was summarized in a review article published in Chem. Rev. 47, 75, 1950 and earned him a D.Sc. in 1949.

In the course of these investigations, in 1947 Dr. Szwarc discovered a new hydrocarbon monomer, p-xylylene (quinomethane), that is formed in the vapor phase at high temperature ($>800°C$) and spontaneously polymerizes upon hitting any surface kept at room temperature. The resultant polymer forms a transparent film conformally encapsulating the object on which it is deposited. The film neither melts nor decomposes until heated above 400°C. Dr. Szwarc determined the structure of this monomer and its polymer. The first publication on this work appeared in 1947 in the Disc. Faraday Soc. 2, 46. Subsequently, a whole class of related quinonoid monomers based on substituted p-xylenes, 1,4-dimethylnaphthalenes, 2,5-dimethylpyrazene, etc., was developed (J. Polymer Sci., 6, 319 (1950)).

In 1952 Dr. Szwarc was appointed professor of Polymer and Physical Chemistry of SUNY in Syracuse N.Y. After coming to this country, he continued the work on various aspects of poly-p-xylyene in collaboration with Lou Errede of M. W. Kellogg Co. Much of the results of this later research was published in a joint review paper in Quarterly Rev. 12, 301 (1958). The work of Bill Gorham of Union Carbide eventually led to commercial development of poly(p-xylylene) marketed under the name PARYLENE.

In Syracuse, in cooperation with Vivian Stannett, Dr. Szwarc began investigating the permeability of gases and vapors through a variety of polymeric films. The most interesting outcome of this work was reported in two papers: "The Permeability of Polymer Films to Gases, a Simple Relationship," published with Stannett in J. Polymer Sci. 26, 89 (1955), and "Permeability of Gases and Vapors through Composite Membranes, Permeability Valves," published with Rogers and Stannett in Ind. Eng. Chem. 49, 1933 (1957). The latter paper describes a new and interesting phenomenon, a rapid permeation in one direction and slow in the other, and quantitatively explains the theory of this behavior. Approximately twenty five other papers on this subject were published by Prof. Szwarc and his colleagues over the period of approximately six years of research.

During the same time period, Professor Szwarc also developed a versatile technique permitting determination of the relative rate constants for addition of small radicals, such as methyl, ethyl, trifluoromethyl, etc. to a large variety of aromatic, ethylenic, acetylenic, etc. compounds. This method allowed him to grade quantitatively the reactivities of various aromatic hydrocarbons, monomers, inhibitors, e.g. quinones, etc. This led to an examination of polar and steric effects in radical reactions. The theory of these additions was developed in a paper presented during the Kekule Symposium in London (1958), and published with Binks in "Theoretical Organic Chemistry" pp. 262-290, (1959), Butterworth. These studies also provided much information on cage reactions and related subjects. The investigations of radical additions continued for approximately twelve years and resulted in over one hundred publications.

In 1955 a study of the cationic polymerization of styrene initiated by trifluoroacetic acid led to discovery of a phenomenon, not appreciated at that time, which became topical in the 70's and 80's. It was found that addition of small amounts of acid (the initiator) to a large amount of styrene (the monomer) resulted in formation of low molecular weight oligomers only, whereas the addition of small amounts of styrene to a large amount of the acid rapidly produced a high molecular weight polymer (MW ~ 15000). This paradoxical result was explained by introducing the concepts of aggregation and homo-conjugation. The relevant paper was published jointly with Stannett in J. Amer. Chem. Soc., 78, 1122 (1956).

In 1956 Professor Szwarc discovered living polymers and electron transfer initiation. Simple and convincing experiments demonstrating the lack of termination and chain transfer were described in his first two papers on this subject entitled: "Polymerization Initiated by Electron Transfer to Monomer, A New Method of Preparation of Block Polymers" with Moshe Levy and Ralph Milkovich published in J. Amer. Chem. Soc., 78, 2656 (1956), and the other published in Nature, 178, 5557 (1956), under the title "Living Polymers." These workers showed how such systems were capable of producing di- and ter-block polymers of predetermined composition and molecular weights. In the paper in Nature, the ramifications of living polymerizations were outlined. The control of molecular weight of each block and the feasibility of functionalization with any desired terminal groups was introduced. Dr. Szwarc deduced a simple relation for difunctional initiators, DP_n = (total monomer)/(1/2 initiator), and the results confirming its validity were described in J. Amer. Chem. Soc., 79, 202 (1957). The first preparation of block polymers of styrene and ethylene oxide was reported in a paper with Richards, Trans. Faraday Soc. 55, 1644 (1959). In the same publication, the influence of endgroups on the properties of otherwise identical polymers was described.

A variety of aspects concerned with living polymers, e.g. effect of impurities, the molecular weight distribution in living polymer systems, monomer-living polymer equilibria, etc., were treated during the years that followed. The problems of termination were reviewed in an article in Adv. Polymer Sci. 2, 275 (1960), and a paper delivered during the Wiesbaden Symposium in 1959 and published in Makromol. Chem. 35, 132 (1960)describes the state of art achieved in this field at that time.

In the 60's, Dr. Szwarc's studies of living polymers shifted toward kinetic problems. Much effort was placed on studying the formation and reaction of complexes of living polymers with aromatic hydrocarbons, kinetics of electron transfer processes, spontaneous isomerization of living polymers, chemistry of radical-anions, etc. Absolute rate constants of propagation were reported in a series of papers published by J. Phys. Chem., 69, 608, 612, 624 (1965). The results revealed that ion-pairs and free ions participate in

2

propagation, but that free ions are the main contributors to the reaction. Continuation of these studies led to a deep understanding of the reactivity of ionic species. The role of tight and loose ion-pairs, of complexation agents, of triple ions, etc. was clarified and quantified. Kinetic studies of homopolymerization were extended to anionic copolymerization. The absolute rate constants for crossover reactions were reported for a series of monomers reacting with living polystyrene, for a series of substituted, living polystyrenes reacting with styrene [J. Amer. Chem. Soc., 85, 1306 (1963)], and extended to the dienes and vinylpyridines. A most unconventional copolymerization of living poly(vinyl naphthalene) with styrene was reported in a paper published in J. Amer. Chem. Soc., 85, 3909, (1963) with Bhasteter and Johan Smid. The results accumulated in that period were published in the monograph "Carbanions, Living Polymers and Electron Transfer Processes" published by Wiley in 1968.

Studies of ionic reactions were extended to processes other than propagation of polymerization. A variety of protonation processes, electron transfers, dimerizations, cis-trans isomerizations, etc., were investigated with emphasis on the role of different types of ionic species on the course of these reactions. The results of these studies were reviewed in two volumes entitled: "Ions and Ion-Pairs in Organic Reactions" edited by Professor Szwarc, who contributed four chapters to this extensive compilation. These books were published by Wiley in 1970 (volume I) and 1972 (volume II).

Directing a research group in the University of Uppsala, Sweden for a period of four years, Professor Szwarc became acquainted with Flash Photolysis. This technique, used in Uppsala and later in Syracuse, allowed quantitative studies of numerous electron transfer reactions. For example, the determination of kinetic parameters governing the electron transfer initiation of polymerization was achieved by applying the flash photolytic technique. Flash photolysis studies were most fruitful, a number of novel and interesting results were obtained, and these were reported in a series of about twenty-five papers.

A great deal of Szwarc's studies of radical anions were performed using the ESR technique. The question of how far an electron can jump as well as the dynamics of flexible polymer chains was investigated. This work was carried out in collaboration with Shimada and published jointly in several papers, e.g., in Chem. Phys. Lett. 28, 540 (1974) and Macromolecules 5, 801 (1972).

Prof. Szwarc also contributed to the understanding of mechanisms of cationic polymerizations. In addition to the previously mentioned study of the cationic polymerization of styrene initiated by trifluoroacetic acid, he developed novel methods of initiation of cationic polymerization, e.g., the initiation of cationic polymerization by transfer of Cl^+ and NO_2^+ ions, and initiation by electron-transfer. In cooperation with deSorgo and David Pepper, he carried out the first stop-flow study of cationic polymerization that demonstrated the formation of the positive polystyryl cation and allowed its spectrum to be recorded. This work was published in J.C.S. Chem. Comm. 419, (1973). He was the first to point out that cationic polymerization induced by ionizing radiation is propagated by free cations [Makromol. Chem. 35a, 123 (1960)].

Professor Michael Szwarc is truly one of the pioneers of polymer science. The members of the ACS Division of Polymer Chemistry recognize his many creative contributions to macromolecular chemistry and his tremendous lasting impact on this field. It is with great pleasure that we honor Professor Szwarc with the Division of Polymer Chemistry award on the occasion of this 1990 Biennial Symposium.

PREPARATION AND POLYMERIZATION OF MACROCYCLIC OLIGOMERS

Daniel J. Brunelle and Thomas L. Evans

GE Research and Development
P. O. Box 8
Schenectady, NY 12301

INTRODUCTION

Ring-opening polymerization reactions (Equation 1) constitute an important class of polymerization techniques. A number of commercial products are prepared via ring-opening polymerization reactions, and the preparation of monomers, studies of catalysis and mechanism, and product development of many of these materials are active areas of academic and industrial research. Ring-opening polymerization chemistry has been extensively reviewed in several monographs,[1] and in many review articles.[2] A wide variety of monomers have been utilized in ring-opening reactions; these polymerizations are commercially utilized for the preparation of polyamides, aliphatic polyesters, silicones, polyalkylenes (via ring-opening methathesis polymerizations), and epoxide thermosets. For this paper, discussion will be limited to three categories of ring-opening polymerization reactions: monomeric strained ring systems, which have high reaction exotherms, monomeric medium ring systems, which have moderate exotherms, and oligomeric large ring systems, which have little or no reaction exotherm.

$$\tag{1}$$

Typically, the ring size of the cyclic monomer has a controlling effect both on the preparation of the monomer, and on its inherent stability and polymerizability. The most reactive systems, those which tend to polymerize completely with little cyclic residue, usually contain strained three- or four-membered rings. Examples of these monomers are ethylene oxide, which has a strained three-membered ring, and pivalolactone, which is a four-membered lactone (equations 2 and 3). Because of the ring strain, the ring-opening polymerization of these materials are generally exothermic; for example, the epoxide opening of equation 2 has a net enthalpy of -104 kJ/mole.[3] The enthalpy of reaction provides a driving force for complete reaction, but may also be a handicap in controlling the polymerization of large volumes.

$$\tag{2}$$

$$\text{(3)}$$

Medium-sized (5-9 members) rings generally have significantly less ring strain, and hence lower reaction exotherms during polymerization. For example, the polymerization of caprolactone (7-membered ring) has an enthalpy of -16.5 kJ/mole (Equation 4)[3]. In the polymerization of such monomers, an equilibrium between cyclic monomer (or oligomers) and linear polymer is achieved during the polymerization. Because the enthalpy for ring-opening is low, and the entropy decreases during polymerization (Δ S for caprolactone = -29 J/° K-mole),[3] these reactions do not proceed to 100% completion, and monomer recoveries of 5-30% are typical. The removal of unreacted monomer may be a limitation if a finished part is to be formed by ring-opening polymerization. Virtually all ring-opening polymerization reactions which involve ring-chain equilibration form measurable quantities of monomeric or oligomeric cyclics.[4]

$$\text{(4)}$$

The ring-opening polymerization of the oligomeric cyclic dimethylsiloxane tetramer seems to be an unusual example, in that the entropy actually increases during polymerization.[5] This increase in entropy is the driving force for the reaction, since the enthalpy is near zero. Nonetheless, the anionic polymerization is an equilibrium reaction, and significant amounts of cyclics remain after polymerization (18.3% for dimethylsiloxane,[6] Equation 5).

$$\text{(5)}$$

The preparation of engineering thermoplastics with aromatic structural units via ring-opening polymerization would be commercially appealing, inasmuch as low molecular weight precursors would lead to high molecular weight polymers without formation of by-products during polymerization. Because of the size of the monomer units, and the distance of chain ends from one another, cyclic *monomeric* precursors for these materials are rare. The cyclic *oligomers* which would be precursors to engineering thermoplastics like polycarbonate or poly(ethylene terephthalate) have been known for some time. Cyclic oligomers are present in many polymers in levels of 0.25-8%,[1,2] and a variety of extraction techniques have been published. The utility of cyclic oligomers of aromatic polyesters or polycarbonates has until now been limited chiefly by the cumbersome and low-yielding techniques for their isolation or preparation. Furthermore, in many cases the discrete oligomers are high-melting crystalline solids, a limitation affecting efficient melt processing.

By conventional thinking, the structure and large ring sizes of oligomeric cyclics in the thermoplastics class would seem to preclude efficient, high yielding cyclization. Even a cyclic carbonate dimer of bisphenol A has a ring size of 24 structural atoms, and each additional monomer unit adds 12 atoms to the ring. It is well known that formation of rings of 5, 6, and 7 structural units are most favored, medium-sized rings (8-11 members) are disfavored, due to transannular repulsions, and beyond ring sizes of 16-20 atoms, the likelihood of cyclic formation becomes minimal.[1,2] This paper reviews work on the use of pseudo-high dilution techniques, which facilitate efficient cyclization reactions, leading to moderate to excellent yields of cyclic oligomeric aromatic carbonates, esters, ethers, amides, imides, and silicones (Equation 6).[7] In all cases, the formation of oligomeric cyclics is selective over formation of linear oligomers, and minimal purification is necessary. The oligomeric cyclic products are mixtures of several ring sizes, and thus have melting points which are depressed relative to the pure discrete cyclic oligomers. For example, the mixture of cyclic oligomeric bisphenol carbonates have a melting range of 190-210° C, compared to melting points of 375° C and 350° C for the discrete cyclic tetramer[8] and trimer[9].

$$\text{(6)}$$

$$n = 2\text{-}10$$

$$m = 50\text{-}500$$

The oligomeric cyclics can be converted to high molecular weight polymers via ring-opening polymerization in the presence of various initiating substances. The most efficient initiators are typically anionic or nucleophilic species such as lithium phenoxide, tetramethylammonium tetraphenylborate, sodium sulfide, or lithium trifluoroethoxide. Because of the large ring size and lack of ring strain, the polymerization reactions are essentially thermoneutral. Similar to the polymerization of cyclic siloxanes, the polymerization reaction is driven by entropy, although apparently to a much greater degree. Equilibrative ring-opening polymerization leads to high molecular weight polymer with little or no detectable cyclic oligomers remaining (<1%). Apparently the ring size of the oligomeric cyclic is such that back-biting on the polymer chain is unlikely.

The technique for preparation of oligomeric cyclics relies upon kinetic reaction control for the formation of cyclics. Fast reactions are necessary for the pseudo-high dilution, intramolecular ring formation. Subsequent polymerization is thermodynamically controlled, with the ring-opening leading to high molecular weight linear polymers, with only very small amounts of cyclic present in the equilibrated mixture. Unlike many ring-opening polymerizations, the reaction has no exotherm, and complete ring-chain equilibration is achieved upon polymerization. Thus, molecular weight distributions approach 2.0.

RESULTS

Cyclic Carbonates

In the mid 1960's, Prochaska was granted several patents for the preparation of monomeric cyclic aromatic carbonates with 7- or 8-membered rings, formed from 2,2'-biphenols (Equation 7).[10] The monomeric cyclic carbonates could be prepared either directly, via phosgenation, or using a vacuum distillation/depolymerization technique similar to that developed by Carothers[11] for the preparation of cyclic aliphatic esters and carbonates. This work was extended by Prochaska to many other 2,2'-bisphenols, and has recently been elaborated upon by Kricheldorf and Jennsen.[12]

$$R = H, CH_3 \tag{7}$$

In 1962, Schnell and Bottenbruch[8] reported the preparation of the cyclic tetrameric carbonate of bisphenol A. Reaction of bisphenol A with an equimolar quantity of bisphenol A bischloroformate in the presence of excess pyridine in a high dilution reaction carried out at 0.05 M concentration led to yields of cyclic tetramer as high as 21% (Equation 8). A variety of bisphenols were similarly converted to their cyclic tetramers. Polymerization at the melting point was also reported. A few years later, Prochaska and then Moody reported preparations of the cyclic trimer of bisphenol A, using similar high dilution techniques.[9]

$$\tag{8}$$

Attempts to optimize these high dilution techniques, with the modern advantage of analysis by high pressure liquid chromatography (HPLC), were unfruitful. The cyclization reaction using pyridine to scavenge the by-product HCl leads mainly to the formation of linear oligomers. Cyclic carbonates are formed as well, and can be isolated by a series of chromatographic or crystallization techniques, but the yields are low. As such, these reactions would not be useful commercially, since the crude reaction products would not be suitable precursors to high molecular weight polymer.

Using triethylamine as the base in the reaction, rather than pyridine gave an increased yield of cyclic carbonates, and use of 4-dimethylaminopyridine[13] gave 40-60% cyclics.[7] Even so, these products remained contaminated by bisphenol A and other linear materials, and polymerization reactions were not explored.

Horbach, et al.[14] reported the preparation of *macro*cyclic polycarbonates (i.e. cyclics with molecular weights of 15,000-30,000), using bisphenol A-bischloroformate in a hydrolysis/condensation reaction catalyzed by triethylamine (Equation 9). The product formed in these reactions has about 15-40% linear polycarbonates, and was generated under special conditions of low temperature, high pH, and high amine concentration. Although no oligomeric cyclics were reported, and the selectivity toward formation of cyclics was low, we were intrigued by this approach. If a method for controlling the

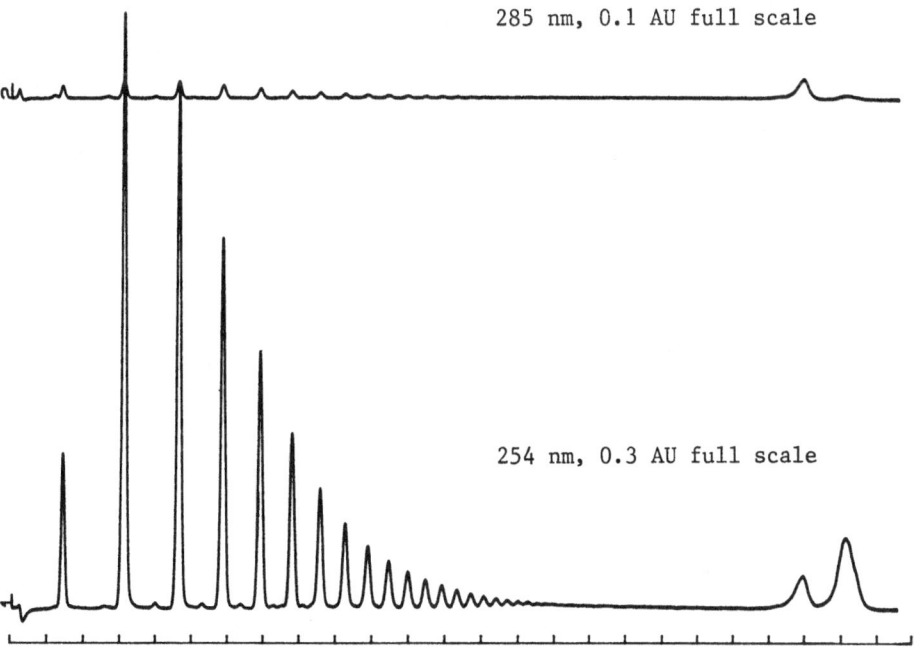

reaction could be found, pseudo-high dilution conditions could be used to form cyclic *oligomers* by addition of a single monomer, rather than trying to balance phenol and chloroformate groups as in previous attempts.

When a reaction was carried out using slow addition of a solution of bisphenol A bischloroformate to an intimately stirred mixture of CH_2Cl_2, triethylamine, and aqueous NaOH, a remarkably selective formation of cyclic oligomers was achieved. The HPLC traces shown in Figure 1 represents the crude products from such a cyclization reaction. The product is composed only of oligomeric cyclics predominantly in the range of dimer to dodecamer, and high molecular weight polycarbonate, with a ratio of cyclics to polymer of 85/15. The level of linear oligomers present, which absorb about 50 times more strongly than cyclics at 285 nm, is estimated to be 0.01-0.03%. The selectivity of cyclic vs linear oligomers is thus about 10,000 to 1.

Figure 1.   HPLC trace of oligomeric cyclics formed via hydrolysis/ condensation using Et_3N.

Careful analysis shows that the range of oligomeric cyclics extends from cyclic dimer to about hexacosamer (n = 26), with the majority of material (>90%) having a degree of polymerization less than 10. Cyclic dimer, never before observed, is formed in 5-10% yield, and is the only strained ring isolated. The ring strain is evident from the FTIR, which shows a shift in the C=O stretch of about 10 cm^{-1}, the X-Ray structure, which shows a cis-trans configuration about the carbonyl, and from the heat of reaction on ring-opening polymerization, which increases with increasing dimer content. The major products from reaction are cyclic trimer, tetramer, pentamer, and hexamer, each formed in 15-25% yields. Gel permeation chromatography shows the wt. avg. MW (relative to polystyrene) to be about 1300, corresponding to a pentamer structure. The major by-product from the reaction is high molecular weight polymer, seen as a broad multimodal peak at long retention time in the HPLC trace. The high MW polymer has wt. avg. MW ranging from 40,000 to 100,000, depending on reaction conditions, and has not yet been characterized as either linear or cyclic material. The presence of high molecular weight polymer has no effect on the subsequent polymerization reaction, other than an increase in the melt viscosity of the cyclics, and a slight moderation of molecular weight. However, the polymer can easily be removed from the reaction product by precipitation of the crude product solution in CH_2Cl_2 into acetone; the cyclics remain soluble while the high molecular weight polycarbonate precipitates, and can be removed by filtration.

The mechanism for the selective formation of cyclics is interesting, and has been the object of detailed studies.[15] In order for bisphenol A bischloroformate to be converted to cyclic oligomeric products, both hydrolysis and condensation reactions must occur. Controlling the ratio of hydrolysis to condensation reactions is crucial, since excessive hydrolysis will lead to recovery of bisphenol A, or to oligomeric linears. Conversely, if hydrolysis is too slow, then the concentration of bisphenol A bischloroformate will increase in the reactor, ultimately leading to conditions which favor intermolecular reactions rather than intramolecular reactions, and forming polymer. Choice of the proper amine catalyst and conditions for reaction which maintain the correct hydrolysis/condensation ratio, while ensuring that reactions occur fast enough to prevent build-up of reactive intermediates, are the keys to successful cyclization.

Interestingly, the structure of the amine catalyst is the primary factor in controlling the selectivity of cyclic vs linear oligomer formation. Replacing triethylamine with pyridine in an otherwise identical reaction leads to selective formation of linear oligomers, to the total exclusion of cyclic oligomeric products as seen in the HPLC trace of Figure 2. Use of other amines, bases, or phase transfer catalysts can give linear oligomers, cyclic oligomers, high molecular weight polymer, no reaction, or mixtures (Table 1). More detailed work on the mechanism of cyclization is presented in an accompanying paper.[15a]

Table 1.   Variation of Products in Attempted Cyclization
Reactions with Various Amines.

Catalyst	% Cyclic Oligomers	Products
Et$_3$N	85	Cyclics, polymer, <.05% linears
Et$_2$NMe	27	polymer, cyclics, linears
pyridine	0	linears and BPA
Me$_2$NBu or quinuclidine	<5	linears and polymer
(i-Pr)$_2$NEt	0	starting materials
n-Bu$_4$OH or Bu$_4$NBr	0	starting materials
n-Bu$_3$N	50	cyclics, polymer, .1% linears

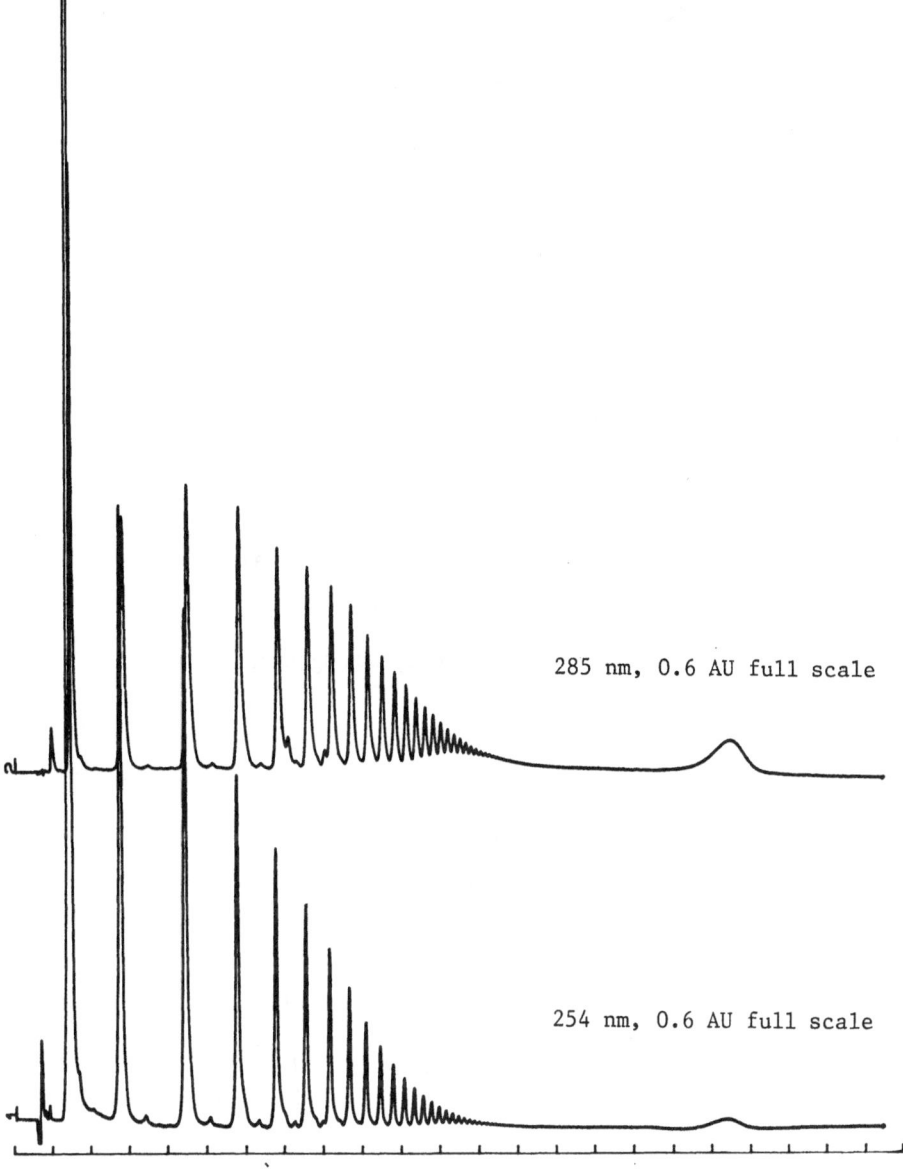

285 nm, 0.6 AU full scale

254 nm, 0.6 AU full scale

Figure 2. HPLC of linear oligomers from hydrolysis/condensation using pyridine.

Due to their low molecular weight, the cyclic oligomeric carbonates have melt viscosities significantly lower than conventionally prepared polycarbonate (Figure 3). This lowered melt viscosity is the property most useful in subsequent applications of the cyclic oligomers. At their melting point, the mixture of cyclic oligomers have significant flow, and have a greater degree of penetration and wetting of fibers in composite applications.

GPC analysis of the cyclic oligomers indicates the M_w to be about 1300 (relative to polystyrene), with a dispersivity of about 1.5. Heating the cyclics to 300° C in glass test tubes under nitrogen in the absence of catalyst gave a modest increase in molecular weight. When the polymerization was initiated by various catalysts, however, very high molecular weight polycarbonates were generated (Table 2). Figure 4 displays the GPC traces

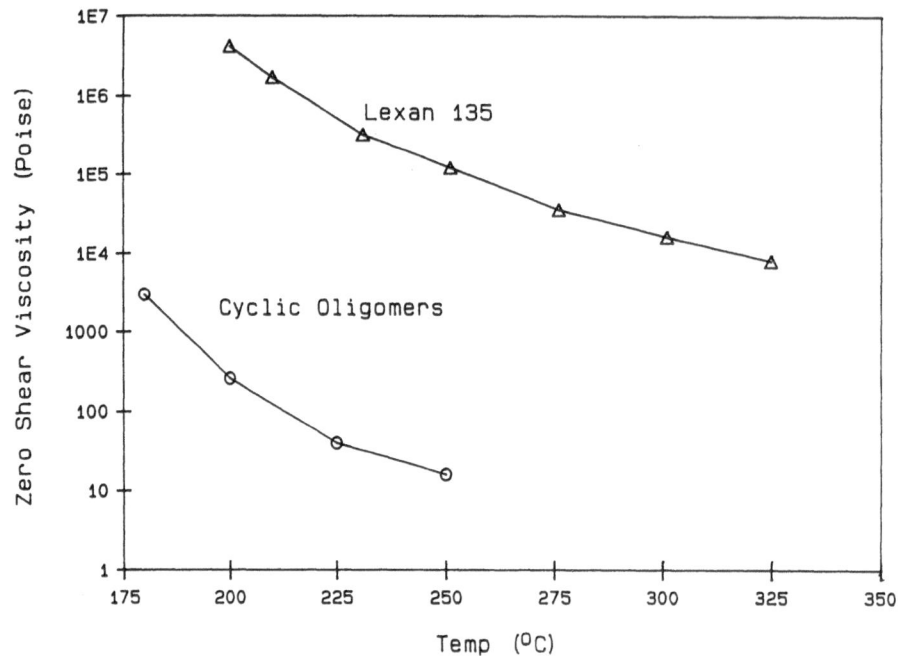

Figure 3. Viscosity comparison of cyclic oligomers to Lexan®.

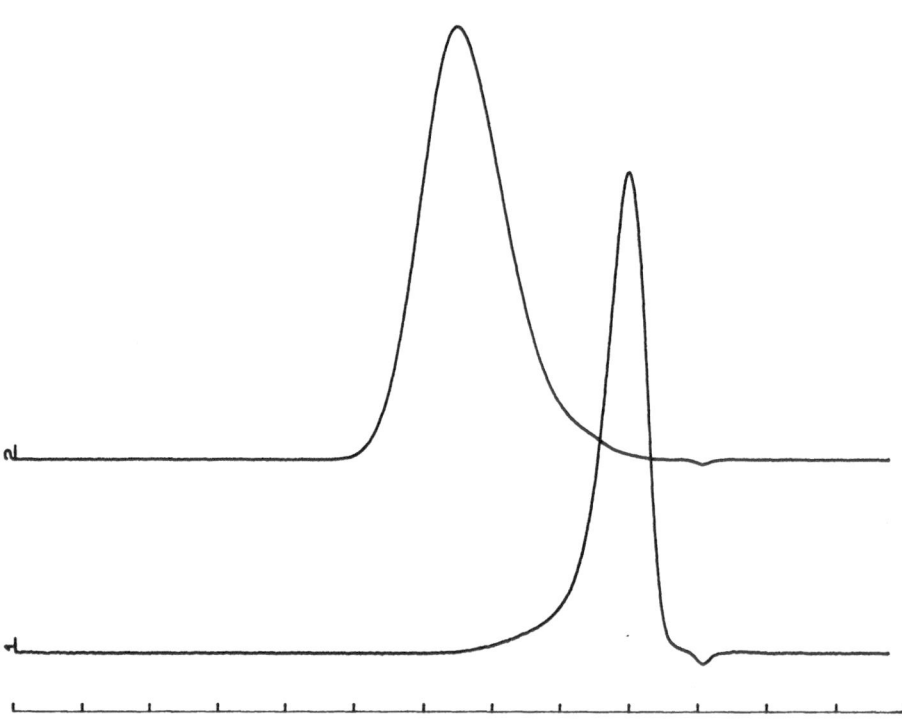

Figure 4. GPC of oligomeric cyclics (bottom) and high Mw polymer (top).

generated from the cyclic oligomers and the resulting high MW polycarbonate. Precipitation of the polycarbonate into acetone and analysis of the soluble portion by HPLC indicated that only about 0.25% cyclics remain after polymerization. The heat of reaction has been measured by differential scanning calorimetry, and has been found to be about -1.2 kJ/mole. This slight exotherm can be correlated to the release of ring strain in opening the cyclic dimer. Thus, the reaction seems to be driven by entropy toward formation of linear polymer. The dispersivity of the product polymer approaches 2.0 (Table 2). This fact, along with the nearly thermoneutral nature of the reaction, indicate that ring-chain equilibration is achieved during the polymerization. The polymerization reaction can be achieved in the melt at 200-300° C, or in solution under various conditions, including reaction in $DMSO/CH_2Cl_2$ at ambient temperature.

Table 2. Melt Polymerization of Cyclic Carbonates (300°C).

Initiator	Initiator Conc. (M %)	Mw	IV (dL/g)
Titanium bis-(isopropoxy)aceto-acetonate	0.05	248,000	2.2
Lithium Phenoxide	0.20	60,190	0.6
Aluminum bis-(ethylacetoacetate)-bis(isopropoxide)	0.10	282,000	2.7
$N(C_4H_9)_4B(C_6H_5)_4$	0.10	280,000	2.7

The molecular weight of the polymer can be controlled by the use of bisphenols or diphenyl carbonate as chain terminators. Addition of various levels of diphenyl carbonate to cyclic oligomer polymerizations provide polycarbonates terminated with phenyl carbonate end groups comparable to commercial polycarbonate grades. The properties of polycarbonates so generated have been measured, and are essentially equivalent to commercial polycarbonate (Table 3). A fundamental study of the polymerization reaction has been reported in a preliminary communication.[16]

Table 3. Mechanical Properties of Polycarbonate Derived from Cyclic Oligomers.

Property	Value
IV (dL/g)	1.0
DTUL (° C)	135.8
Tensile Yield (psi)	9,024
Tensile Break (psi)	10,230
% Elongation	74
Flexural Yield (psi)	13,640
Flexural Modulus(psi)	317,400
Notched Izod (ft-lbs/in)	19.1

The cyclization and polymerization chemistry developed and optimized for bisphenol A is fairly general and has been used for a variety of bisphenols. Various functionalized polycarbonates have been prepared by incorporation of other functional groups such as ester, amide, ketone, sulfone, and urethane groups into a bisphenol monomer. Conversion to the bischloroformate and cyclization of these materials normally leads to good yields of cyclic carbonate oligomers. Subsequent polymerization under similar conditions to those described above afford the functionalized high molecular weight polycarbonate copolymers (Equation 10; Table 4).

$$X = \text{ketone, sulfone, amide, urethane, ester}$$

Cyclic Aromatic Esters

Discovery of an efficient means of preparing cyclic oligomeric carbonates sparked the interest in applying this methodology to the preparation of other systems. Extension of the method to cyclic arylates (aryl aryl esters) seemed to be straightforward, since the aryl acid chlorides should have reactivities similar to aryl chloroformates. Surprisingly, little is known in the literature about cyclic arylates. Tyuzyo, *et al.* have reported a cyclic 2 + 2 arylate from reaction of phthaloyl chloride with bisphenol A,[17] but no similar cyclics have been reported from either iso- or terephthaloyl chloride. Although the cyclic alkyl aryl esters based on iso- or terephthalic acid and aliphatic diols are known,[18] their direct synthesis has never been demonstrated. A method for the direct formation of cyclic arylate oligomers via direct phase-transfer catalyzed reaction of isophthaloyl chloride with bisphenols has now been developed (Equation 11).[19]

Addition of a solution of isophthaloyl chloride in CH_2Cl_2 and a solution of bisphenol A disodium salt in water to a well-stirred reactor containing a phase transfer catalyst (PTC) over 30-60 minutes affords cyclic arylate oligomers in 60-80% yield (Table 5). Use of 1-3% Adogen (methyl trialkyl ammonium chloride) as the PTC gave the best results. The mixture of cyclic oligomers contained predominantly the 3 + 3 adduct, with decreasing amounts of higher oligomers. The 2 + 2 oligomer is highly crystalline and sometimes precipitates from solution. The cyclics are formed selectively over linear oligomers, with the balance of material being high molecular weight polymer.

Table 4. Mixed Functionality Cyclics and Polymers

Cyclic	Cyclic Yield	Polymerization
	80% (75% ester; 25% carb)	$M_w = 67,000$ $M_w/M_n = 2.21$ $T_g = 167°$ C
	95%	$M_w = 48,000$ $M_w/M_n = 2.15$ $T_g = 154°$ C
	90%	$M_w = 67,000$ $M_w/M_n = 2.12$ $T_g = 154°$ C
	90-%	$M_w = 48,000$ $M_w/M_n = 2.35$ $T_g = 169°$ C
	92%	$M_w = 88,000$ $M_w/M_n = 2.06$ $T_g = 165°$ C
	88%	$M_w = 48,000$ $M_w/M_n = 2.14$ $T_g = 128°$ C

15

Table 5. Effects of Experimental Conditions on Formation of Cyclic Arylates[a].

Solvent	Catalyst (eq)	Temp	Cyclics Yield
CH_2Cl_2	Et_3N (0.2)	25° C	15%
CH_2Cl_2	Adogen (0.10)	25° C	25%
CH_2Cl_2	Adogen (0.10	40° C	35%
$CHCl_3$	Adogen (0.10)	61° C	50%
$CHCl_3$	Bu_4NBr (0.10)	61° C	40%
CH_2Cl_2	BU_4NBr (0.02)	40° C	85%[b]
CH_2Cl_2	Adogen (0.01)	40° C	65%[c]

[a] All reactions used equimolar amounts of bisphenol A disodium salt and isophthaloyl chloride in reactions at 0.2 M, except where noted.
[b] Reaction using spirobiindane bisphenol **1**.
[c] Reaction at 0.02 M.

Other bisphenols and diacid chlorides can also be used to prepare cyclic arylates. The use of the spirobiindane bisphenol **1** in reaction with isophthaloyl chloride gives significantly higher yields of cyclic oligomers. This propensity to form cyclics has been seen in other systems using bisphenol **1**.[20] Spirobiindane bisphenol **1** can be charged into the reaction flask along with two equivalents of NaOH and the phase transfer catalyst, followed by slow addition of isophthaloyl chloride over 1/2 to 1 hour. If bisphenol A disodium salt is used as a co-monomer, slow addition of the salt is necessary.

1

Guggenheim, et al demonstrated that the cyclic arylates can be polymerized at elevated temperature (360° C) in the presence of an anionic initiator. The polymerization of a cyclics/polymer mixture, which has a lower melting point, can be carried out at a somewhat lower temperature.[21] The individual cyclics melt at about 385° C with polymerization occurring, even in the absence of catalyst. Polymerization leads to polyarylates with wt. avg. MW of about 40-60,000, and the expected glass transition temperatures (bisphenol A polyarylate, T_g = 167° C, spirobiindane polyarylate, T_g = 242° C).

Ether and Thioether Imides, Sulfones, and Ketones

As seen in the preparation of cyclic arylate oligomers, the spirobiindane moiety conveys a propensity to form cyclics. Cella, Fukuyama, et al have prepared a number of aromatic ether and thioether imides, sulfones, and ketones in cases where the spirobiindane has been built into the structure of one of the monomers.[22] Using bisphenol **1** as a synthetic precursor to dianhydride **2** or diamine **3** has enabled the preparation of a variety of ether polyimide structures via subsequent reaction with various amines or dianhydrides (Table 6).

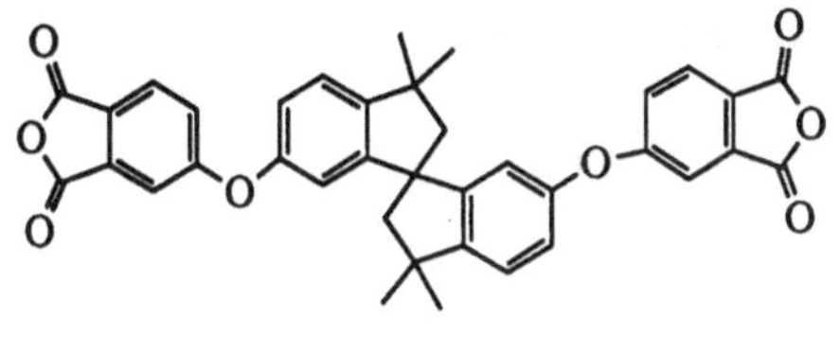

2

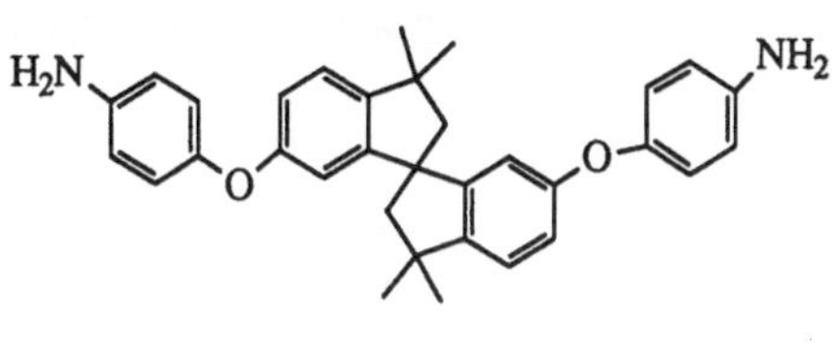

3

Table 6. Formation of Cyclic Oligomeric Etherimides.

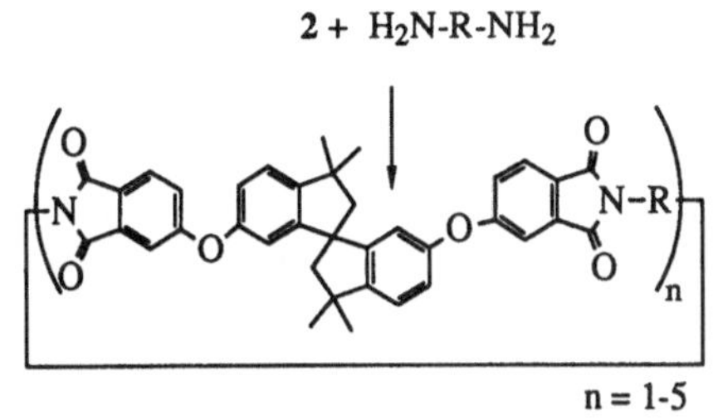

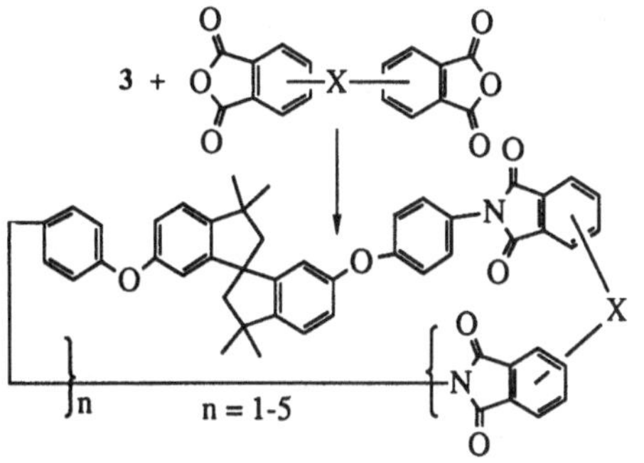

2 + H₂N-R-NH₂

n = 1-5

3 +

n = 1-5

R	Cyclics Yield	-X-		Cyclics Yield
(m-phenylene)	77	-O-	(4,4)	40
(diphenyl ether)	50	-O-	(3,3)	30
(diphenyl methane)	25	-S-	(4,4)	40
		-S-	(3,3)	75
(spirobiindane)	64	*(isopropylidene diphenyl)*		45
		(hexafluoroisopropylidene diphenyl, CF₃/CF₃)		40

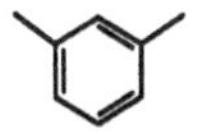

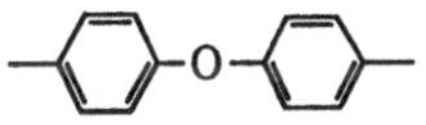

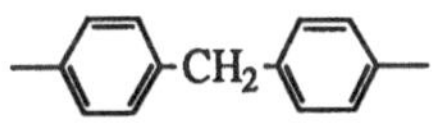

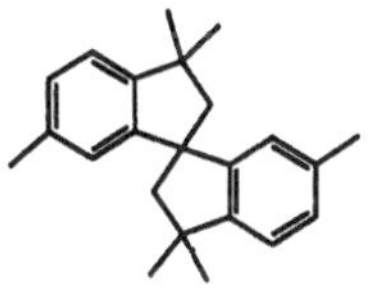

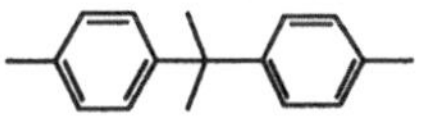

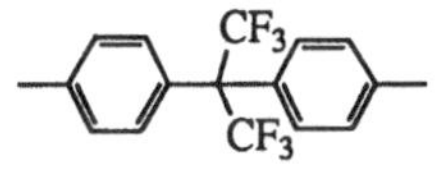

Ring-opening polymerization of these ether-imide structures via a transetherification reaction has been achieved. A survey of several potential catalysts has shown that sulfur nucleophiles, such as sodium sulfide or sodium thiophenoxide are effective initiators for the polymerization reaction. Although model studies have indicated that thioetherimides are significantly more reactive than etherimides toward transetherification, the thioetherimide cyclics have melting points too high to consider melt polymerization. Reaction of a random ether/thioetherimide at 200° C with sodium sulfide in DMAC affords very high molecular weight polymer (M_w ca 140,000), as well as some low oligomers. Further heating as a thin film provides a film with good integrity and with $T_g = 230°$ C.

Cyclic ethersulfone and etherketone oligomers have been prepared by Cella and Fukuyama using high dilution techniques during reaction of bisphenols with bis-4-fluorophenyl sulfone or with 4,4'-difluorobenzophenone in dipolar aprotic solvents such as dimethyl sulfoxide.[22] Again, the presence of the spirobiindane moiety affords significantly higher yields of cyclics. Using final product concentrations of 0.01-0.15 M, yields of cyclic oligomers varied from 25-75% (Table 7).

Table 7. Cyclic Etherketones and Ethersulfones.

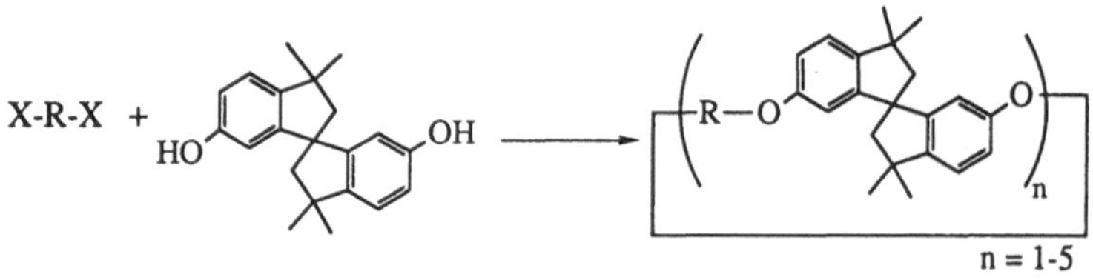

X-R-X	Cyclics Yield (%)
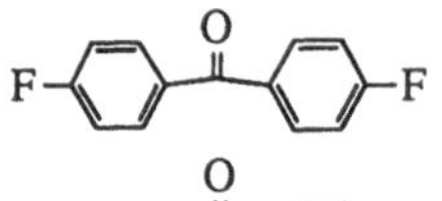 F—⟨⟩—C(=O)—⟨⟩—F	47
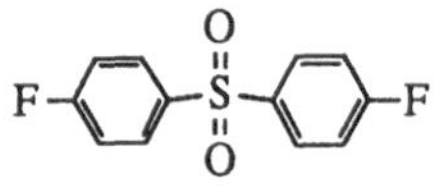 F—⟨⟩—S(=O)₂—⟨⟩—F	45
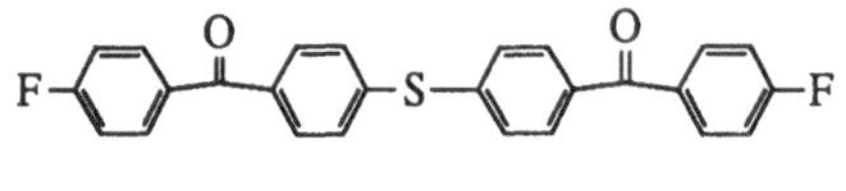 F—⟨⟩—C(=O)—⟨⟩—S—⟨⟩—C(=O)—⟨⟩—F	48
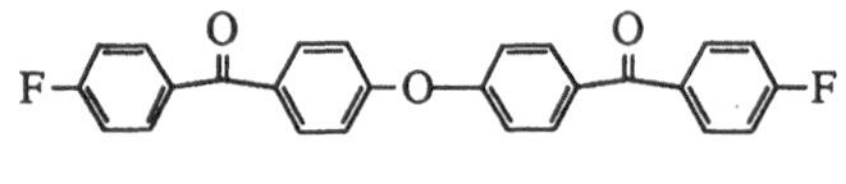 F—⟨⟩—C(=O)—⟨⟩—O—⟨⟩—C(=O)—⟨⟩—F	40
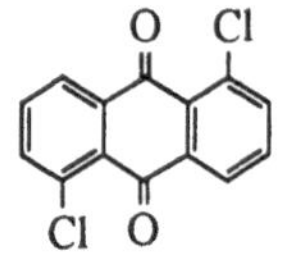 dichloroanthraquinone	52

Like the ether and thioetherimides, the ethersulfone and etherketone cyclics are activated toward ring-opening polymerization via transetherification. Heating of the cyclic oligomer mixture at 380-400° C in the melt with 1.0 mol% of bisphenol A disodium salt produced a polyethersulfone having a wt. avg. MW of about 80,000. Polymerization of the etherketone cyclics has not yet been reported.

CONCLUSION

Discovery of an efficient means for preparation of cyclic oligomeric carbonates, has made these materials available for fundamental polymerization studies. The work on cyclic carbonates has been extended to include many other systems, which may or may not contain the carbonate functionality. Cyclic oligomers are useful precursors to engineering thermoplastics, because no volatiles or by-products are generated during their polymerization. Additionally, the low melt viscosity of the cyclic oligomers, compared to the final polymers, make these materials useful for a variety of applications which the conventionally-prepared polymers would not be suited.   These ring-opening polymerizations are unusual in that the ring-chain equilibrium dramatically favors the linear polymer form, and only low levels of cyclic materials are present after polymerization, even though the reaction is not enthalpically driven. The preparation and polymerization of cyclic oligomers continues to be an active area of research.

ACKNOWLEDGEMENTS

We thank T. G. Shannon, C. B. Berman, J. C. Carpenter, D. Y. Choi, and D. A. Williams for technical assistance, H. M. Relles and J. W. Verbicky for their support, and E. P. Boden, J. A. Cella, T. L. Guggenheim and J. M. Fukuyama for discussion.

REFERENCES

1. (a) For example, see: (a) Ivin, K. J. and Saegusa, T., Eds, *Ring-opening Polymerization, Vols 1-3*; Elsevier Applied Science: London, 1984; (b) McGrath, J. E., Ed., *Ring-opening Polymerization: Kinetics, Mechanisms, and Synthesis*; ACS: Washington, D.C., 1985; (c) Semlyen, J. A., Ed., *Cyclic Polymers*, Elsevier Applied Science: New York, 1986; (d) Schill, G., *Catenanes, Rotaxanes, and Knots*, Academic Press: New York, 1971.

2. For example, see Chapters 31-37 and Chapters 45-53 in Volume 3 of *Comprehensive Polymer Science*, Allen, G.; Benington, J. C., Eds., Eastmond, G. C.; Ledwith, A.; Russo, S.; Sigwalt, P., Volume Eds., Pergamon Press: Oxford, 1989.

3. Ivin, K. J., Chapter II-8 in *Polymer Handbook, 2nd Ed.*, Brandrup, J. and Immergut, E. H., Eds, Wiley: New York, 1975.

4. Penczek, S.; Kubisa, P.; Matyjaszewski, K., *Adv. Polym. Sci.*, **1985**, *68/69*, "Cationic Ring-Opening Polymerization", 35.

5. Lee, C. L.; Johannson, O. K., *J. Poly. Sci.: A-1*, **1966**, *4*, 3013.

6. Wright, P.V. and Semlyen, J. A., *Polymer*, **1970**, *11*, 462.

7. (a) Brunelle, D. J.; Boden, E. P.; Shannon, T. G.; *et al.*, *Polym. Prep.*, **1989**, *30 (#2)*, 569; (b) Brunelle, D. J.; Boden, E. P.; Shannon, T. G., *Jour. Amer. Chem. Soc.*, **1990**, *112*, 2399; (c) Brunelle, D. J.; Shannon, T. G., *Macromolecules*, **1991**, *24*, in press.

8. Schnell, H.; Bottenbruch, L., *Ger. Pat.* 1,229,101 1966; (b) Ibid, *Belg. Pat.*, 620 620, 1960; (c) Ibid., *Macromolecular Chem.*, **1962**, *57*, 1.

9. (a) Prochaska, R. J., U.S. Pat. 3,274,214, 1966; (b) Moody, L. S., U.S. Pat. 3,155,683, 1964.

10. (a) Prochaska, R. J., U.S. Pats. 3,220,980, 3,221,025 (1965), 3,274,214 (1966), and 3,422,119, (1969).

11. (a) Hill, J. W. and Carothers, W. H., *J. Am. Chem. Soc.*, **1933**, *55*, 5031; (b) Spanagel, E. W. and Carothers, W. H., *Ibid.*, **1935**, *57*, 929.

12. Kricheldorf, H. R. and Jenssen, J., *Eur. Polym. Jour.*, **1989**, *25*, 1273.

13. Hoftle, G.; Steglich, W.; Vorbruggen, H., *Angew. Chem. Int. Ed. Eng.*, **1978**, *17*, 569.

14. (a) Weirauch, K.; Horbach, A.; Vernaleken, H., U.S. Pat. 4,229,948, 1981; (b) Horbach, A., *Polymer. Prep.*, **1980**, *21*, 185.

15. (a) Boden, E. P.; Brunelle, D. J., this volume; (b) Brunelle, D. J. and Boden, E. P., *Polym. Prep.*, **1991**, *31*, in press.

16. Evans, T. L.; Berman, C. B.; Carpenter, J. C.; Choi; D. Y; and Williams, D. A., *Polym. Prep.*, **1989**, *30* (2), 573.

17. Tyuzyo, K.; Harada, Y.; Suzuki, J., *Polymer Lett.*, **1964**, *2*, 43.

18. (a) Wick, G.; Zeitler, H., *Angew. Makromol. Chem.*, **1983**, *112*, 59; (b) Davis, A. C., *Polymer*, **1977**, *18*, 305; (c) East, G. C.; Girshab, A. M., *Polymer Commun.*, **1982**, *23*, 323; (d) Ha, W. S.; Choun, Y. K., *J. Poly. Sci: Poly. Chem. Ed.*, **1979**, *17*, 2103.

19. (a) Brunelle D. J. and Shannon, T. G., U.S. Pat. 4,829,144, 1989; (b) Brunelle, D. J.; Guggenheim, T. L.; Boden, E. P.; Shannon, T. G.; Guiles, J. W., U.S. Patent 4,696,998, 1987; (c) Guggenheim, T.L.; McCormick, S.J.; Kelly, J. J.; Brunelle, D. J.; Colley, A. M.; Boden, E. P.; and Shannon, T. G., *Polym. Prep.*, **1989**, *30* (2), 579.

20. (a) Brunelle, D. J.; Evans, T. L.; Shannon, T. G., U.S. Patent 4,736,016, 1988; (b) Brunelle, D. J.; Guggenheim, T. L.; Cella, J. A., *et al.*, U.S. Patent 4,980,453, 1990.

21. Brunelle, D. J.; Evans, T. L.; Shannon, T. G., U.S. Patent 4,775,741, 1988.

22. (a) Cella, J. A.; Talley, J. J.; Fukuyama, J., *Polym. Prep.*, **1989**, *30* (#2), 581; (b) Cella, J. A.; Fukuyama, J.; Guggenheim, T. L., *Ibid.*, **1989**, *30* (#2), 142.

MECHANISM FOR THE AMINE-CATALYZED

FORMATION OF POLYCARBONATE CYCLICS

Eugene P. Boden and Daniel J. Brunelle

GE Research and Development
Schenectady, NY 12301

INTRODUCTION

The importance of cyclic precursors for the preparation of polyamides, aliphatic polyesters, silicones, and epoxide thermosets by ring opening polymerization has been well documented.[1] However, use of cyclic aromatic polyesters or polycarbonates has been limited not only by low-yielding procedures for their preparation,[2] but also by the high melting points of the individual cyclics produced in these reactions.[3] Recently, we have reported several high yielding procedures for preparing oligomeric cyclic polycarbonates, polyarylates, and polyimides for use in ring opening polymerizations.[4] The low melt viscosity of these oligomeric cyclics together with the very high molecular weights which are achievable upon polymerization promise to make these polymer intermediates very useful in a variety of applications. We report herein mechanistic and process studies which have led to the high yielding procedure for preparing bisphenol A cyclic polycarbonates under interfacial reaction conditions. The role of the tertiary amine catalyst will be addressed in detail. Many of the concepts developed in this work are applicable to interfacial polycarbonate forming reactions in general and should provide utility in both novel and traditional polycarbonate syntheses. Although several procedures already exist for preparing cyclic polycarbonates, the interfacial process described herein is the only one known which *selectively* affords cyclics with almost total exclusion of low MW linear oligomers (< 0.1% of oligomers with molecular weights < 5000). The presence of linear oligomers must remain low to ensure the formation of high MW polymer upon ring opening polymerization.

GENERAL REACTION PARAMETERS

In a typical cyclization reaction, a methylene chloride solution of BPA-*bis*-chloroformate[5] is continuously added to a well stirred mixture of methylene chloride, water, sodium hydroxide and a tertiary amine catalyst (see Scheme 1). The incoming BPA-*bis*-chloroformate must first undergo hydrolysis of one chloroformate moiety followed by condensation of the resulting phenolate with another chloroformate moiety. However, torsional strain prevents the formation of a cyclic monomer so a bimolecular condensation reaction must occur next. The resulting dimer *bis*-chloroformate can then undergo partial hydrolysis followed by intramolecular condensation to form a cyclic dimer. Since an *oligomeric* distribution of cyclics is necessary to ensure that the mixture possesses a depressed melting point, both oligomerization and condensation reactions in addition to hydrolysis reactions are necessary. Ideally, the incoming BPA-*bis*-chloroformate should quickly undergo partial hydrolysis followed by condensation reactions to form cyclic oligomers, which are stable to the reaction conditions, prior to the addition of more BPA-*bis*-chloroformate. Under these conditions, referred to as *pseudo*-high dilution reaction conditions, the reactive intermediates remain very dilute to minimize intramolecular reactions, while the cyclic products continue to increase in concentration. Product concentrations of 0.5 to 1.0 molar (in BPA carbonate repeat units) have routinely been prepared even on large scale.

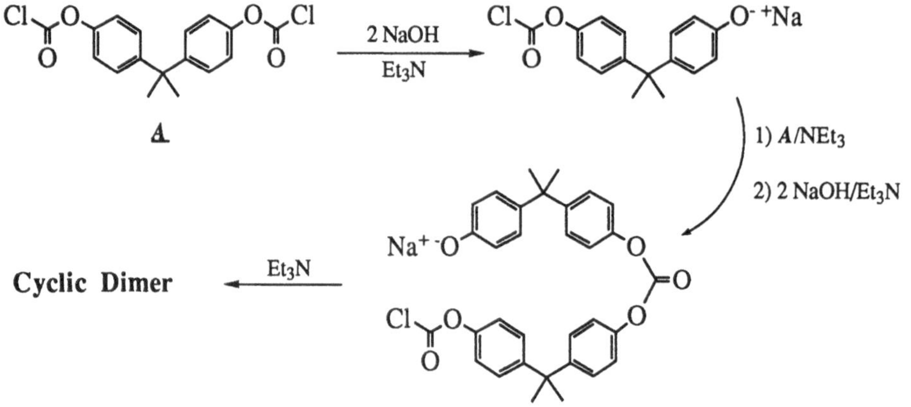

Scheme 1. Conversion of BPA-*bis*-chloroformate to cyclic dimer.

An advantage to using BPA-*bis*-chloroformate under interfacial conditions over classical anhydrous conditions is that the stoichiometry does not become an issue. A second advantage is that sodium hydroxide can be used as an acid scavenger instead of a tertiary alkyl amine. Under anhydrous conditions, adding precisely the same levels of both BPA-*bis*-chloroformate and bisphenols is difficult. Additional phenol present in the product limits the MW achievable upon polymerization of the cyclics, and residual chloroformate produce melt instability of the polymer. Fortunately, as the phenols are liberated by hydrolysis of chloroformate under interfacial reaction conditions, condensation reactions quickly occur to form carbonate linkages. Thus, there is no need for a slight excess of either reagent. Under interfacial conditions, the rate of condensation reactions must be significantly faster than the rate of hydrolysis reactions to prevent low MW linear oligomers from being produced instead of cyclics.

Thus, in order to control the formation of cyclics, a number of variables needed to be optimized. For instance, factors such as phase ratios, base strength, reaction temperature, interfacial stirring, chloroformate addition rates, catalyst concentration and catalyst structure all played an important role in yielding a fully optimized reaction. In contrast, the preparation of linear polycarbonates under interfacial conditions generally requires much less carefully controlled conditions, the most important parameter being the level of chain stopper used to regulate the molecular weight of the polymer. Another important difference is that phosgenation of an interfacial reaction mixture of bisphenols producing high MW linear polymer requires no hydrolysis of chloroformates which are produced in-situ. The abundant supply of bisphenols present in the reaction mixture readily lead to bimolecular condensation reactions and preclude significant levels of chloroformate hydrolysis.

In developing a viable process for the preparation of cyclics on large scale and the elucidation of a mechanism for this process, we have relied heavily on product analysis of cyclics reactions and on model reactions. Product analysis is particularly useful since there is a wide range of products possible as reaction conditions are varied. Any of the following products can be formed and observed by HPLC analyses: 1) cyclics; 2) high MW polymer; 3) low MW linear oligomers; 4) bisphenol A; and 5) unreacted starting material. Within this text, reference to high yields of cyclics refers not only to a good distribution of cyclics, but also to a low level of high MW polymer and nearly total exclusion of low MW linears including bisphenol A. In terms of mass balance nearly all reactions produce very high yields.

The results of optimizing most of these variables will be presented separately, but a summary of these results will be used as a starting point for this paper. For instance, we have found that phase ratios and interfacial mixing are inter-dependent and can have a dramatic effect on the reaction outcome. By minimizing the level of water to the point of sodium chloride saturation in the aqueous phase, very intense mixing can be used while achieving high yields of cyclics. As additional water is added to the reaction, more casual stirring must be maintained to prevent excessive hydrolysis of the chloroformate. Although we have looked at a variety of rates for BPA-*bis*-chloroformate addition and of reaction temperatures, including a variety of solvents, these parameters are not crucial to the outcome, allowing one to make some arbitrary assignments. Continuous addition of chloroformate to the reaction mixture over 30 minutes is sufficiently slow to achieve good yields of cyclics and yet fast enough for convenient large scale preparation of cyclics. The hydrolysis/condensation of BPA-*bis*-chloroformate to form cyclics is very exothermic. To prevent changes of individual reaction rates from variations in temperature over the course of the reaction, all experiments were performed in refluxing methylene chloride. Although seemingly straightforward, a substantial amount of effort has been directed at the level and type of base used for scavenging the HCl formed from the hydrolysis/condensation of chloroformate. A variety of $M^+(^-OH)_n$ bases can be used with only minor differences in product distributions observed. However, the *level* of base is very important. In principle, only 2.0 eq. of sodium hydroxide/eq. of BPA-*bis*-chloroformate are necessary to achieve the desired hydrolysis/condensation reactions (see Scheme 2). Under very basic conditions, the CO_2 formed by hydrolysis can be trapped to form either Na_2CO_3 or $NaHCO_3$ depending on the pH of the reaction. We have found that using 3.25 - 3.75 eq. of sodium hydroxide/eq. of pure BPA-*bis*-chloroformate produces the highest yields of cyclics. This level of base produces a final pH of the aqueous phase of ~ 10.5 - 11.0. Ideally, one would use pH-controlled caustic addition to maintain the desired pH range, but due to the low levels of water used in the reaction, pH measurements are not generally reliable.

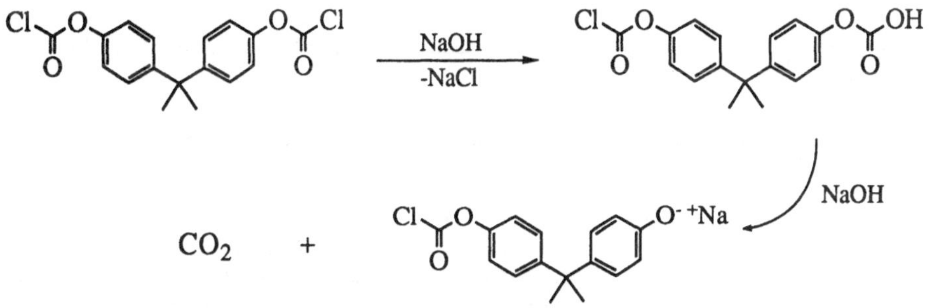

Scheme 2. Hydrolysis of a chloroformate to form carbon dioxide and a phenoxide.

ROLE OF AMINE CATALYSTS

The last variable, the nature and concentration of the amine catalysts used, requires a more detailed analysis to understand its importance. Most of the early work on preparing cyclics was based on the use of triethylamine since much of the literature on the preparation of polycarbonates has centered around this catalyst. As the importance of the catalyst became more fully realized, a variety of other amines were tested to help determine the role of the catalyst. Ultimately, it was desirable to determine if the catalyst functioned as a general base catalyst or as an acylation catalyst similar to the hyper-nucleophilic amine 4-dimethylaminopyridine[6] (see Scheme 3). If general base catalysis were the sole function of the amine, then classical phase-transfer catalysts and very basic amines should prove to be effective at promoting cyclics formation. If the amine functions solely as an acylation catalyst, then sterically unhindered tertiary amines which are very nucleophilic should be ideal catalysts. However, the amine could act in one mode to effect hydrolysis of the chloroformate and in another mode to promote condensation, complicating the mechanism and possibly requiring multiple catalysts or a unique catalyst.

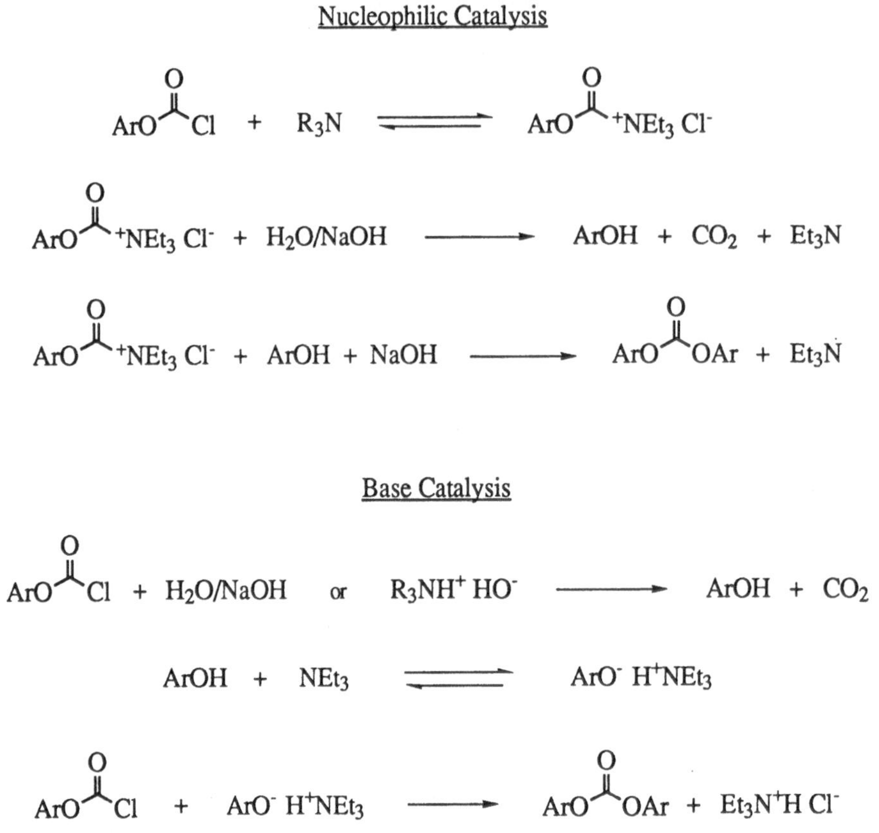

Nucleophilic Catalysis

$$AroO-C(=O)-Cl \; + \; R_3N \; \rightleftharpoons \; AroO-C(=O)-^+NEt_3 \; Cl^-$$

$$AroO-C(=O)-^+NEt_3 \; Cl^- \; + \; H_2O/NaOH \; \longrightarrow \; ArOH \; + \; CO_2 \; + \; Et_3N$$

$$AroO-C(=O)-^+NEt_3 \; Cl^- \; + \; ArOH \; + \; NaOH \; \longrightarrow \; AroO-C(=O)-OAr \; + \; Et_3N$$

Base Catalysis

$$AroO-C(=O)-Cl \; + \; H_2O/NaOH \quad or \quad R_3NH^+ \; HO^- \; \longrightarrow \; ArOH \; + \; CO_2$$

$$ArOH \; + \; NEt_3 \; \rightleftharpoons \; ArO^- \; H^+NEt_3$$

$$AroO-C(=O)-Cl \; + \; ArO^- \; H^+NEt_3 \; \longrightarrow \; AroO-C(=O)-OAr \; + \; Et_3N^+H \; Cl^-$$

Scheme 3. Amine functioning as an acylation catalyst or as a base catalyst.

If the amine acts as an acylation catalyst, the potential exists for an unwanted side reaction to occur. It is well known that under anhydrous conditions acyl ammonium salts derived from the reaction of a chloroformate and a tertiary amine can decompose to form urethanes.[7] We have recently reported a rate constant (1.3 ± 0.2 min^{-1}) for the case involving triethylamine and phenyl chloroformate in methylene chloride at 39°C (see Scheme 4).[8] Formation of urethanes in this manner under normal interfacial conditions to form high MW polymer would have minimal effects on the properties of the polymer and may not even be observed. However, urethane formation in the cyclization reaction would produce a capped oligomer which cannot cyclize, therefore leading only to polymer. In this manner, a fraction of the amine used in the reaction could produce a significant level of unwanted high MW polymer. Evidence for the formation of urethanes during the preparation of cyclics will be presented later in this paper.

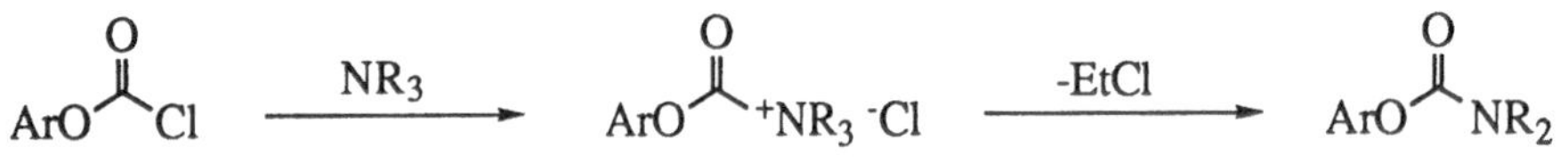

$$AroO-C(=O)-Cl \; \xrightarrow{NR_3} \; AroO-C(=O)-^+NR_3 \; ^-Cl \; \xrightarrow{-EtCl} \; AroO-C(=O)-NR_2$$

Scheme 4. Formation of a urethane from a tertiary amine and a chloroformate.

Very early experiments revealed that the amine was absolutely critical in the first step of the reaction, hydrolysis. In the absence of any amine catalyst, chloroformate could be recovered virtually unreacted from the interfacial mixture even when pH's approached 14. If base catalysis were the dominant means of chloroformate hydrolysis, then typical phase-transfer catalysts in the presence of sodium hydroxide should at least promote hydrolysis of chloroformate to phenols. However, a variety of phase-transfer catalysts, including n-Bu$_4$N$^+$ $^-$OH, produced little or no reaction of the bischloroformate during the time frame of a normal cyclization reaction. Under homogeneous conditions or very long reaction times, the chloroformate can be consumed to produce primarily linears and polymer with negligible levels of cyclics.

In contrast, the condensation reaction of a chloroformate moiety with a phenol under interfacial conditions proceeds very rapidly in the presence of either tertiary amines or quaternary ammonium compounds. The rates are very fast and are difficult to quantify. In fact, samples which are removed during the course of a cyclization reaction, quenched rapidly with acid to stop further reaction, and analyzed by HPLC reveal only carbonate linkages. No chloroformate or phenolic species are observed.

Similar results are observed in model studies if phenyl chloroformate is added to an interfacial reaction mixture containing caustic and a tertiary amine followed by systematic quenching of samples taken over the course of the reaction with a secondary amine. Diphenyl carbonate is produced to the exclusion of phenol or the phenyl dialkyl urethane which results upon quenching with amine. If phase-transfer catalysts are used instead of tertiary amines, then diphenyl carbonate is not produced and urethanes are the major product formed upon quenching samples with the secondary amine. However, if equal molar amounts of phenyl chloroformate and a phenol are added to an interfacial mixture containing either a tertiary amine or a phase-transfer catalyst, then diphenyl carbonate is produced very rapidly. Thus, at least *the hydrolysis portion of the cyclization reaction requires a tertiary amine acylation catalyst.*

These results are consistent with theories developed for phase-transfer catalyzed reactions.[9] Phase-transfer catalyzed hydrolysis of chloroformate would necessitate the transfer of hydroxide from the aqueous phase to the organic phase since the chloroformates are not soluble in the aqueous phase. It is well known that only extremely lipophilic phase-transfer catalysts can achieve this transfer, and that standard catalysts such as n-Bu$_4$N$^+$ X$^-$ generally cannot. If triethylamine were acting to carry hydroxide to the organic phase, then a species such as Et$_3$NH$^+$ $^-$OH would need to exist. This species would be less lipophilic than n-Bu$_4$N$^+$ $^-$OH and should not transfer. Also, under the high pH reaction conditions used in the preparation of cyclics, the existence of Et$_3$NH$^+$ $^-$OH seems highly unlikely.

Early studies indicated that approximately 0.2 eq. of triethylamine/eq. BPA-*bis*-chloroformate were necessary to ensure a high yield of cyclics, compared with approximately 0.005 - 0.02 eq. for a normal interfacial polymerization. Somewhat surprisingly, other very similar tertiary amines, such as tripropylamine and N-ethylpiperidine, produced much higher levels of polymer at the same molar concentrations. It was later found that the *stoichiometry* of triethylamine vs BPA-*bis*-chloroformate was not so important, but rather that the triethylamine *concentration* in methylene chloride controlled the cyclics/polymer ratio in the cyclization reaction. In early cyclization reactions, the amine was either entirely added at the start of reaction or was added during the reaction in 3-5 equal portions. Since half the organic volume of the reaction was added as a chloroformate solution in methylene chloride over 30 minutes, the triethylamine concentration decreased by half over the course of the reaction in reactions where it was added at the start. In reactions where the amine was added in portions, the amine concentration increased steadily through the reaction. A key observation revealed that the ratio of polymer to cyclics was higher at the start of reaction than at the end. By adding half the amine initially and adding the remaining half over the course of reaction, the amine concentration remained constant throughout the reaction, and the ratio of polymer/cyclics also remained constant throughout the reaction.

Using this latter approach, cyclization reactions were carried out at varying amine concentrations, and it was found that polymer levels varied depending on the concentration of triethylamine. The optimum amine concentration appeared to be approximately 0.012 mL/mL methylene chloride (0.087 M) for 0.5 molar cyclization reactions. If triethylamine was most effective at a specific concentration, then other amines may also have individual optimal concentrations. Thus, comparable studies were performed for several similar amines (see Figure 1). It was found that these amines worked nearly as well as triethylamine for forming high yields of cyclics but at *different* optimal amine concentrations. The fact that the less hindered amine (N-ethylpiperidine) was effective at lower molar concentration and that the more bulky amine (tripropylamine) was effective at higher molar concentration suggested that ease of formation of the acyl ammonium salt or of its subsequent reactions might be an important consideration. These results prompted a further investigation of the nature of acyl ammonium salt formation by a series of NMR experiments.

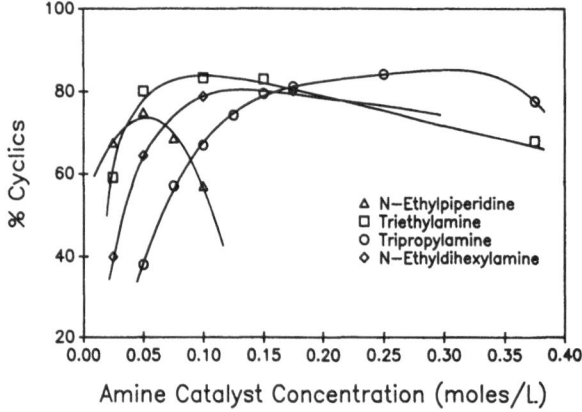

Figure 1. Concentration dependence of tertiary amines.

NMR EXPERIMENTS WITH ACYL AMMONIUM SALTS

The reaction of a chloroformate with a tertiary amine to form an acyl ammonium salt is potentially a reversible reaction (see Equation 1).[10]

$$NEt_3 \quad + \qquad\qquad\qquad\qquad \rightleftharpoons \qquad\qquad\qquad\qquad \tag{1}$$

If so, an equilibrium mixture of the acyl ammonium salt and the starting materials would exist, and the relative proportions in solution would depend on the concentration of the reagents and the nature of the particular amine. In fact, H^1NMR can be used to easily distinguish unreacted triethylamine from the corresponding acyl ammonium salt. In addition, ethyl chloride and the urethane that results from the decomposition of the acyl ammonium salt can also be differentiated from the other species present. The equilibrium constant for acyl ammonium salt formation from triethylamine and phenyl chloroformate was determined by measuring the relative amounts of acyl ammonium salt and triethylamine present when phenyl chloroformate and triethylamine were mixed in CD_2Cl_2 at several temperatures. Of course, the exact concentration of the starting materials must be known, and the loss of material to urethane formation must be taken into account. We found that the equilibrium constant for reaction of triethylamine with phenyl chloroformate to form an acyl ammonium salt at 30°C is approximately 39. Results of the NMR study of the acyl ammonium salt equilibria at several temperatures with triethylamine, along with comparison to amounts of acyl ammonium salts formed using other amines, are summarized in Table 1.

Table 1. Equilibrium Constants Determined From NMR Experiments.

Entry	Amine	Temperature	K_{eq}
1	Et_3N	-10.0	8990
2	Et_3N	0.0	1570
3	Et_3N	10.0	408
4	Et_3N	15.0	296
5	Et_3N	20.0	125
6	Et_3N	30.0	39.0
7	Bu_3N	15.0	39.7
8	N-Ethylpiperidine	15.0	625
9	$MeNEt_2$	15.0	> 50000
10	Me_2NBu	15.0	> 50000
11	4-DMAP	15.0	> 50000
12	$(iPr)_2NEt$	15.0	< 0.001
13	Pyridine	15.0	< 0.001

NMR chemical shifts relative to TMS for the Et_3N case are: acyl ammonium salt, 4.23 (q, J=7.3), EtCl, 3.59 (q, J=7.3), diethyl urethane, 3.43 (q, J=7.2) and 3.35 (q, J=7.1), and Et_3N, 2.46 broad or unresolved quartet.

In order to force the equilibrium toward starting material and to determine more accurately the equilibrium constant, the concentration of each reagent was reduced to as low as 0.05 M. At very low levels of reagents, adventitious water in the CD_2Cl_2 (unless meticulous precautions were taken in distilling the CD_2Cl_2 from P_2O_5) or on the glassware can result in hydrolysis/condensation of the chloroformate producing Et_3N-HCl (and diphenyl carbonate) and resulting in an absorption at 3.00 ppm. In the NMR experiments, the acyl ammonium salt (4.23 ppm) and the decomposition product (diethyl urethane, 3.43 and 3.35 ppm) produced well resolved quartets. Curiously, the expected quartet for the triethylamine methylenes was not present. Instead, the triethylamine methylenes were seen as a broad, unresolved absorption at 2.5 ppm. Initially, peak broadening was thought to be caused by rapid exchange with the amine-hydrochloride. However, the triethylamine methylenes remained unresolved even when very dry conditions were used. Furthermore, the presence of excess triethylamine did not significantly increase the resolution. At present, we have no rationale to explain the peak broadening other than some unknown type of complexation of the amine with the chloroformate or with the acyl ammonium salt.

The other amines listed in Table 1 also displayed unique resonances in the NMR. Although no acyl ammonium salt was observed for the very hindered N-ethyl diisopropylamine (entry 12), peak broadening of the methylene protons and the methine protons was again observed. The sterically unhindered amine, MeNEt$_2$, reacts completely with phenyl chloroformate to form the acyl ammonium salt, and produces two non-equivalent sets of methylenes (multiplets at 4.98 ppm and 4.11 ppm) indicative of a single conformational isomer, unlike the triethylamine case.

Use of these various amines in cyclization reactions produced results which correlated well with the NMR data. The extremely sterically hindered N-ethyl diisopropylamine, which produced no acyl ammonium salt by NMR, left the chloroformate virtually unreacted under standard cyclization reaction conditions. Although tributylamine does form acyl ammonium salts, it has an equilibrium constant nearly an order of magnitude lower than triethylamine when compared at the same temperatures (Table 1, entries 7 and 4). This amine produced cyclics, but required much higher catalyst loadings, similar to tripropylamine which was graphically represented in Figure 1. The equilibrium constant for N-ethylpiperidine was twice the value for triethylamine at similar temperatures (Table 1, entries 8 and 4), and affords an optimum yield of cyclics at half the amine concentration required for triethylamine.

The extreme sensitivities of the cyclization reaction on the structure of the amine catalyst together with the NMR equilibrium data strongly suggest the need for maintaining a specific concentration of acyl ammonium salt in the reaction mixture. The relatively high levels of amine are not necessary simply to convert all the chloroformate to acyl ammonium salts since too much amine can lead to increased levels of polymer (refer to Figure 1). These effects were accentuated in the sterically unhindered methyl amines (Table 1, entries 9 and 10). Only the acyl ammonium salts were observable in these examples as reflected in the equilibrium constants (> 50,000). Using these catalysts in cyclization reactions at normal amine concentrations produced a variety of products, especially low MW linear oligomers and polymer. However, *by reducing catalyst concentrations by more than two orders of magnitude*, reasonable levels of cyclics were produced which contained low levels of low MW linears. The results of cyclization reactions employing a variety of catalysts are contained in Table 2.

The results suggest that although rapid conversion of the incoming BPA-*bis*-chloroformate to an acyl ammonium salt is important, conversion to *bis*-acyl ammonium salts may lead to polymer and low MW oligomers. Thus, the amine concentration must be properly balanced to sequentially form a *mono*-acyl ammonium salt followed by hydrolysis to form a *mono*-chloroformate capable of cyclizing after the remaining chloroformate end-group is converted to an acyl ammonium salt. Formation of *bis*-acyl ammonium salts from the *bis*-chloroformate oligomers may produce an extremely water soluble moiety which would pass into the aqueous phase and suffer complete hydrolysis to form linear oligomers.

Having established the necessary existence of an acyl ammonium salt intermediate for hydrolysis reactions and its likely existence for condensation reactions, and also the premise that the ratio of hydrolysis to condensation will control the relative proportions of cyclic and linear oligomers, the working model for cyclics formation under pseudo high

Table 2.   Interfacial Hydrolysis/Condensation Reactions of Bisphenol A
Bis-chloroformate using Various Catalysts[a].

Catalyst	Cat. Conc.	% Cyclics	Products
Et$_3$N	0.1[b]	85	cyclics and polymer
"	0.005	<5	polymer and cyclics
Pyridine	0.1	0	linears
"	0.5	0	Bisphenol A
N-ethylpiperidine	0.1	58	cyclics and polymer
"	0.05[b]	75	cyclics and polymer
n-Pr$_3$N	0.1	68	cyclics and polymer
"	0.25[b]	84	cyclics and polymer
Quinuclidine	0.1	<1	36% linears + polymer
"	0.005[b]	34	polymer, cyclics, <1% linear
Et$_2$NMe	0.1	27	linears, cyclics, polymer
"	0.005[b]	64	cyclics, polymer, <1% linears
EtNMe$_2$, Me$_3$N, DABCO[c]	0.1	<1	linears
Proton Sponge[d], i-Bu$_3$N, Et$_4$NOH	0.1	0	No reaction
n-Bu$_4$NBr, PEG 600[e]	0.1	0	No reaction
4-Dimethylaminopyridine	0.1	80	cyclics and polymer

[a] Final product concentrations in all cases were 0.5 molar in BPA-carbonate repeat units.
[b] Optimum concentration for cyclic formation.
[c] 1,4-Diazabicyclo[2.2.2]octane.
[d] 1,8-bisdimethylamino)naphthalene.
[e] Polyethylene glycol, Avg. MW = 600.

dilution reaction conditions can now be presented (see Scheme 5). The BPA-*bis*-chloroformate is virtually insoluble in the aqueous phase and most of the tertiary amine resides in the organic phase due to the high pH of the interfacial mixture. Thus, the formation of the acyl ammonium salt *B* must occur in the organic phase. It has been established that chloroformate hydrolysis occurs via intermediate *B*. Although the hydrolysis reaction may result from dissolved water in the organic phase, it is attractive to think of this charged intermediate migrating to the interface where hydrolysis would then occur. The resulting phenoxide may reside at the interface while the remaining chloroformate moiety reacts with triethylamine. This charged acyl ammonium salt may then be attracted to an oppositely charged sodium phenoxide which has already formed. Subsequent condensation most likely occurs via intermediate *D*, again via reaction at or near the interface. In order for a pseudo high dilution reaction to occur, the concentration of all the intermediates (A, B, B', C and D) must be very low to minimize bimolecular oligomerization reactions. After intermediate D undergoes an intramolecular condensation to form a cyclic, it becomes essentially inert to the reaction medium and has no further effect on the overall reaction outcome.

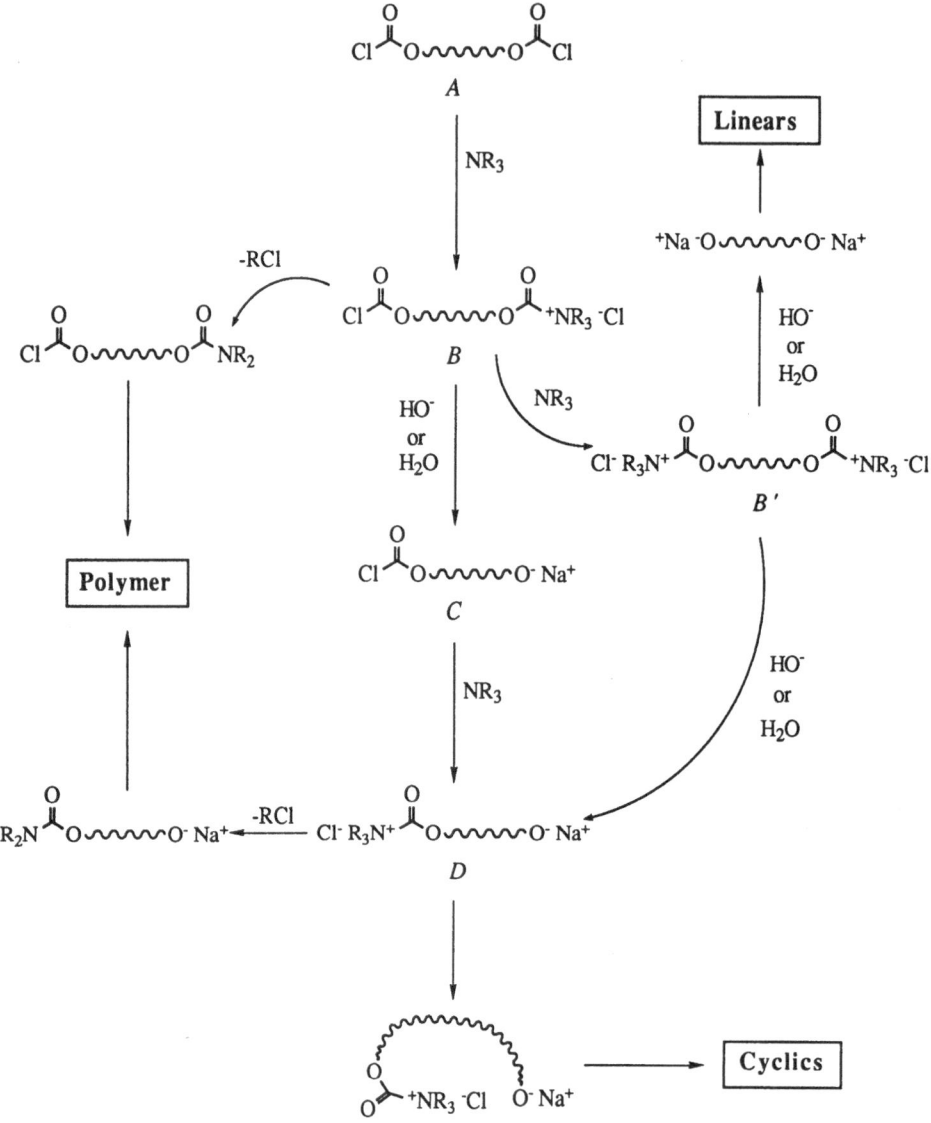

Scheme 5. Tertiary Amine Catalyzed Formation of Cyclics, Linears, and Polymers.

The rates and/or equilibria in each step which control the concentration of each reactant must be properly balanced to ensure that none of the intermediates increase in concentration during the reaction. Since the concentration of the incoming BPA-*bis*-chloroformate or an oligomeric bischloroformate present in the reaction mixture must remain very low, the conversion of the chloroformate to an acyl ammonium salt must be very fast. Maintenance of a low level of chloroformate A will require a sufficiently high concentration of triethylamine to ensure that the chloroformate A is consumed very quickly.

However, care must be taken not to produce high levels of the *bis*-acyl ammonium salt *B'*, which may lead to extensive hydrolysis.

The amine catalyst has two very important roles. First, the amine must be nucleophilic enough and present in sufficient concentrations to ensure rapid formation of acyl ammonium salts from the available chloroformate. Second, the amine must form an acyl ammonium salt which can differentiate between phenoxides and either water or hydroxide. If hydrolysis reactions are competitive with carbonate forming reactions, then excessive hydrolysis could occur which would lead to low MW linear oligomers. In fact, if hydrolysis rates were comparable with rates for carbonate formation, then measurable quantities of linears and very high levels of polymer should be formed. It should be noted that under *homogeneous* reaction conditions, phenoxides and hydroxides react at approximately the same rates with acyl halides.[11] Fortunately, most of the tertiary alkyl amines tested under interfacial conditions are able to discriminate well, and the formation of linears caused by hydrolysis is not a problem. Methyl amines appear to be one of the few exceptions but only at relatively high amine concentrations. However, virtually all pyridine based amines produce significant quantities of linears.

The NMR data suggests that since pyridine forms no observable acyl ammonium salt (Table 1, entry 13), BPA-*bis*-chloroformate should be recoverable virtually unreacted when it is used in a cyclization reaction. However, the BPA-*bis*-chloroformate was completely consumed producing primarily low MW linear oligomers and essentially no cyclic oligomers. The reaction was almost as selective for linear oligomers as the triethylamine-catalyzed reaction was for cyclics. Pyridine is a poorer nucleophile than the trialkyl amines, as evidenced by the NMR experiments. Due to the nature of the pyridyl moiety, an acyl ammonium salt derived from pyridine would be more sterically accessible to nucleophiles than are the acyl ammonium salts derived from any of the tertiary alkyl amines. Since pyridine is five orders of magnitude less basic than triethylamine, it should also be a far better leaving group in reactions of its acyl ammonium salt. Thus, any acyl ammonium salt which does form should be more reactive, and should be indiscriminating in its subsequent reaction with water, hydroxide or phenoxide. Nonetheless, an acyl ammonium salt *must* form because reaction occurred. Results with pyridine are probably not the result of a change in mechanism to acid-base chemistry, since 2-methylpyridine, which is a somewhat stronger base than pyridine, does not react at all. The fact that pyridine reacts, yet is a far weaker base than the hindered amines mentioned earlier, seems to further preclude acid-base chemistry as the source of these results. Both the nucleophilicity of the amine and the reactivity of its acyl ammonium salt toward nucleophiles are important considerations in determining the outcome of the reaction. As expected, high concentrations of pyridine led to a low oligomeric distribution of linears while low levels of pyridine produced a higher distribution of linears and polymer. We found that a variety of pyridyl amines produce either linears and polymer or no reaction at all depending on the steric constraints of the amine. Pyridines with 2-substitution (e.g. 2-picoline or 2,2'-bipyridine) apparently are not nucleophilic enough to form an acyl ammonium salt to any significant degree under the conditions of the reaction and return starting materials under normal conditions. Lipophilic pyridines which have alkyl or aryl substituents in the para position as well as a polymer bound pyridine also produce similar distributions of linears.

A notable exception to the behavior of the pyridine class was seen in the case of DMAP. Unlike other pyridyl amines, DMAP behaved similarly to triethylamine producing a good yield of cyclics (65-75%) with exclusion of low MW linears. These results suggested that both acyl ammonium salt formation and differentiation between water/hydroxide and phenoxide of the acyl ammonium salt were better than with pyridine. It was clear from the NMR that an acyl ammonium salt formed readily with DMAP (Table 1, entry 11). This acyl ammonium salt would have a less electropositive carbonyl carbon due to donation of electrons from the p-dimethylamino moiety. A more stable acyl ammonium salt would be less susceptible to nucleophilic attack and should differentiate between nucleophiles better.

Use of DMAP as a catalyst for the preparation of cyclics has been very useful in understanding another aspect of the cyclization reaction. As described earlier, the acyl ammonium salt derived from triethylamine can decompose to form urethanes and ethyl chloride. Although this reaction is slow ($k = 1.3$ min^{-1}) relative to hydrolysis and condensation reactions, low levels of urethanes could account for the formation of polymer since a urethane capped oligomer cannot cyclize. The polymer formed in a typical cyclization reaction has a number average MW of approximately 18 K corresponding to an average oligomer size of approximately 70 BPA units. If each chain has one urethane end-group, then only 1.0% of the amine used in a cyclization reaction would need to be consumed to form 10% polymer. Obviously, if the polymer unit is *bis*-capped, then twice as much amine would need to be consumed. Since there is no mechanism for decomposition of an acyl ammonium salt derived from DMAP, we reasoned that the polymer formed in a DMAP catalyzed reaction should be different from that formed when triethylamine is used as a catalyst. In fact, sampling over the course of a 30 minute cyclization reaction revealed that the MW of polymer formed with triethylamine did not change over the course of the reaction, but the MW of polymer formed using DMAP steadily increased (see Table 3).

Table 3. Comparison of Polymers Formed Using Et$_3$N vs DMAP Catalysis.

Catalyst	Time (min)	% Polymer	Polymer Peak MW[a]
Et$_3$N	5	15.4	46,000
Et$_3$N	15	18.9	47,000
Et$_3$N	30	18.0	47,000
DMAP	5	35.8	47,000
DMAP	15	36.3	86,000
DMAP	30	35.7	102,000

[a] Molecular weights determined by GPC using polystyrene standards.

Thus, in an optimized cyclization reaction using triethylamine, the incoming BPA-*bis*-chloroformate was very quickly converted to a mixture of cyclics and polymer of a fixed composition. When DMAP was used as a catalyst, the polymer formed was not capped and proceeded to grow over the course of the reaction. Analyzing the crude cyclics with polymer present from a triethylamine catalyzed reaction reveals that significant levels of urethanes are present. In fact, for a typical lab reaction which yielded 15% polymer, the level of urethanes detected corresponded to 1.5 urethanes/chain. It has not yet been determined whether the formation of urethanes is the *cause* of polymer formation or the *result* (i.e. the polymer may be formed first followed by conversion to urethane after the likelihood of intramolecular reaction becomes vanishingly small). Certainly, urethane formation is not the only mode of polymer formation since optimized DMAP catalyzed cyclization reactions still form significant levels of polymer.

Although it is difficult to conclusively determine the role that different amines play in a cyclization reaction, the experimental evidence has allowed a mechanism for formation of cyclics to be suggested which satisfactorily explains the different products which arise from both major and minor changes in chemical structure of the amine catalyst. These trends are summarized in Table 4.

Table 4. Effects of Various Amines on Cyclization Reactions.

Amine	Products
EtNR$_2$ (R = n-alkyl)	optimum catalyst for Cyclics (15-50 mole%)
MeNR$_2$	polymer, linears, cyclics (20 mole%) cyclics & polymer (0.5-1 mole%)
Me$_2$NR	polymer and linears (1-20%)
4-dialkylaminopyridines	cyclics and polymer
pyridine & 3 or 4 subst. pyridines	BPA and low linears (20-40 mole%) linears and polymer (1-10 mole%)
2-substituted pyridines	no reaction
branched tertiary alkylamines	no reaction
Quaternary ammonium salts	no reaction

CONCLUSION

Relying significantly on product analysis, we have attempted to rationalize a mechanism for the formation of BPA polycarbonate cyclics under interfacial conditions which can explain the observed results. The importance of maintaining *pseudo*-high dilution reaction conditions via use of reactive, nucleophilic amines has been delineated. The effects of amine structure and concentration on the outcome of the reaction have been addressed in detail. It has been concluded that the formation and subsequent reactions of acyl ammonium salts control the effectiveness of cyclization reactions in this interfacial reaction. The data suggesting that acyl ammonium salt formation is required for hydrolysis of chloroformate seems conclusive. Although the data concerning the intermediacy of acyl ammonium salts in the condensation reaction seems valid, invocation of such intermediates cannot be made unambiguously. Certainly, general base catalysis plays little if any role in hydrolysis/condensation reactions leading to cyclics under interfacial conditions. Finally, it has been shown that decomposition of acyl ammonium salts to carbamates is a means of producing unwanted linear polymer.

ACKNOWLEDGMENTS

We thank T. G. Shannon and R. A. Sawyers for technical assistance and K. R. Stewart for helpful discussions.

REFERENCES

1.   For recent reviews on ring-opening polymerizations, see: (a) J. E. McGrath, Ed., "Ring-opening Polymerization: Kinetics, Mechanisms, and Synthesis", American Chemical Society, Washington, D.C. (1985). (b) K. J. Ivin and T. Saegusa, Eds, "Ring-opening Polymerization", Elsevier Applied Science, London, Vols. 1-3 (1984).

2. (a) H. Schnell and L. Bottenbruch, Ger. Pat. 1,229,101 (1966). (b) Ibid, Belg. Pat. 620,620 (1962). (c) Ibid., Macromolecular Chem., 57:1 (1962). (d) R. J. Prochaska, U.S. Pat. 3,274,214 (1966). (e) L. S. Moody, U.S. Pat. 3,155,683 (1964).

3. The melting point for Bisphenol A cyclic trimer (350°C) may be found in ref 2d and for cyclic tetramer (375°C) in ref 2c.

4. (a) D. J. Brunelle, E. P. Boden, and T. G. Shannon, J. Am. Chem. Soc., 112:2399 (1990). (b) E. P. Boden, D. J. Brunelle, and T. G. Shannon, Polym. Prepr., 30(2), 571 (1989). (c) D. J. Brunelle, E. P. Boden, and T. G. Shannon, U.S. Patent 4,727,134 (GE, 1988). (d) D. J. Brunelle and T. G. Shannon, U.S. Patent 4,644,053 (GE, 1987). (e) T. L. Guggenheim et al, Polym. Prepr., 30(2), 579 (1989).

5. Preparation of BPA-*bis*-chloroformate described in Brit. Pat. 613,280, 1948 (Wingfott Corp.). Use of diethylaniline instead of dimethylaniline allows reaction at °C and provides purer product.

6. G. Holfe, W. Steglich, and H. Vorbruggen, Angew. Chem. Int. Eng. Ed., 17:569 (1978).

7. (a) J. D. Hobson and J. G. McCluskey, J. Chem. Soc. C., 2015 (**1967**). (b) J. A. Campbell, J. Org. Chem., 22:1259 (1957). (c) W. B. Wright and H. J. Brabander, J. Org. Chem., 26:4057 (1961). (d) T. Kometani, S. Shiotani, and K. Mitsuhashi, Chem. Pharn. Bull., 24(2), 342 (1976).

8. P. G. Kosky and E. P. Boden, J. Poly. Sci.: Part A: Poly. Chem., 28:1507 (1990).

9. (a) M. Halpern, Y. Cohen, Y. Sasson, and M. Rabinovitz, Nouv. J. Chim., 8:443 (1984). (b) M. Halpern, Y. Sasson, and M. Rabinovitz, Tetrahedron, 38:183 (1982). (c) M. Halpern, Y. Sasson, I. Willner, and M. Rabinovitz, Tetrahedron Lett., 22:1719 (1981).

10. We would like to thank Dr. J.A. King III for sharing with us information he had on this equilibrium.

11. (a) R. F. Hudson and M. Green, J. Chem. Soc., 1055 (1962). For similar examples in reactions with esters, see: W. P. Jencks and M. Gilchrist, J. Am. Chem. Soc., 90:2622 (1968). (b) N. L. Bender and W. A. Glasson, J. Am. Chem. Soc., 81:1590 (1959).

POLY(ENAMINONITRILES):

NEW MATERIALS WITH INTERESTING PROPERTIES

J. A. Moore

Polymer Science and Engineering Program
Rensselaer Polytechnic Institute
Troy, NY 12180-3590

INTRODUCTION

The thermal stability of polymers is influenced by both chemical and physical factors.[1,2] The chemical factors are primary bond strength, secondary or van der Waals' bonding forces, hydrogen bonding, resonance stabilization, mechanism of bond cleavage, molecular symmetry, purity, crosslinking and branching. The physical factors include molecular weight, molecular weight distribution and crystallinity. The primary bond strength contributes the most to the thermal stability of polymers. The bond dissociation energy[4] of a carbon - carbon single bond is 350 kJ/mol, and that of a carbon - carbon double bond is 610 kJ/mol. A double bond in an aromatic system is further strengthened by resonance stabilization, adding 164 to 287 kJ/mol to the bond dissociation energy. Almost all thermally stable polymers contains aromatic carbocyclic or heterocyclic rings. The bond dissociation energy of a carbon - fluorine bond of 430 kJ/mol is increased to 504 kJ/mol by a second fluorine atom attached to the same carbon atom. Therefore, perfluorinated materials are more thermally stable. Inorganic polymers which have inherent higher bond strengths (e.g. B-N, Si-N, Ti-O) are often susceptible to chemical attack such as hydrolysis. As a consequence, few useful polymers containing inorganic backbones have been developed.[5]

Secondary and van der Waals' bonding forces provide additional strength and thermal stability. An additional 25 to 41 kJ/mol are contributed to molecular stability by dipole - dipole interaction and hydrogen bonding. Therefore, heat resistant polymers often contain polar groups such as carbonyl, which are able to form strong intermolecular associations. Polymers containing electron withdrawing groups as connecting groups are generally more stable than those containing electron donating groups.[2]

Contemporary Topics in Polymer Science, Vol. 7., Edited by
J.C. Salamone and J. Riffle, Plenum Press, New York, 1992

The thermal decomposition pathways of high temperature polymers are very complicated processes. In the case of aromatic heterocyclic polymers, the degradation occurs mainly through random chain scission. Crosslinking improves the thermal stability of polymers primarily because more bonds must be broken before weight loss or reduction in mechanical properties occurs. Branching in a polymer leads to lower thermal stability by decreasing the crystallinity of polymers.

Molecular weight and molecular weight distribution not only affect thermal stability, but also many other physical properties such as mechanical strength. The higher molecular weight polymers are more stable than are lower molecular weight polymers because they can tolerate more chain cleavage without significant property reduction and have higher T_g values. However molecular weight effects are not significant beyond a certain molecular weight limit.

Polymers with High Heat Resistance

Considering structural effects on the thermal stability of polymers, poly(p-phenylene) (I) would appear to have an ideal structure as a thermally stable organic polymer.

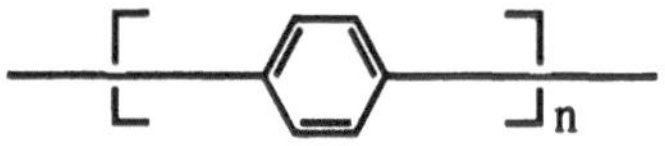

I

Indeed, poly(p-phenylene) is stable over 500 °C in air[6], but this material is insoluble and infusible. Generally, the types of chemical structure which impart heat resistance have a tendency to lead to insolubility and infusibility. Therefore, structural modification must be made so that fabrication is possible. Introducing flexible linking groups into a polymer backbone is a good example of such structural modifications, even though it may decrease the thermal stability. Of all the flexible linking groups that have been used, it is reported[5] that the following have the least effect on thermal stability.

-CO-, -COO-, -CONH-, -S-, -O-, -C(CF$_3$)$_2$-

For example, Nomex^{TM} (II) and Kevlar^{TM} (III) have been successfully commercialized by Du Pont.

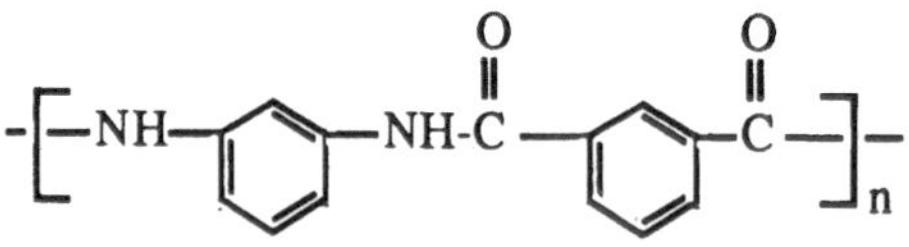

II

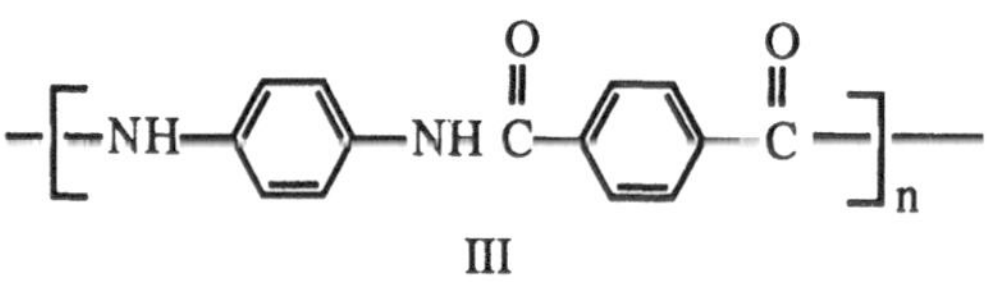

III

These materials have very good thermal stability, but have very limited solubility and they are not melt processable.[7]

The most successful approach to improve processability is the so-called "two step process". First, a soluble, high molecular weight prepolymer or oligomeric material is prepared. These prepolymers are often soluble in organic solvent or have low melting points so that fabrication from solution or the melt is possible. After the material is formed into its final shape, it is cured by heating, resulting in rigid stable ring structures which are insoluble and infusible. Maleimide endcapped polyimide (IV)[8] is an example of an oligomeric material. Curing of this material can be achieved by the reaction of end groups with a diamine, which undergoes Michael addition. However, oligomeric materials often fail to form good films.

The poly(amic acid) (V) prepared from pyromellitic dianhydride and 4,4'-diaminophenyl ether is a typical example of a high molecular weight prepolymer. The initial condensation product, poly(amic acid), is soluble in organic solvents and can be cast to films, which are subsequently cyclized to the insoluble polyimide (VI).[9]

This material, commercialized by Du Pont as Kapton^{TM} (VI), exhibits excellent thermal stability and good dielectric behavior. The poly(amic acid) solution in N-methyl pyrrolidone is also commercialized by Du Pont as Pyralin^{TM} which has found an application as an interlayer dielectric material in the fabrication of integrated circuits.[10] The principle drawback of the poly(amic acid) approach is the release of a volatile by-product, water, during curing. Volatile by-products which can cause voids cannot be tolerated in composite structures and thin film applications. Poly(amic acids) are also susceptible to hydrolysis because the carboxylic acid moieties adjacent to amide bonds of poly(amic acids) catalyze the hydrolysis of amide bonds.

Polyimide in Electronic Applications

Organic polymers have attracted much interest in electrical and electronic applications because of their electrical insulating nature. Polyimides have gained much attention because of their excellent thermal stability and low dielectric constant. Polyimides have found applications in matrix resins for circuit boards, encapsulants, adhesives, passivation coatings, alpha particle barriers, ion implant masks and interlayer dielectrics.[10,11]

Since polyimide film was first introduced as a dielectric layer in multilevel interconnection in Very Large-Scale Integration (VLSI) by Sato et al.[12], much attention has been focused on the use of polyimides as an interlayer dielectric materials due to their excellent thermal stability, low dielectric constant (3.0-3.5), high dielectric strength (typically 10^6 volts/cm), chemical inertness and planarizing property.[13] The fluid characteristics of polymer solutions provide an ideal flatness (planarization) which reduces the problem of amplified topography that arises from previous metallization and permits good step coverage in overlying conductors. Pyralin[TM] from Du Pont and PIQ[TM] from Hitachi[13e] are commercially available semiconductor grade poly(amic acid) solutions in NMP. The polyimides are formed by imidization of spin-coated films of poly(amic acid). As mentioned before, the imidization process generates water which can cause microvoids and, in the worst case, film defects. Any trapped moisture is also a potential source of corrosion of nearby metals.

One approach to overcome the problem caused by volatile by-products is to make soluble, fully cyclized polyimides by incorporating flexible linking groups into polymer backbones. Polyimides containing hexafluoro isopropylidene (Hoechst Celanese, Sixef[TM])[14], carbonyl and indane (Ciba-Geigy, Probimide 200[TM])[15], and siloxane (GE, SPI-100[TM])[16] are commercially available. However, flexible linking

groups generally cause a reduction in the thermal stability of polymers and lower glass transition temperature of polymers.

Isomerization Polycyclization

The ideal material for an interlayer dielectric layer of VLSI is a soluble prepolymer with suitable functional groups which can undergo cyclization through a rearrangement reaction without the evolution of volatile by-products resulting in thermally stable rigid polymer. A recent review on polyheteroarylene chemistry by Korshak[17] described a process called isomerization polycyclization, i.e. curing the polymer by isomerization without generating volatile by-products. Poly(imino-quinazolyne) and poly(iminobenzoxazine-dione) are synthesized by reacting di[(o-amino) nitrile] (X=NH) (VII) and di[(o-oxi) nitrile] (X=O) (VIII) with diisocyanates followed by cyclization. Another example of isomerization polycyclization is the poly(iminoimide) (IX) obtained by cyclization of the condensation from the reaction product of a dinitrile dicarboxylic acid dichloride with a diamine.[18]

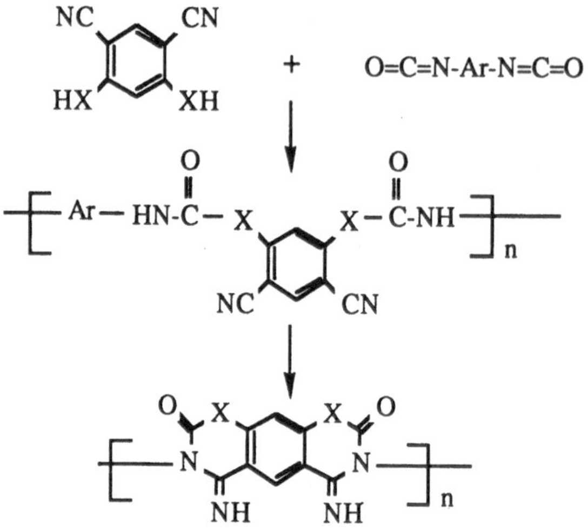

VII   X = NH
VIII X = O

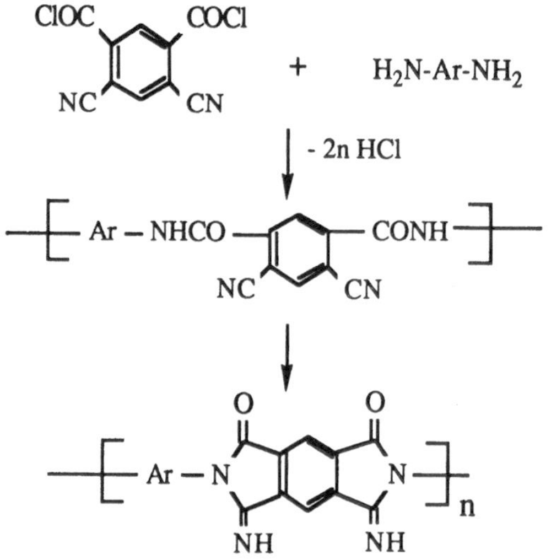

IX

It was claimed that the thermal and mechanical properties of this material are comparable with those of polyimides.

Moore and Mitchell[19] prepared soluble poly(enaminoester) (X) from a,a'-bis(carbomethoxy) diacetyl benzene and aromatic diamines. The polymers were subsequently thermally cyclized (Conrad-Limpach reaction) to poly(quinolines) (XI), which is analogous to the conversion of poly(amic acids) to polyimides in terms of converting a soluble prepolymer to a stable polymer of cyclized, rigid ring structures with the formation of volatile by-products. The cyclized product however, was infusible and insoluble.

It was felt that, in addition to the rigidity of the partially cyclized backbone, the extensive hydrogen bonding capability of the quinacridone segments would have to be modified if soluble systems which could be processed were to be obtained. Other considerations included preventing the evolution of volatile byproducts with suitably chosen cure chemistry and stability (shelf life) of the intermediate polymer.

RESULTS AND DISCUSSION

An analogy between dicyanomethylidine and carbonyl groups was pointed out by Wallenfells[23] in 1966. Because of the strong electron withdrawing characteristics of nitrile groups, (the electronegativity of -CN substituted carbon atoms resembles that of N, O, and F atoms), a $C=C(CN)_2$ group may be viewed as the structural equivalent of a carbonyl group. The groups have similar inductive and resonance effects, and have close parallels in many reactions. The following dicyanomethylidine derivatives may also be considered as structurally equivalent to the corresponding carbonyl derivatives.

NC, CN — Cl / O, Cl   NC, CN — OR / O, OR

NC, CN — OH / O, OH   NC, CN — N— / O, N—

Poly(enaminonitriles)

If we could prepare a bifunctional molecule containing chlorovinylidene cyanide groups we might be able to prepare a new class of polyaramide-like polymers which would be thermally stable and soluble until cured. Further, there is ample precedent in the literature for heterocyclic syntheses utilizing the nitrile group as the acylation unit.

By replacing the ester or acyl chloride groups with vinylidene cyanide groups, we were able to prepare a soluble poly(enaminonitrile) (XII) by allowing 1,4-bis(1-chloro-2,2-dicyanovinyl) benzene to react with 4-aminophenyl ether[20]. This new class of polymers which we called poly(enaminonitrile) was cured to an insoluble 4-aminoquinoline structure (XIII) without any volatile by-products. It should be noted that 4-iminodihydroquinoline which is presumably the structure of the thermal product formed first tautomerizes to the fully aromatic 4-aminoquinoline.[21] This material has excellent thermal stability, but the thermal expansion coefficient of this polymer is three times higher than that of polyimides and the dielectric constant is about 6 before curing and 5 after curing, which is about twice that of typical polyimides.[22]

Extensive synthetic efforts have revealed that any aromatic acid chloride which we have tried can be converted to the corresponding chlorovinylidene cyanide[24]. Almost all aromatic diamines can be used as long as they are not substituted in such a manner that the nucleophillicity of the amine nitrogen atom is sterically hindered or electronically diminished by electron withdrawing groups. Aliphatic diamines are sufficiently nucleophillic that they can be used with bis (chlorovinylidene cyanides) to produce high molecular weight (intrinsic viscosities > 1.0 dl/g) poly (enaminonitriles) by interfacial polycondensation[25]. By approprriate structural variations it is possible to prepare poly(enaminonitriles with glass transition temperatures ranging from 100 to almost 500 °C.

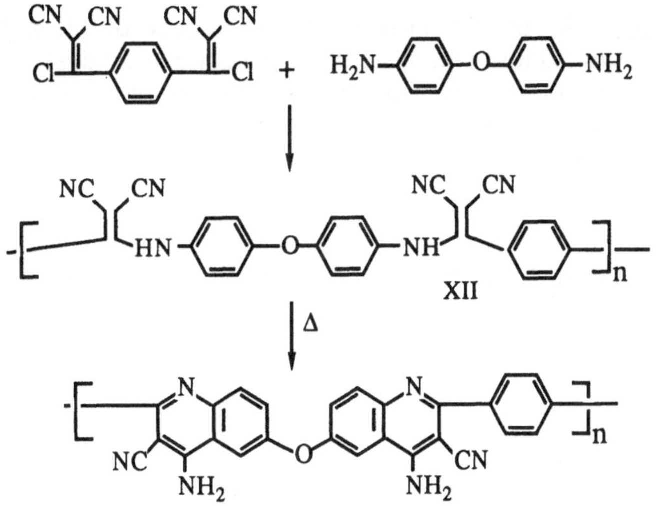

XII

XIII

PEAN Blends : The solubility properties of these new polymers are most intriguing and are entirely a function of the enaminonitrile functionality. The "soft" or polarizable character of the vinylidene cyanide was hypothesized to impart a more pronounced tendency for these polymers to interact well with solvents. It was observed early in this effort that solutions of 1,3-ODA PEAN (i.e., the polymer derived from 1,3-phenylene-bis(chlorovinylidene cyanide) and oxydianiline) in diemethoxyethane became turbid on heating to about 35 °C but became homogeneous again on cooling. The temperature at which this phenomenon is observed has been termed the lower critical solution temperature and is commomly observed when hydrogen bonding provides the driving force for solubilization of both small and large molecules. Further investigation revealed that higher homologs of dimethoxy ethane [$CH_3(-O-CH_2-CH_2-)_nOCH_3$] were not only solvents but they also exhibited LCST behavior with the difference that the critical temperature for phase separation shifted to higher temperature as n increased from 1 to 11, leveling asymptotically to about 140 °C. This behavior sugggested that these polymers might form compatible blends with poly(ethylene oxide). It was found that 1,3-ODA PEAN does, in fact, form a compatible blend with PEO up to 50 wt.-% and exhibits a completely reversible phase separation at about 160 °C[26]. Other strong hydrogen bond-accepting polymers such as poly(N-vinyl pyrrolidone), poly(ethyl oxazoline), poly(N,N-dimethyl methacryl- amide) and poly(4-vinyl pyridine) form compatible blends over the entire composition range but do not undergo thermally induced phase separation up to the decomposition temperature of the blend component[27].

<u>Polypyrazoles</u>: In an attempt to expand the scope of this approach, we reacted bis hydrazines with bis(chlorovinylidene cyanides) in the hope that we would be able to isolate the intermediate poly(enhydrazino- nitriles) and subsequently effect curing at lower temperature than with the PEAN system[28].

To date, we have seen no evidence of the intermediate and have only been able to isolate the polypyrazole form. High molecular weight samples (intrinsic viscosity > 1.0 dl/g in H_2SO_4) are generally difficultly soluble in polar, aprotic solvents but lower molecular weight samples dissolve in DMF or DMSO. Interestingly, these materials become insoluble when heated above 150 °C. Once cured they are as thermally stable as PEANs and are surprisingly resistant to degradation by atomic oxygen.

CONCLUSIONS AND PROSPECT

As a part of a program to prepare high molecular weight processable polymers that cure thermally without evolution of small molecules, we have successfully exploited a vinylic nucleophilic substitution pathway to prepare novel polyenaminonitriles and polypyrazoles with a new array of properties including high thermal stability while at the same time the polymers retain unusually good solubility and can be cast readily to tough films. We are continuing our efforts to develop synthetic methodology which will allow the incorporation of the vinylidene cyanide moiety and its analogs into a variety of other thermally stable but difficult to process polymers.

ACKNOWLEDGMENT

The tireless effort of my coworkers who are listed in the reference section is noted with satisfaction and gratitude. The work described here has been supported by the Office of Naval Research.

REFERENCES

1. Cassidy, P. E. Thermally Stable Polymers; Marcel Dekker: New York,1980.
2. Hergenrother, P. M. In Encyclopedia of Polymer Science and Technology, 2nd ed.; Mark, H. F.; Bikales, N. M.; Overberger, C. G.; Menges, G.; Kroschwitz, J.I., Eds.; Wiley-Interscience: New York, 1987; Vol. 7, p 639.
3. Wendlandt, W. W.; Gallagher, P. K. In Thermal Characterization of Polymeric Materials Turi, E. A., Ed.; Academic Press: New York, 1981; p 1.
4. Cottrell, T. L. The Strength of Chemical Bonds, 2nd ed.; Butterworths: London, 1958.
5. Critchley, J. P.; Knight, G. J.; Wright, W. W. Heat Resistant Polymers; Plenum Press: New york, 1983.
6. Speight, J. G.; Kovacic, P.; Koch, F. W. J. Macromol. Sci., Rev. Macromol. Chem. 1971, 5, 295.
7. Preston, J. G. In Encyclopedia of Polymer Science and Technology, 2nd ed.; Mark, H. F.; Bikales, N. M.; Overberger, C. G.; Menges, G.; Kroschwitz, J. I., Eds.; Wiley-Interscience: New York, 1988; Vol. 1, p 381.
8. St. Clair, A. K.; St. Clair, T. L. Polym. Eng. Sci. 1982, 22, 9.
9. Sroog, C. E. J. Polym. Sci., Macromol. Rev. 1976, 11, 161.
10. Harvey, J. A. In Proceedings of the 3rd International SAMPE Electronic Materials and Process Conferences; Los Angeles, CA, 1989; p 124.
11. (a) Burggraaf, P. Semicond. Int. 1988, 11(3), 58.
    (b) Senturia, D. S. In Polymers for High Technology; Bowden, M. J.; Turner, S. R., Eds.; ACS Symposium Series 346; American Chemical Society: Washington, DC, 1987; p 428.
    (c) Wong, C. P. Adv. Polym. Sci. 1988, 84, 63.
    (d) Lee, Y. K.; Craig, J. D. In Polymers for Electronic Applications; Feit, E. D.; Wilkins, C. W., Eds.; ACS Symposium Series 184; American Chemical Society: Washington, DC, 1982; p 108.
    (e) Adduci, J. M. In Polyimides; Mittal, K. L., Ed.; Plenum Press: New York,1984; p 1023.
    (f) Lacombe, R. H.; Greenblatt, J. Ibid, p 647.
12. Sato, K.; Harada, S.; Saiki, A.; Kimura, T.; Okubo, T.; Mukai, K. IEEE Trans. Parts. Hybrids and Packaging 1973, Vol PHP-9, 176.
13. (a) Wilson, A. N. Thin Solid Film 1981, 83, 145.
    (b) Economy, J. In Contemporary Topics in Polymer Science; Vandenberg, E. J., Ed.; Plenum Press: New York, 1984; p 351.
    (c) Rohde, O.; Riedieker, M.; Schaffner, A.; Bateman, J. Solid State Technol.1986, 29(9), 109.

(d) Jensen, J. In <u>Polymers for High Technology;</u> Bowden, M. J.; Turner, S. R., Eds.; ACS Symposium Series 346; American Chemical Society: Washington, DC,1987; p 466.

(e) Saiki, A.; Mukai, K.; Harada, S.; Miyadera, Y. In <u>Polymers for Electronic Applications;</u> Feit, E. D.; Wilkins, C. W., Eds.; ACS Symposium Series 184; American Chemical Society: Washington, DC, 1982; p 123.

14. Khanna, D. N.; Mueller, W. H. In <u>Proceedings of the 3rd International SAMPE Electronic Materials and Process Conferences;</u> Los Angeles, CA, 1989; p 905.

15. Ciba-Geigy, "Probimide^{TM} 200 Series"; technical report; 1986.

16. John, T.; Valenty, V. B. In <u>Proceedings of the 3rd International Conference on Polyimides;</u>SPE: Ellenville, New York, 1988; p 36.

17. Korshak, V. V.; Rusanov, A. L. In <u>Polymer Yearbook 3;</u> Pethrick, R. A.; Zaikov, G. E., Eds.; Harwood: Chur, 1986; p 115.

18. Voznesenskaya, N. N.; Yarosh, V. N.; Flerova, A. N.; Zaitseva, E. L.; Teleshov, E. N.; Pravednikov, A. N. <u>Vysokomolek. Soyed.</u> 1978, B20, 196.

19. Moore, J. A.; Mitchell, T. D. <u>J. Polym. Sci., Polym. Chem. Ed.</u> 1980, 18, 3029.

20. (a) Moore, J. A.; Robello, D. R. <u>Macromolecules</u> 1989, 22, 1084.
    (b) Moore, J. A.; Robello, D. R. <u>Macromolecules</u> 1986, 19, 2667.

21. Jones, G. <u>Quinolines;</u> Wiley: London, 1977.

22. Moore, J. A.; Robello, D. R.; Mehta, P. G. In <u>Proceedings of the 3rd International SAMPE Electronic Materials and Process Conferences;</u> Los Angeles, CA, 1989.

23. (a) Wallenfels, K. <u>Chemia</u> 1966, 20, 203; <u>Chem. Abstr.</u> 1967, 66, 2109.
    (b) Wallenfels, K.; Friedrich, K.; Rieser, J.; Ertel, W.; Thieme, K. <u>Angew. Chem. Int. Ed. Eng.</u> 1976, *15*, 261.

24. Moore, J. A.; Mehta, Parag. G. <u>Proc. Polym. Sci. Mat. Eng.</u> 1989, 60, 74.

25. Moore, J. A.; Mehta, Parag. G. <u>Proc. Polym. Mat. Eng. Sci.</u> 1990, 62, 351.

26. Moore, J. A.; Kim, Ji-Heung; Seidel, Peter R. <u>Chem. Mat.</u> , in press.

27. Moore, J. A.; Kim, Ji-Heung <u>Macromolecules,</u> submitted for publication.

28. Moore, J. A.; Mehta, Parag. G. <u>Macromolecules</u> 1988, 21, 2644.

PERFECTLY ALTERNATING SEGMENTED POLYIMIDE SILOXANE COPOLYMERS

M. E. Rogers, D. Rodrigues, A. Brennan*, G. L. Wilkes
and J. E. McGrath**

Departments of Chemistry and Chemical Engineering
NSF Science and Technology Center: High Performance Polymeric
Adhesives and Composites
Virginia Polytechnic Institute and State University
Blacksburg, Virginia 24061-0212

ABSTRACT

Perfectly alternating segmented, fully cyclized, polyimide siloxane copolymers were synthesized and characterized as new materials. The reaction utilized the transimidization of aminopropyl terminated polydimethyl siloxane oligomers with aminopyridine capped polyimide oligomers based on oxydiphthalic anhydride and bisaniline P. The reaction was conducted in refluxing chlorobenzene at about 130°C. Transimidization processes with the aliphatic amine end group appear to be quite rapid under these conditions, and high molecular weight copolymers were achieved in times as short as 10 minutes. The resulting copolymers could be cast or melt pressed into transparent films whose thermal and mechanical properties reflected the composition and block molecular weights. In general, the reactions proceeded smoothly under homogeneous conditions in chlorobenzene and produced materials with glass transition temperatures somewhat depressed from the glass transition temperatures of the starting polyimide oligomers. The method allows excellent control of the microphase morphology, which is a function of block size and interaction parameters. The synthesis and characterization of these materials will be discussed.

1. INTRODUCTION

Since their development in the 1960's, polyimides have become an increasingly important class of materials finding a wide range of applications in the aerospace and microelectronics industry (1,2). Contributing to their success are the many desirable properties of polyimides including good thermooxidative stability and excellent mechanical properties (3). However, polyimides are often insoluble and intractable resulting in processing difficulties which limit their applications. These processing problems can be overcome by incorporating flexible copolymers such as polysiloxanes (4) or polyarylene ethers (5) in the polyimide backbone. The polyimide siloxane segmented copolymers are of particular interest since incorporation of amine functionalized polysiloxanes into polyimides results in improved solubility and processability, decreased water absorption, atomic oxygen resistance (4), lower dielectric

---

* University of Florida, Department of Material Science and Engineering, 317 MAE, Gainseville, FL 32611

** To whom all correspondance should be addressed

constants (6) and enhanced adhesion (7). The unique properties of the polyimide siloxane segmented copolymers make them especially attractive for microelectronic applications (8).

The polyimide siloxane copolymers are generally synthesized, as shown in Scheme 1, from a combination of a dianhydride and diamine monomer and an aminopropyl terminated polysiloxane to form the segmented polyamic acid. The polyamic acid is then cyclodehydrated by either thermal or solution imidization to give randomly segmented polyimide siloxane copolymers (4). This is indeed the classical approach of reacting one flexible oligomer with two monomers which form the "hard" segment (9).

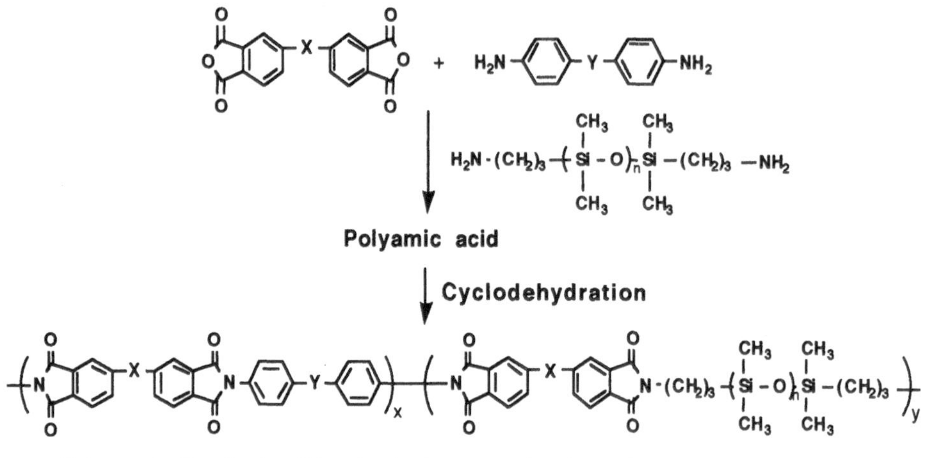

Scheme 1. Synthesis of Randomly Segmented Polyimide Siloxane Copolymers

Perfectly alternating segmented polyimide siloxane copolymers made by combining preformed polyimide oligomers and polysiloxanes would have properties similar to those of the randomly segmented polyimide siloxane copolymers. However, the perfectly alternating segmented copolyimides should have higher structural regularity giving better defined microphase separation (10). Thus, depending on their composition, well defined perfectly alternating segmented copolyimides may afford higher upper glass transition temperatures, improved tensile strength and modulus as compared to analogous randomly segmented copolyimides.(9)

Takekoshi, et al. (11), demonstrated that an amine-imide exchange or transimidization method could be employed for the synthesis of polyetherimide homopolymers by reacting bisphthalimide monomers capped with 2-aminopyridine and appropriate diamine monomers in the presence of a transition metal catalyst. Thus, polyimide oligomers endcapped with 2-aminopyridine were postulated to also be reactive toward amine terminated oligomers through a transimidization route. Amino alkyl (and possibly aryl) terminated oligomers such as polydimethylsiloxane and polyarylene ethers of controlled molecular weight combined with polyimides of a predetermined molecular weight and endcapped with 2-aminopyridine will react to afford perfectly alternating segmented polyimide copolymers. Aspects of the synthesis and characterization of perfectly alternating segmented polyimide siloxane copolymers were investigated and will be discussed in this paper.

2. EXPERIMENTAL

2.1 Synthesis

2.1.1 Oligomer Synthesis: The polyimides reported herein were largely based on oxydiphthalic anhydride (ODPA), and the aromatic diamine bisaniline P (Bis P). The ODPA was provided by Occidental Chemical Company and the Bis P was provided by Air Products and Chemicals, Inc. Both were of high purity and used as received. 2-aminopyridine (2AP) was purchased from Aldrich and recrystallized from a mixture of 75% (by volume) chloroform and 25% petroleum ether. The polymerization solvents were N-methylpyrrolidone (NMP), o-dichlorobenzene (DCB) and chlorobenzene (ØCl). They were distilled over phosphorous pentoxide and stored in sealed flasks under nitrogen until use. The poly(amic acid) preparation was performed in a four-necked flask equipped with a mechanical stirrer, nitrogen inlet and a condenser with drying tube. Thus, the dianhydride was dissolved in NMP with slight heating, then cooled to ambient temperature. A calculated amount of the reactive endcapping agent 2-aminopyridine was added to the reaction mixture and allowed to react with the ODPA while stirring for 15 to 20 minutes. The diamine was then added as a powder and rinsed with NMP to bring the final reaction solids content to ~17-20 percent. The reaction was allowed to proceed for 20 hours to allow for the generation of the required molecular weight and for equilibration to a most probable molecular weight distribution.

The imidization was conducted as previously described.(12) A reverse Dean Stark trap with a condenser filled with DCB was fitted to the flask. An additional amount of DCB was added to the reaction mixture to bring the solvent ratio to 80%NMP/20%DCB. The reaction mixture was heated to 165°C by immersion in a hot silicone oil bath. The reaction was allowed to heat for a total of 24 hours to ensure complete imidization. The solution was cooled to ambient temperatures, and precipitated in methanol in a high speed blender. Upon cooling, the ODPA-Bis P solutions became turbid. The polyimide oligomers were collected and dried in a vacuum oven for 18 hours at 200°C and for 1 hour at 300°C.

The synthesis of amine terminated polysiloxane (PSX) has been reported elsewhere (4). Aminopropyl terminated polydimethylsiloxanes with <Mn>'s of 1070 and 2670 g/mole, as determined by potentiometric titration, were used. The synthesis of high molecular weight, fully cyclized, soluble polyimides endcapped with phthalic anhydride (PA) has also been reported (12).

2.1.2 Copolymer Synthesis: The copolymer synthesis was carried out in a three neck round bottom flask equipped with a mechanical stirrer, nitrogen inlet, thermometer and a condenser with drying tube. The stoichiometry was offset by using an excess of the polyimide oligomer. In every case, the calculated <Mn> was 40,000 g/mole. The polyimide oligomer was dissolved in chlorobenzene and heated to ~125°C. The amine terminated oligomer was added slowly and rinsed with chlorobenzene to bring the final reaction solids content to 15 percent. The reaction was stirred at ~125°C for 30 minutes. The solution was cooled to ambient temperatures and cast into films. The films were heated slowly in a vacuum oven and dried for at least 3 hours at 250°C.

2.2 Characterization

2.2.1 Intrinsic viscosity measurements: Intrinsic viscosity measurements were performed in chloroform at 25°C using a Cannon-Ubbelohde viscometer.

2.2.2 Proton NMR Analysis: All proton spectra were measured on a Varian Unity 400 MHz NMR. Samples were dissolved in deuterated chloroform. TMS was used as an internal reference for the polyimide oligomers and chloroform was used as the reference for the copolymers.

2.2.3 Potentiometric titration: The <Mn>'s of the aminopropyl terminated polydimethylsiloxanes were determined by potentiometric titration with HCl on a MCI Automatic Titrator GT-05.

2.2.4 Thermal Analysis: Glass transition temperatures, Tg's, were obtained by differential scanning calorimetry on a Seiko DSC 210. Scans were run at 10°C per minute and the reported values were obtained from a second heating after quick cooling. Dynamic mechanical thermal analysis (DMTA) was also carried out on the polyimide siloxane copolymer films using a Seiko Model 200 DMS at 1hz. The thermooxidative stability was determined by thermogravimetric analysis in air on a Perkin-Elmer TGA-7 at 10°/minute heating rate.

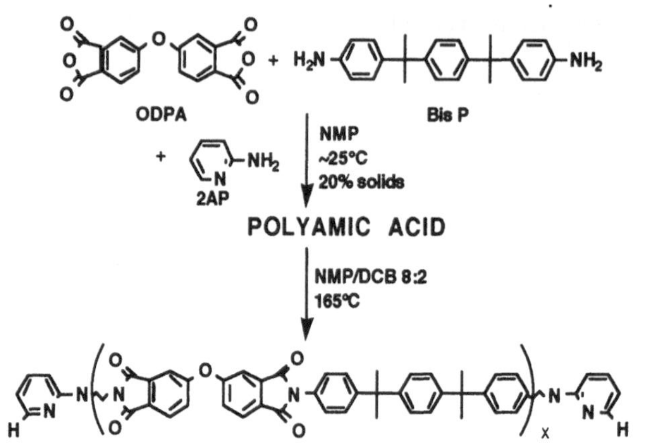

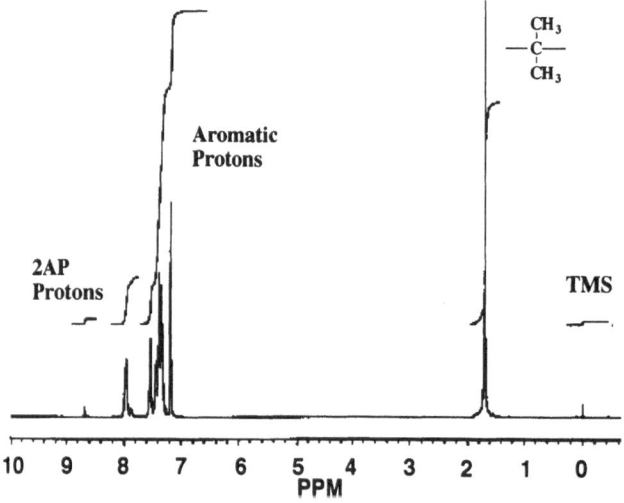

Scheme 2. Polyimide Oligomer Synthesis

3. RESULTS AND DISCUSSION

The synthesis of the polyimides (PI) is illustrated in Scheme 2. The polymerizations were conducted in a single "one pot" reactor, which minimizes solution transfer steps. Number average molecular weights, <Mn>, of the oligomers were evaluated by end group analysis using ^{1}H-NMR from the integral ratio of the proton for 2AP, at 8.7ppm, to the Bis P methyl protons at 1.7 ppm, see Figure 1. Table 1 shows the comparison of the theoretical <Mn> to the <Mn> determined by ^{1}H NMR. The low intensity of the 2AP proton at 8.7 ppm limits the accuracy of the integration, but nevertheless, the <Mn> determined by ^{1}H NMR is in reasonable agreement with the theoretical <Mn>.

The copolymer synthesis was carried out as shown in Scheme 3. Two series of polyimide siloxane copolymers were investigated. In the PI-PSX1070 series, an aminopropyl terminated polydimethylsiloxane oligomer having a molecular weight of 1070 g/mole was reacted with three polyimide oligomers of ODPA-Bis P-2AP with theoretical molecular weights of 4000, 6000, and 8000 g/mole. In the PI-PSX2670 series, an aminopropyl terminated polydimethylsiloxane oligomer with a molecular weight of 2670 g/mole was reacted with the same polyimide oligomers used in the PI-PSX1070 series. In each polymerization, the stoichiometry was offset by using an excess of the polyimide oligomer. The calculated <Mn> was 40,000 g/mole.

Figure 1. ^{1}H NMR of ODPA-Bis P-2AP Polyimide with a theoretical <Mn> = 4000 g/mole

Table 1. ODPA-Bis P-2AP Polyimide Oligomers

<Mn> Theoretical	<Mn> By ^{1}H NMR	[η] dl/g
4000	4900	0.21
6000	7300	0.23
8000	9700	0.33

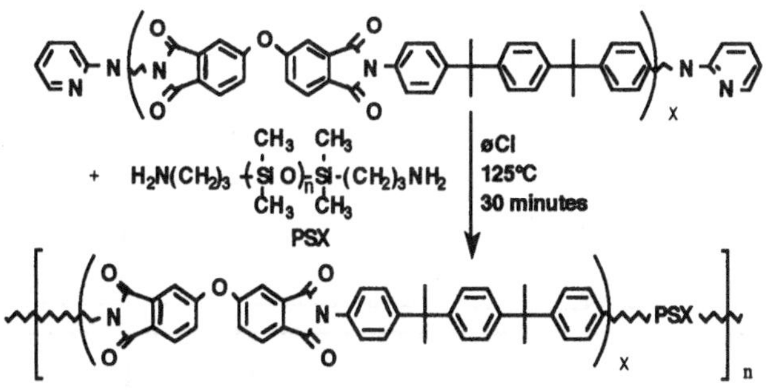

Scheme 3. Perfectly Alternating Segmented Copolymer Synthesis by the Transimidization Route

Intrinsic viscosity data of the PI-PSX1070 and PI-PSX2670 series is given in Table 2. For a relatively short reaction time of 30 minutes, high molecular weight was achieved as evidenced by the increase in intrinsic viscosities of the copolymers over intrinsic viscosities of the polyimide oligomers. Compared with the ODPA-Bis P-PA homopolymer having a theoretical <Mn> of 40,000 g/mole and an intrinsic viscosity of 0.54 dl/g, the polyimide-siloxane copolymers all showed higher intrinsic viscosities.

Figure 2 shows a representative ^{1}H NMR of a polyimide siloxane copolymer with 31 wt% polysiloxane. The theoretical molecular weight of the polyimide oligomer is 4000 g/mole and the polysiloxane had a number average molecular weight of 2670 g/mole. The disappearance of the 2AP proton at 8.7 ppm indicates that the reaction went to near complete conversion.

Table 2. Intrinsic Viscosities

[η] of PI Oligomer* (dl/g)	[η] of PI-PSX1070 Copolymer (dl/g)	[η] of PI-PSX2670 Copolymer (dl/g)
1. 0.21	0.71	0.64
2. 0.23	0.87	0.71
3. 0.33	1.05	0.73

*The theoretical <Mn> of the PI oligomers in 1, 2 and 3 are 4000, 6000 and 8000 g/mole respectively.

DSC data of the two polyimide siloxane series and a high molecular weight ODPA-Bis P-PA homopolymer is presented in Table 3. All of the copolymers exhibited two Tg's indicating phase separation in the copolymers. The Tg's of the polyimide oligomers range from 219°C to 240°C and the high molecular weight polyimide homopolymer shows a Tg of 267°C. The upper Tg's, which result from the polyimide phase, of the polyimide siloxane copolymers range from 201°C to 233°C for the PI-PSX1070 series and from 212 to 235°C for the PI-PSX2670 copolymers. In both series, the Tg of the copolymers is less than the Tg of the corresponding polyimide oligomers indicating that there is some mixing between the polyimide and PSX phases. The depression of the Tg is even more pronounced when the

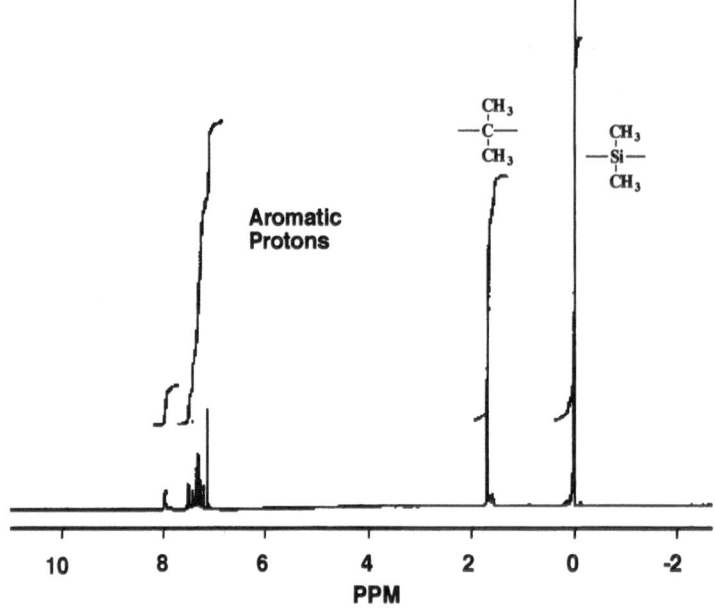

Figure 2. ^{1}H NMR of a ODPA-Bis P Polyimide Siloxane Containing 31 wt% PSX

lower molecular weight polyimide oligomer (4000 g/mole) is used. The lower molecular weight PSX oligomer used in the PI-PSX1070 series results in a greater depression of the upper Tg's than the higher molecular weight PSX used in the PI-PSX2670 series. The depression of the upper Tg is not dependent on the weight % PSX incorporated but depends on the size of the polyimide and PSX oligomers with the lower molecular weight oligomers giving lower upper Tg's and thus increased phase mixing.

Dynamic mechanical analysis at 1Hz from -150°C to 300°C displays trends which are analogous to those found in the DSC results. In Figure 3. the dynamic storage modulus for the three copolymers in the PI-PSX1070 series has a value of ~4 GPa at -150°C and decreases as the temperature is raised. The decrease in the storage modulus is greatest for the copolymer with a 4000 g/mole polyimide and is least for the 8000 g/mole polyimide copolymer indicating that miscibility is dependent on the molecular weight of the oligomers. The sharp decrease in the storage modulus of the copolymers at the glass transition temperature also depends upon the oligomer molecular weight with the Tg increasing with increasing molecular weight of the polyimide oligomers.

Table 3. Thermal Analysis

Theoretical <Mn> of PI Oligomer (g/mole)	Tg of PI oligomer (°C)	wt% PSX Theoretical	Upper Tg of PI-PSX(°C)	5% wt. loss by TGA (°C)
PI-PSX2670 Series				
4000	219	31	212	445
6000	232	22	229	453
8000	240	17	235	468
PI-PSX1070 Series				
4000	219	16	201	472
6000	232	11	227	498
8000	240	8	233	511
ODPA-Bis P-PA				
40,000	267	--	--	544

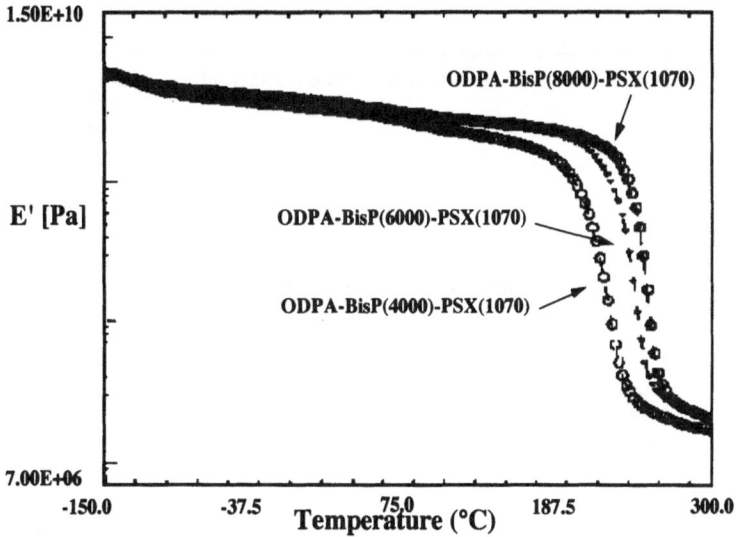

Figure 3. Stiffness-Temperature Curves for the PI-PSX(1070) Series

The two phase nature of the copolymers is shown most clearly in the tan δ curves given in Figure 4. The tan δ versus temperature curves of the PI-PSX1070 copolymers series show relaxations due to both the polyimide and polysiloxane phases. The relaxation due to the polysiloxane occurs between -150°C and -90°C. The peak position does not depend on the polyimide molecular weight and remains constant between 120°C and 125°C. The origin of the peaks at 75°C is unclear and is being further investigated. The relaxations at temperatures above 200°C are due primarily to cooperative segmental motion of the polyimide blocks. The peak temperature increases with increasing molecular weight and the breadth of these dispersions reflect the effect of mixing which, as discussed above, is dependent on the oligomer molecular weight.

Also included in Table 3, is the 5% weight loss measured by TGA. The thermooxidative stability of the copolymers is less than the polyimide homopolymer due to the presence of the PSX oligomers. The 5% weight loss is dependent on the wt% PSX incorporated since the thermooxidative stability of the copolymers decreases as greater weights of the PSX are incorporated into the copolymer. Two thermograms from the PI-PSX2670 series are shown in Figure 5. The copolymer with the higher weight percent PSX begins to degrade first but it also gives a greater char yield.

4. CONCLUSIONS

Controlled molecular weight polyimides oligomers endcapped with 2-aminopyridine were successfully synthesized by a convenient "one pot" method. Utilizing transimidization chemistry, polyimides endcapped with 2-aminopyridine were reacted with aliphatic aminopropyl terminated polydimethylsiloxane oligomers to afford perfectly alternating segmented copolymers of high molecular weight. These copolymers could be transformed into tough, transparent films. The copolymers exhibit two Tg's due to the microphase separation of the polyimide and polysiloxane segments. However, phase mixing was evident from the DSC and DMA results with the degree of mixing dependant on the polyimide and polysiloxane block lengths.

5. ACKNOWLEDGEMENTS

We appreciate the support of this research by the NSF Science and Technology Center for High Performance Polymeric Adhesives and Composites at Virginia Tech under contract DMR8809714. An IBM fellowship provided to Martin E. Rogers is also gratefully acknowledged.

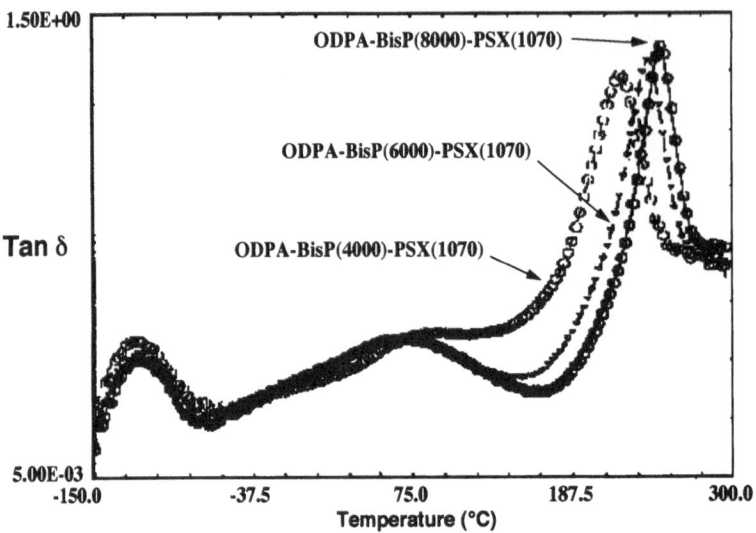

Figure 4. Tan δ - Temperature Curves for the PI-PSX(1070) Series

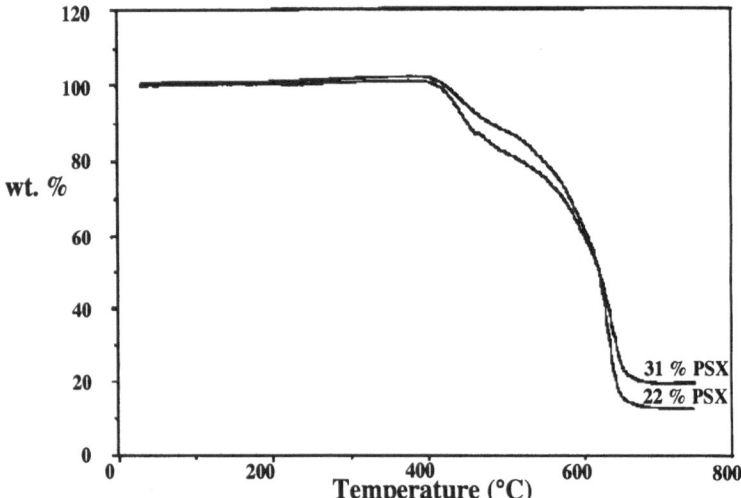

Figure 5. Thermogram of Polyimide Siloxane Copolymers

6. REFERENCES

1. K. L. Mittal, Editor, "Polyimides", Volumes 1 and 2, Plenum Press (1984); C. Feger, M. M. Khojasteh and J. E. McGrath, "Polyimides: Chemistry Materials, and Characterization", Elsevier (1989).
2. M. Bowden and S. R. Turner, Editors, "Polymers for High Technology, Electronics and Photonics", ACS Symp. Series 46 (1987).
3. D. Wilson, P. Hergenrother, and H. Stenzenberger, Editors, "Polyimides", Chapman and Hall (1990).
4. C. A. Arnold, J. D. Summers, Y. P. Chen and J. E. McGrath, Polymer (London), 30(6), 986 (1989); J. D. Summers, Ph.D. Thesis, VPI and SU ,1987; C. A. Arnold, Ph.D. Thesis VPI and SU (1989).
5. D. L. Wilkens, C. A. Arnold, M. J. Jurek, M. E. Rogers and J. E. McGrath, Journal of Thermoplastic Composite Materials, 3, 4 (1990).
6. C. A. Arnold, Y. P. Chen, D. H. Chen, M. E. Rogers and J. E. McGrath, Mat. Res. Soc. Symp. Proc., Vol. 154, 149 (1989).
7. T. Yoon, C. A. Arnold and J. E. McGrath, J. Adhesion, (accepted 1991).
8. 5th International SAMPE Electronics Conference, vol. 5 (1991).
9. A. Noshay and J. E. McGrath, "Block Copolymers Overview and Critical Survey", Academic Press (1977).
10. G. York, Ph. D Thesis, VPI and SU, 1991; publication pending.
11. T. Takekoshi, J. L. Webb, P. P. Anderson and C. E. Olsen, IUPAC 32nd International Symposium on Macromolecules, 464 (1988); T. Takekoshi, in "Polyimides", D. Wilson, P. M. Hergenrother and H. Stenzenberger Editors, Chapman and Hall (1990).
12. R. O. Waldbauer, M. E. Rogers, C. A. Arnold, G. A. York, Y. Kim and J. E. McGrath, Polymer Preprints, 31(2), 430 (1990); Makromol. Chem. Series, accepted (1992).

HIGH MOLECULAR WEIGHT AROMATIC BIPHENYLENE POLYMERS

BY NICKEL COUPLING OF ARYL DICHLORIDES

G. T. Kwiatkowski and I. Colon

Amoco Performance Products, Inc., Alpharetta, Georgia 30202
and Union Carbide Corporation, Bound Brook, New Jersey 08805

Introduction

A number of methods are available for the preparation of aromatic polymers. The majority of these methods involve the formation of a chemical bond between a carbon atom and a heteroatom. Typical of these are, for example, aromatic polycarbonates; the latter are prepared by the reaction of a dihydric phenol with phosgene or derivative thereof.[1-7] The polymerization proceeds via formation of a carbon–oxygen bond. A similar situation is encountered with the class of polyarylates -- the polyesters from dihydric phenols and aromatic diacids.[8-11]

The oxidative polymerization of substituted phenols to poly(phenylene oxide)s is another example where a polymeric chain is formed by carbon–heteroatom coupling.[12-17] Thus, the reaction of 2,6–dimethylphenol with oxygen, in the presence of a copper–amine complex, yields high molecular weight poly(2,6–dimethyl–1,4–phenylene oxide).

The nucleophilic substitution reaction of an activated benzenoid halide with a phenoxide anion is currently the method used worldwide for the preparation of aromatic poly(aryl ether sulfones)[18-22] and poly(aryl ether ketones).[18,23,24] Amoco Performance Products' UDEL Polysulfone and RADEL Polyphenylsulfone are made commercially via this route. The subject substitution reactions are also used by ICI to produce their Victrex Poly(ether sulfone) (PES) and Victrex Poly(ether ether ketone) (PEEK).

In another approach, poly(aryl ether sulfones) were synthesized by the electrophilic Friedel–Crafts reactions of sulfonyl halides with aromatic hydrocarbons.[22] The critical step in these polymerizations is the formation of the carbon–sulfur bond. High polymers were obtained, though they were not always completely linear. Carbonyl aryl carbon–carbon bonds are created in Friedel Craft reactions leading to poly(aryl ketones).

Polymerizations which involve the formation of aromatic carbon–carbon bonds are of high interest. One would expect such polymerizations to afford novel and unique polymer structures that cannot be prepared via the known methods. Moreover, alternative and more economical routes to existing polymers that are now based on unavailable and/or expensive starting materials may be at hand.

There is a significant number of references[23-36] describing efforts to make high polymers via aromatic carbon–carbon bond formation. These efforts were unsuccessful. In one case, however, high polymer was prepared by an elegant oxidative coupling (the Scholl reaction). High molecular weight polymer was obtained upon treating 4,4'–di(1–naphthoxy)diphenyl sulfone and 4,4'–di(1–naphthoxy)benzophenone with ferric chloride. The method requires equimolar amounts of the Lewis acid and is limited to systems that are capable of undergoing the Scholl reaction.

The nickel coupling of aryl chlorides was studied in–depth in our laboratories. It was demonstrated[38] that the corresponding biaryls can be produced in quantitative yields. Hence, it was felt that the method should be appropriate for the preparation of high polymers. Our studies, as herein reported, do indeed show that materials of excellent quality can be prepared via this novel route. Also, proper choice of the starting monomers allows for the synthesis of an almost infinite variety of polymers.[39] In general, these polymers contain biphenyl moieties which contribute excellent thermal, mechanical, impact and alloying characteristics.

Experimental

The polymerization technique and the characterization of the obtained polymers are described in a recently published article.[40]

Results and Discussion

(a)   Biphenyl and Terphenyl–Based Polymers

The nickel coupling route which results in the formation of an aromatic carbon–carbon bond, leads to polymers which contain, in the simplest case, a biphenyl group. This is illustrated in an idealized way in equation (I):

The formation of polymers derived from higher polyphenyl homologs is easily envisioned, equation (II):

Clearly, the method is uniquely suited for the preparation of high polymers based on polyphenyl–containing monomers. Note, that these polyphenyl monomers are expensive: their preparations are often tedious multi–step procedures. Inspection of equations (I) and (II), on the other hand, indicates that polymers (3), (4) and (6) should require comparatively inexpensive starting materials.

Biphenyl– and terphenyl–based polymers generally display outstanding combinations of properties. Biphenol–based polymers have high T_g's, may be crystalline and have good environmental stress–crack resistance; they are very tough and show improved stability to acids, bases and oxidizing agents.

Thus, comparison of polymers (7) and (8)

$$\left[O\!-\!\!\bigcirc\!\!-\!\!\underset{CH_3}{\overset{CH_3}{C}}\!\!-\!\!\bigcirc\!\!-\!O\!-\!\!\bigcirc\!\!-\!SO_2\!-\!\!\bigcirc \right]$$

(7)

$$\left[O\!-\!\!\bigcirc\!\!-\!\!\bigcirc\!\!-\!O\!-\!\!\bigcirc\!\!-\!SO_2\!-\!\!\bigcirc \right]$$

(8)

shows that replacement of the bisphenol–A moiety in (7) by the biphenol moiety leads to a polymer with a number of improved features. The introduction of the biphenyl group results in excellent mechanical properties (see next section) as well as in

o   a significantly higher glass transition temperature, i.e. 215–220°C for (8) versus 190°C for (7);

o   a significantly higher notched Izod impact strength, i.e. 12–14 ft. lbs./in. for (8) versus about 1.3 ft. lbs./in. for (7); the high impact strength is maintained after high temperature cyclic exposures;

o   a significantly better chemical and environmental stress–crack resistance; and

o   improved alloying characteristics. Poly(aryl ether sulfones) having the biphenyl group were found to be miscible with selected polyimides and poly(amide imides).[41] Moreover, good mechanical compatibility was observed upon alloying the biphenyl–containing poly(aryl ether sulfones) with other poly(aryl ether sulfones), as well as with poly(aryl ether ketones), although the blends had only limited solubility.[42]

Biphenol–based poly(aryl ether sulfones) are also uniquely suitable for medical applications; they combine toughness and transparency with the ability to withstand numerous sterilization cycles with steam containing morpholine, at pressures of 500 psi or higher.

The poly(aryl ether) (9),

$$\left[O\!-\!\!\bigcirc\!\!-\!\!\underset{CH_3}{\overset{CH_3}{C}}\!\!-\!\!\bigcirc\!\!-\!O\!-\!\!\bigcirc\!\!-\!\!\bigcirc \right]$$

(9)

prepared via the Ullman condensation of bisphenol–A with 4,4'–dibromodiphenyl[43] illustrates the toughness achievable with this class of materials. Polymer (9) [RV in chloroform = 1.11 dl/g] had a T_g of 175°C, a tensile modulus of 240,000 psi, a tensile strength of 11,000 psi, an elongation of 150% and a pendulum impact strength of greater than 700 ft. lbs./in.[3]

Poly(aryl ether sulfones) derived from the dihalide (10) were made by condensing (10) with various diphenols.[44,45,46] They displayed good mechanical properties and high glass transition temperatures; and were particularly useful in composite applications.[47]

X—〈◯〉—SO₂—〈◯〉—〈◯〉—SO₂—〈◯〉—X

(10)

The use of monomers having additional p–phenylene rings, i.e. terphenyl or quaterphenyl derived, leads to polymers with increased rigidity and increased degree of crystallinity. Thus, terphenyl diol (11) was condensed with a number of dihalobenzenoid compounds.[48]

HO—〈◯〉—〈◯〉—〈◯〉—OH

(11)

The condensation product with 4,4'–dichlorodiphenyl sulfone had a glass transition temperature of 251°C and a melting point of 389°C. One would expect it to have excellent toughness, comparable to that of polymer (8). Another polymer derived from (11), i.e. (12) which was made via the Ullman reaction, had a T_g of 235°C, a tensile strength of 11,500 psi, an elongation of 140% and a pendulum impact of greater than 590 ft. lbs./in.[3]

$$\left[O—〈◯〉—〈◯〉—〈◯〉—O—〈◯〉—\underset{CH_3}{\overset{CH_3}{C}}—〈◯〉 \right]$$

(12)

Crystalline high melting (380–400°C) and high T_g (260–310°C) poly(aryl ether sulfones) were obtained by the condensation of (13) with hydroquinone, 4,4'–biphenol and 4,4"–bis (4–hydroxyphenylsulfonyl)–p–terphenyl.[48,49] Interestingly, however, the condensation of (13) with

Cl—〈◯〉—SO₂—〈◯〉—〈◯〉—〈◯〉—SO₂—〈◯〉—Cl

(13)

4,4'–dihydroxybenzophenone or with 4,4'–dihydroxydiphenyl sulfone yielded amorphous products with T_g's of about 280°C.

The foregoing brief discussion clearly shows that polymers containing the biphenyl or the terphenyl unit generally present very attractive combinations of properties. These polymers range from amorphous and very tough to the highly crystalline, totally insoluble materials. As indicated earlier, the preparations of the monomers required to make these polymers frequently necessitate multistep tedious procedures with the result that the starting materials are often prohibitively expensive.

The nickel coupling approach has the advantage [equations (I) and (II)] of using comparatively inexpensive starting materials; most of the structures discussed above can be prepared using this approach. The nickel coupling route is definitely one of the preferred routes for the synthesis of biphenyl– or terphenyl–based condensation polymers; as long, of course, as these polymers are soluble in aprotic solvents which are the preferred solvents for the coupling reaction (vide infra).

(b)   Preparation of High Molecular Weight Poly(Biphenylene Ether Sulfone)

Our initial investigation was centered around the preparation of high molecular weight poly(biphenylene ether sulfone) (8); this material can be prepared by the nucleophilic polycondensation shown in equation (III):

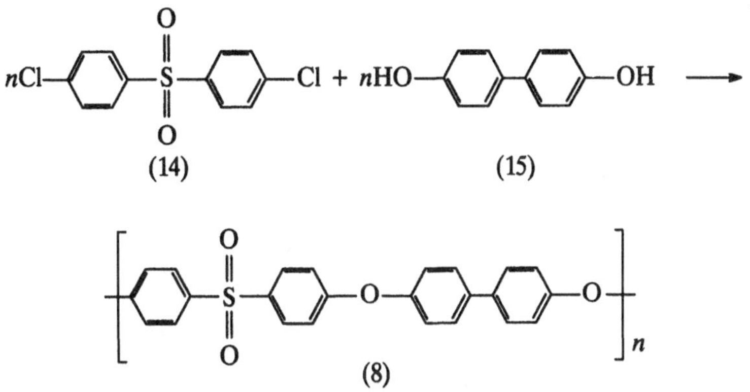

As discussed in the previous section, polymers containing biphenyl moieties display a unique property profile, in particular, high toughness and thermal–oxidative stability. The required biphenyl monomers, however, are either expensive or not available commercially. Accordingly, the following alternative route to polymer (8), involving a nickel coupling step, was investigated [equation (IV)]:

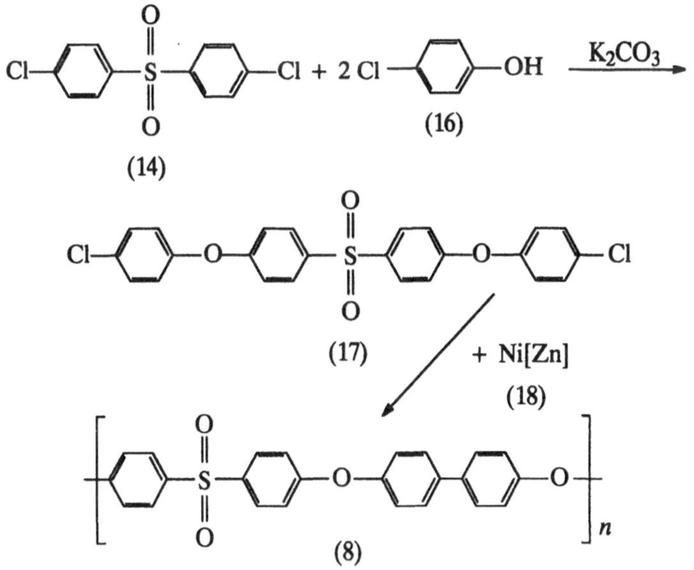

The reactions depicted in equation (IV) were studied in detail. The first reaction, i.e. the condensation of p-chlorophenol (16) with 4,4'-dichlorodiphenyl sulfone (14) is a classical nucleophilic aromatic substitution; it proceeds smoothly in polar aprotic solvents (e.g., DMAC) and yields the dichloro intermediate (17) in quantitative yield. A detailed study was necessary, however, to determine the critical parameters which control the formation of high polymer via the second reaction i.e. via the nickel coupling of (17) to (8).

As indicated in a previous paper[38] near quantitative yields of biphenyl derivatives can be obtained from select chloroaromatic compounds using zero valent nickel–triphenylphosphine complex in the presence of zinc metal. The conditions employ a catalytic amount of nickel–triphenylphosphine complex in the presence of zinc metal in dry, dipolar aprotic solvents at 60 to 80°C. The reaction is complete in a few minutes.

$$2R\!-\!\!\langle\bigcirc\rangle\!\!-\!Cl + Zn \xrightarrow[\text{DMAc, N}_2,\ 80°C]{\text{NiCl}_2,\ \text{PPh}_3} R\!-\!\!\langle\bigcirc\rangle\!\!-\!\!\langle\bigcirc\rangle\!\!-\!R + ZnCl_2 \quad (V)$$

$$(19) \qquad\qquad (18) \qquad\qquad\qquad\qquad (20) \qquad\qquad (21)$$

$$>98\%$$

A number of nickel salts can be used to generate the catalyst. Nickel chloride and bromide were the most active. Zinc, magnesium and manganese in combination with the nickel catalyst gave high yields of coupled products from chloroaromatic derivatives. Triarylphosphines were the best ligands.

Acidic substituents such as phenolic hydroxyl and carboxylic acids cause reduction reactions to the corresponding arene. Water and other protic sources must also be avoided. Aryl chlorides with electron–withdrawing or weakly electron–donating substituents give high yields of biaryl products. Monomers with strong electron–donating substituents such as methoxyl, produce some reduction products. This side reaction can be suppressed by using 2,2'-bipyridine as the ligand.

Initial attempts at polymerizing monomer (17) resulted in very low molecular weight product, despite the fact that great care was taken in purifying and drying the monomer and the N,N–dimethylacetamide solvent used in the coupling reaction. End–group analysis of the resulting polymer revealed that there was virtually no chlorine in the final polymer, and that hydrogen or nickel were the only end–groups. Based on the results of the water analysis in the initial polymerization mixture, there was not enough water present to account for the amount of apparent reduction taking place. Previous mechanistic studies[38] indicated that the level of nickel catalyst would have to be kept low in order to produce high polymer. Yet, in spite of that, even low catalyst levels did not result in high polymer. The complexity of the problem prompted us to use statistical experimental design; thus a sixteen experiment, two–level fractional factorial design was used to study the effect of nine variables on polymer molecular weight. The nine variables examined were: the amount of nickel chloride; the amount of excess zinc; the amount of solvent; the reaction temperature; the amount of triphenylphosphine (TPP); the amount of bipyridine (a ligand found to be useful in the coupling of aryl chlorides); the reaction time; the methods used to dry the solvent; and length of time between active catalyst (Ni°) generation and reaction with monomer. Interestingly enough, of the nine variables, only three had a truly significant effect (>95% confidence level) on the molecular weight, as measured by the reduced viscosity (RV) of the polymer (Table I). The most significant variable was the amount of triphenylphosphine (estimated effect = 0.232 on the reduced viscosity of the polymer in going from low to high TPP levels). The result was surprising, since studies with aryl chloride couplings had not shown a great triphenylphosphine dependence. The apparent discrepancy may be due to the fact that in monomer chemistry a yield of 95% is considered excellent; while yields must be 99% or greater, if high polymer is desired. The variable with the next largest effect was the reaction temperature; its effect was negative. This was consistent with previous studies that showed that the amount of side reactions increased with temperature. As had been anticipated, there was also a negative effect due

TABLE I

Significant Variables in Nickel Coupling to Produce Poly(aryl ether sulfones)
as Determined by Experimental Design

Variable	Effect on RV	Confidence Level[a]
NiCl$_2$	−0.084	>95%
Zn	0.061	>90%
Reaction temp.	−0.103	>99%
TPP	0.232	¯100%

[a]Based on the precision of replicated runs.

to the level of nickel. Although not as statistically significant as the other three, there also appeared to be a real positive effect due to the amount of zinc. Using the above information, it was a simple task to define conditions which resulted in high polymer (RV >0.5 dl/g).

Comparison of properties of polymer (8) made via the conventional nucleophilic displacement route and by nickel coupling is presented in Table II. The only difference observed was in the impact resistance and this was probably due to high levels of zinc chloride contaminating the polymer made by the coupling process. Nonetheless, the impact strength of both polymers was outstanding. Simply slurrying the polymer in hot water allowed to decrease the amount of zinc chloride in it, and to increase its impact toughness.

TABLE II

Properties of Poly(aryl ether sulfone) (8) Produced by
Conventional Process and Nickel Coupling

Property	Conventional	Nickel Coupling
Reduced Viscosity (dl/g)	0.57	0.81
T$_g$ (°C)	215	215
1% Secant Modulus (psi)*	217,000	238,000
Tensile Strength (psi)	9,730	9,650
Yield Elongation (%)	7.0	7.0
Pendulum Impact (ft. lb/in^3)	252	166

*Moduli were determined using crosshead travel to determine strain.

Having successfully prepared high polymer via the nickel coupling route, we turned our attention toward developing an integrated two–step process [see equations (IV)] where the isolation and purification of the intermediate (17) [4,4'–di(p–chlorophenoxy)diphenyl sulfone] would not be required. The monomer hitherto used was prepared by the reaction of 4,4'–dichlorodiphenyl sulfone (14) with two equivalents of p–chlorophenol (16) in N,N–dimethylacetamide/toluene with potassium carbonate as the base. The monomer was isolated by coagulation in methanol/water, purified by triple recrystallization from isopropanol, and dried in a vacuum oven prior to polymerization. In principle, an integrated process involving monomer formation with azeotropic removal of the water of reaction, filtration to remove the potassium salts (excess K_2CO_3 and KCl), and reaction with Zn and Ni° to form polymer, appeared possible. The economic advantage of such an integrated process is readily apparent.

Pertinent experimentation has quickly determined that the most important factor for obtaining high molecular weight polymer is the removal of water, formed during the preparation of the intermediate (17). Keeping this important variable in mind it was possible to prepare polymers, without isolation of the intermediate dichlorodiphenoxy derivative, having reduced viscosities as high as 1.17 dl/g.

(c)   Scope

Further studies have shown that the nickel coupling polymerization is very general and allows for the preparation of a practically infinite number of polymers. This was first demonstrated by copolymerizing the dichloro diphenoxy derivative (17) with a series of comonomers [equation (VI)]:

$$xCl-\underset{(17)}{\text{⬡}-O-\text{⬡}-\overset{\overset{O}{\|}}{\underset{\underset{O}{\|}}{S}}-\text{⬡}-O-\text{⬡}-Cl} + yCl-Ar-Cl + Zn \xrightarrow{[Ni]} \text{Copolymer} \qquad (VI)$$

(17)   + yCl-Ar-Cl + Zn   → Copolymer
              (22)   (18)

A number of high molecular weight polymers were obtained. They are listed in Table III.

Table III

Properties of High Polymers Produced by the Copolymerization
of 4,4'-Bis(p-Chlorophenoxy)diphenylsulfone and Aryl Dichlorides

No.	Aryl Dichloride	Wt %	RV (dl/g)	T_g (°C)	1% Secant Modulus (psi)**	Tensile Strength (psi)	Pendulum impact (ft-lbs/in^3)
1	None		0.81	215	238,000	9,650	166
2	4,4'-Dichlorodiphenylsulfone	15	0.69	225	232,000	9,870	138
3	4,4'-Dichlorodiphenylsulfone	20	0.55	230	255,000	10,200	120
4	p-Dichlorobenzene*	10	0.56	218	234,000	9,840	179
5	4,4'-Dichlorobenzophenone	10	0.58	210	227,000	9,620	147
6	Ethyl bis(p-chlorobenzoate)	10	0.54	200	243,000	10,300	147
7	2,5-dichlorothiophene	10	0.40	---	---	---	---

*The use of 15 weight percent (36.2 mole %) of p-Dichlorobenzene resulted in a partially soluble polymer in NMP.

**Moduli were determined using crosshead travel to determine strain.

It is of interest to consider the structure of some of the copolymers listed in Table III. Thus, the copolymerizations with p–dichlorobenzene are expected to yield a polymer (No. 4) containing terphenyl groups along with smaller amounts of higher polyphenyl groups. Paradichlorobenzene (an aryl chloride having an electron withdrawing group) is more reactive toward zero–valent nickel than the diphenoxy derivative (17) [see preceding section (b)]. However, due to the limited solubility of the copolymer, amounts of p–dichlorobenzene comonomer were limited to about 10 weight percent (26.3 mole %). The condensation

$$\sim O\text{—}\langle\text{—}\rangle\text{—}Cl \;+\; Cl\text{—}\langle\text{—}\rangle\text{—}Cl \;+\; Cl\text{—}\langle\text{—}\rangle\text{—}O\sim$$

(23)　　　　　(24)　　　　　(23)

will lead to terphenyl groups; formation of

$$Cl\text{—}\langle\text{—}\rangle\text{—}\langle\text{—}\rangle\text{—}Cl$$

(25)

however, is also expected, at least in the initial stages of the reaction. Cross–coupling of (25) with (23) will give rise to quaterphenyl units. It is conceivable that higher polyphenyl groups are also formed, albeit probably in small amounts.

The structure of the copolymer from the dichlorodiphenoxy monomer (17) and 4,4'–dichlorodiphenyl sulfone is intriguing. Again, the comonomer (4,4'–dichlorodiphenyl sulfone) is an aryl chloride having an electron withdrawing group. Hence, formation of biphenyl units (26), flanked on both sides by an SO_2 group, is expected to predominate during the initial stages of the reaction. As the reaction proceeds, formation of units (27) and finally of units (28) will take place:

(26)

(27)

(28)

Thus, the reaction leads to a polymer containing three types of biphenyl units. Comparison of the glass transition temperatures (T_g's) of polymers 1, 2 and 3 (Table III) shows that as the amount of 4,4'–dichlorodiphenyl sulfone comonomer increases, so does also the T_g of the final polymer. It is due to the contribution of units (26) and (27); higher weight proportions of these units ($ArSO_2$– versus ArO–) lead to polymers with higher T_g's.

Note, that the experiments were performed by simultaneously charging all of the monomers into the reaction mixture. As pointed out previously, both p-dichlorobenzene and 4,4'-dichlorodiphenyl sulfone have electron withdrawing substituents and react faster with Ni° than monomer (17). At increasing monomer contents progressively longer sequences of the aryl dichloride homopolymer are, therefore, expected to form under our conditions. Highly crystalline polyphenyl segments, formed when p-dichlorobenzene was used, led to premature polymer precipitation at comonomer contents greater than about 20 weight percent (44 mole %). Even at 15 weight percent of p-dichlorobenzene (36.2 mole %), the copolymer was only partially soluble in NMP. Essentially, the more crystalline the homopolymer, the less of the corresponding monomer could be tolerated under our reaction conditions. This is also the reason why 4,4'-dichlorodiphenyl sulfone, even though faster reacting than p-dichlorobenzene, could be used in higher amounts (Table III). The above limitation could be overcome by gradually adding the aryl dichloride over the course of the reaction so as to minimize the amount of homopolymerization of the more crystalline monomer. Gradual addition of the comonomer would improve the reaction efficiency and yield a more random type of copolymer.

In another approach to demonstrate the breadth of the scope of the nickel-catalyzed polymerizations, the monomer used was prepared as in equation (IV) except that the mole ratio of p-chlorophenol (16) to 4,4'-dichlorodiphenyl sulfone (14) was less than 2. A particularly interesting case is depicted in equation (VII) where the mole ratio of the two monomers is 1:1.

Monomer (29) has two chlorine atoms that differ considerably in their reactivity toward zero-valent nickel. The chlorine in the para position to the sulfone group should react first; it is, therefore, expected that the final polymer will have structure (30). Polymer (30) can also be synthesized via the nucleophilic route by polycondensing 4,4'-bis(p-chlorophenylsulfonyl) biphenyl (31) with 4,4'-biphenol (32) [equation (VIII)]:

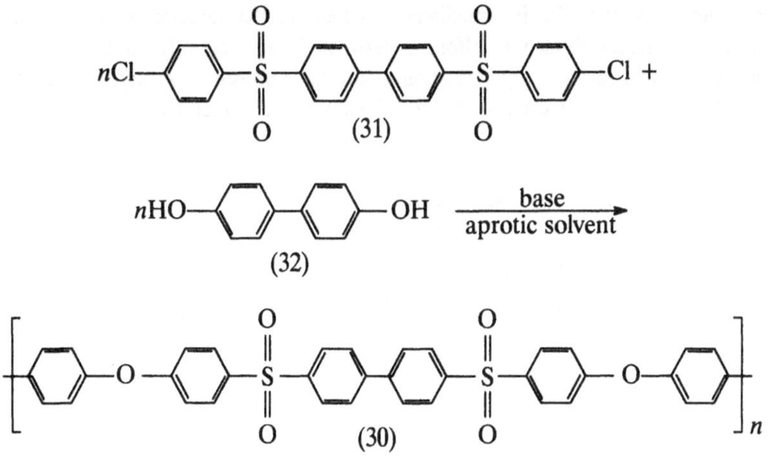

$$n\text{Cl}-\text{C}_6\text{H}_4-\overset{\overset{\displaystyle O}{\|}}{\underset{\underset{\displaystyle O}{\|}}{S}}-\text{C}_6\text{H}_4-\text{C}_6\text{H}_4-\overset{\overset{\displaystyle O}{\|}}{\underset{\underset{\displaystyle O}{\|}}{S}}-\text{C}_6\text{H}_4-\text{Cl} \ +$$

(31)

$$n\text{HO}-\text{C}_6\text{H}_4-\text{C}_6\text{H}_4-\text{OH} \xrightarrow[\text{aprotic solvent}]{\text{base}}$$

(32)

(VIII)

(30)

The data in Table IV show that the properties of the two materials, i.e. (30) made via the nickel coupling route [equation (VII)]; and (30) made via the nucleophilic route [equation (VIII)] are practically identical.

The T_g of polymer (30) (p-chlorophenol/4,4'-dichlorodiphenyl sulfone mole ratio of 1:1) is 265°C; the T_g of polymer (8) (p-chlorophenol/4,4'-dichlorodiphenyl sulfone mole ratio of 2:1) is 215°C. Accordingly, any T_g between 215 and 265°C and greater should be achievable by simply varying the p-chlorophenol/4,4'-dichlorodiphenyl sulfone mole ratio within the range of from about 2:1 to about 1:1 and above. Such is indeed the case as further illustrated by the data of Table IV.

Table IV

Properties of Polymers Obtained by Varying the PCP:DCDPS Ratio
in the Monomer Reaction

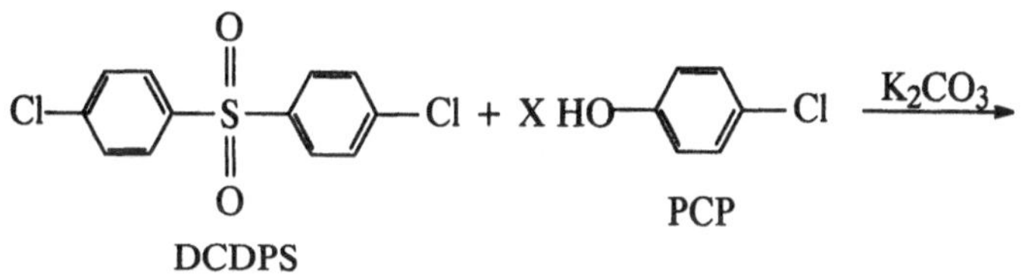

Property	Polymer, X = 2	Polymer, X = 1	Polymer, X = 1.5	Control[a]
Reduced viscosity	0.81	0.69	0.60	0.78
T_g (°C)	215	265	235	265
1% Secant modulus (psi)[b]	238,000	229,000	248,000	240,000
Tensile strength (psi)	9,650	10,800	11,000	10,500
Yield elongation (%)	7.0	8.0	7.7	8.0
Pendulum impact (ft-lbs/in³)	166	176	126	150

[a]Polymer made via the nucleophilic route, equation (VIII).
[b]Moduli were determined using crosshead travel to determine strain.

It is noteworthy that the intermediates obtained in the reaction where the ratio of p-chlorophenol to 4,4'-dichlorodiphenyl sulfone is between 2 and 1 are oily product mixtures. Their isolation, purification and further polymerization would have presented serious problems had it not been for the possibility of performing the reactions via the integrated process (see preceding section).

The integrated process was also used to prepare another group of polymers, equation (IX):

Representative examples are summarized in Table V.

The broad scope of the reaction was further demonstrated by a series of exploratory scanning experiments using a variety of monomers. The results are listed in Table VI.

Table V

Ar	RV(dl/g).
$-C_6H_4-C(CH_3)_2-C_6H_4-$	0.43
$-C_6H_4-$	0.31
$-C_6H_4-SO_2-C_6H_4-$	0.38

TABLE VI

Exploratory Scanning Experiments

Monomer	Results
Br—⟨C₆H₄⟩—O—⟨C₆H₄⟩—Br	Insoluble powder, m.p. >260°C.
Cl—⟨C₆H₄⟩—S(=O)₂—⟨C₆H₄⟩—Cl	Insoluble in NMP. RV (p–chlorophenol) = 0.18 dl/g.
Cl—⟨C₆H₄⟩—S(=O)₂—⟨C₆H₄⟩—Cl + Br—⟨C₆H₄⟩—O—⟨C₆H₄⟩—Br (47:53 by wt.)	Insoluble powder
Cl—⟨C₆H₄⟩—C(=O)—⟨C₆H₄⟩—Cl	Insoluble powder
Cl—⟨C₆H₄⟩—C(=O)OCH₂CH₂OC(=O)—⟨C₆H₄⟩—Cl	RV = 0.1 dl/g.
Cl—⟨C₆H₄⟩—C(=O)—O—⟨C₆H₄⟩—C(CH₃)₂—⟨C₆H₄⟩—O—C(=O)—⟨C₆H₄⟩—Cl	RV = 0.18 dl/g.
Cl—⟨thiophene⟩—Cl	Dark red, insoluble powder; readily absorbed iodine to give a black material which could be compacted to shiny, highly conducting pellets.

It is readily apparent that one can produce an almost infinite variety of polymers using the nickel coupling route. The possibility of using inexpensive monomers, such as p–dichlorobenzene, is particularly worth noting.[50] It should be remembered, however, that the reactions are performed in homogeneous solutions, generally in polar aprotic solvents (e.g., DMAC), preferably at temperatures of about 70°C. Hence, in order to obtain high molecular weight material, the monomers, and more particularly the obtained polymers, must be soluble in the polymerization medium under the above rather mild conditions. Attempts at applying the nickel coupling polymerization to the preparation of high molecular weight crystalline materials (Table VI) failed because of lack of polymer solubility. Attempts by others[51] to prepare via this route the highly crystalline, insoluble poly(aryl ether ketones) also resulted in low molecular weight products. On the other hand, cyclic monomers which yielded high molecular weight poly(aryl ether ketones) upon ring–opening polymerization, were successfully prepared using the nickel coupling route.[52]

(d)   Important Reaction Parameters

Several factors with regard to this polymerization are worthy of note and are discussed below.

1.   **Water removal.** This is without question the most important aspect of the entire reaction sequence. It is crucial because in the presence of Ni(O) water reduces aryl halides to the arene, acts as a polymer chain terminator and can deactivate the catalyst.[53] This is the reason that the post reaction azeotrope step has such a large effect on the reduced viscosity (RV) of the polymer produced via the integrated process.

2.   **Inert atmosphere.** While relatively small amounts of oxygen can be tolerated, especially in the presence of large quantities of TPP, oxygen is a powerful deactivator of the catalyst. Even though the reaction medium is highly reducing in nature, it has been found that nickel oxide is not reduced under these conditions. For this reason it is important that an inert atmosphere as devoid of oxygen as possible be maintained during the nickel coupling reaction.

3.   **Zinc purity.** This is another critical factor for the successful production of high polymer. Not only should the zinc have a low level of oxide, but it should have a high surface area. Most cases for failure to produce high polymer are attributable to poor quality zinc. The catalyst employed in this reaction has the advantage of having a built–in indicator (color). The proper color for the reaction mixture is red–brown to brown. Although the reaction mixture will turn greenish on the addition of the monomer, it should begin to display streaks of its characteristic color within a short while. A persistent greenish color (especially a deep green) is an indication that nickel is not being reduced quickly in the reaction medium, while a grayish color indicates total catalyst deactivation.

(e)   Mechanistic Considerations

Previous studies in our laboratories[38] have established the following key steps for the nickel coupling of aryl chlorides (L represents a ligand, e.g., triphenylphosphine):

70

$$\text{Ni}^\circ\text{L}_3 + \text{ArCl} \xrightarrow{\ -\text{L}\ } \text{ArNi}^{\text{II}}\text{ClL}_2 \qquad\qquad (1)$$

$$\text{ArNi}^{\text{II}}\text{ClL}_2 + 1/2\ \text{Zn} \xrightarrow{\ \text{L}\ } \text{ArNi}^{\text{I}}\text{L}_3 + 1/2\ \text{ZnCl}_2 \qquad\qquad (2)$$

$$\text{ArNi}^{\text{I}}\text{L}_3 + \text{ArCl} \xrightarrow{\ -\text{L}\ } \text{Ar}_2\text{Ni}^{\text{III}}\text{ClL}_2 \qquad\qquad (3) \qquad\qquad (\text{X})$$

$$\text{Ar}_2\text{Ni}^{\text{III}}\text{ClL}_2 \xrightarrow{\ \text{L}\ } \text{Ar-Ar} + \text{Ni}^{\text{I}}\text{ClL}_3 \qquad\qquad (4)$$

$$\text{Ni}^{\text{I}}\text{ClL}_3 + 1/2\ \text{Zn} \longrightarrow \text{Ni}^\circ\text{L}_3 + 1/2\ \text{ZnCl}_2 \qquad\qquad (5)$$

Step (1), the oxidative addition of ArCl to zero–valent nickel, is very fast and yields a Ni^{II}– terminated species. This latter species undergoes reduction [step (2)] which is rate–determining during the early stages of the reaction, when the aryl chloride concentration is high. Toward the end of the reaction, however, step (3) becomes rate–determining. Step (4), which results in the formation of the biaryl, is very fast and yields $\text{Ni}^{\text{I}}\text{ClL}_3$; the latter is then reduced [step (5)] to regenerate the Ni°.

In the absence of aryl chloride, the Ni^{I} species of step (2) undergoes a very slow process of biaryl formation as shown:

$$2\text{ArNi}^{\text{I}}\text{L}_3 \xrightarrow{\ -\text{L}\ } \text{Ar}_2\text{Ni}^{\text{II}}\text{L}_2 + \text{Ni}^\circ\text{L}_3$$

$$\Big\downarrow \text{L}$$

$$\text{Ar-Ar} + \text{Ni}^\circ\text{L}_3 \qquad\qquad (\text{XI})$$

Clearly, the use of low levels of nickel will be beneficial if rapid formation of high molecular weight polymer is desired.

Steps (2) and (3) are both accelerated by electron–withdrawing substituents on the aryl group so that aryl chlorides with such groups tend to react faster than those with electron–donating substituents. In order to avoid excessive block formation of the more reactive monomer, its relative concentration must be kept in inverse proportion to its reactivity.

In addition to being slower reacting, aryl groups with electron–donating substituents also promote side reactions that act as polymer terminating steps. The most important of these involves the triphenyl phosphine ligands. Triphenylphosphine works so well as a ligand in this reaction, because it provides a good balance of electron donation to nickel to promote oxidative addition of aryl chloride and because it is bulky enough to also ensure sufficient amount of coordinative unsaturation of nickel. A delicate balance with respect to the environment in the vicinity of the nickel atom must be maintained. On the one hand, there must be enough coordinative unsaturation to promote oxidative addition of the aryl chloride; on the other hand, if the coordinative unsaturation is too high phenyl transfer from the triphenyl phosphine ligand will take place (vide infra). Substituents on the aryl ring play a major role. The equilibrium will be shifted to the right when Y is electron donating because the electronic requirements of the nickel are being met by the aryl substituent rather than by the coordinated triphenylphosphine. This, in turn, increases the coordinative unsaturation of the nickel. At higher levels of coordinative unsaturation the nickel fills its coordination requirements by phenyl transfer from the ligand:

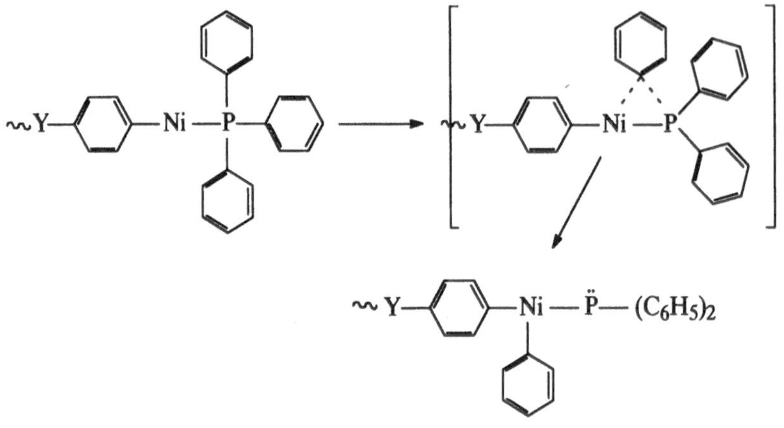

(XIII)

Our studies have demonstrated that intramolecular transfer of a phenyl group involving ipso substitution of nickel for phosphorus takes place. An alternative ortho–metalation mechanism was ruled out by using substituted triarylphosphines.

Reductive elimination from the rearranged species yields a low molecular weight biphenyl–terminated material:

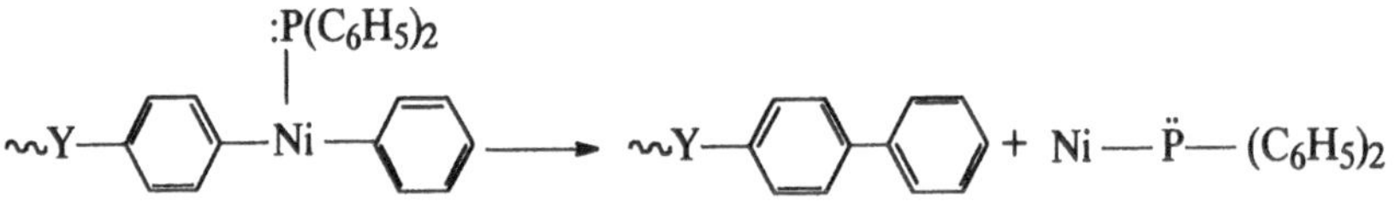

Electron donating groups promote the phenyl transfer reaction by stabilizing the highly coordinatively unsaturated nickel species and by stabilizing the transition state involved in phenyl transfer from phosphorous to nickel. The need for relatively high concentrations of triphenyl-phosphine to produce high molecular weight polymer is readily explained by this mechanism. The high triphenylphosphine levels are needed to shift the equilibrium of equation (XII) to the left; and, accordingly, to minimize the amounts of highly coordinatively unsaturated nickel species which lead to premature termination via phenyl group transfer.

Conclusions

A new method for the formation of high molecular weight aromatic biphenylene polymers via the nickel coupling of aromatic dichlorides is described. Aromatic biphenylene polymers display a number of highly attractive features which are discussed. The novel reaction is performed in a dry aprotic solvent (e.g., in DMAC) using catalytic amounts of zero–valent nickel, triphenyl-phosphine ligand and zinc metal. The reactions must be performed under an inert atmosphere and in the absence of water. In order to obtain high polymer one must use low amounts of nickel, high triphenylphosphine/nickel ratios, excess zinc metal, and moderate temperatures (70°C). The method is general and allows for the preparation of an almost infinite variety of polymers. The two prerequisites for a successful preparation of high molecular weight product are solubility of the monomer, and more particularly of the obtained polymer, in the reaction medium; and the necessity that any functional groups present in the monomer(s) (and in the resulting polymer) be inert toward the zinc/nickel system.

Critical features of the polymerization mechanism are also reviewed.

REFERENCES

1. H. Schnell, *Angew.Chemie*, **68**, 633–640 (1956).
2. H. Schnell, *Chemistry and Physics of Polycarbonates*, Interscience, New York, 1964.
3. D. W. Fox, in *Kirk–Othmer Encyclop. Chem. Technol.*, 3rd ed., M. Grayson and D. Eckroth, Eds., Wiley, New York, 1982, Vol. 18, pp. 479–494.
4. J. Ferguson, *Macromol. Chem. (London)*, **2**, 49–68 (1982).
5. J. Ferguson, *Macromol. Chem. (London)*, **3**, 76–92 (1984).
6. D. C. Clagett and S. J. Shafer, *Polym. Eng. Sci.*, **25**(8), 458–461 (1985).
7. S. K. Sikdar, *Chemtech*, **17**(2), 112–118 (1987).
8. A. J. Conix, *Ind. Chim. Belge*, **22**, 1457 (1957).
9. V. V. Korshak and S. V. Vinogradova, *Polyesters*, Pergamon, Oxford, 1965.
10. G. Bier, *Polymer*, **15**, 527 (1974).
11. L. M. Maresca and L. M. Robeson, in *Engineering Thermoplastics: Properties and Applications*, J. M. Margolis, Ed., Dekker, New York, 1985, p. 255.
12. A. S. Hay, *Fortschr. Hochpolym.-Forsch.*, **4**(4), 496–527 (1967).
13. A. S. Hay, *Macromolecules*, **2**(1), 107–108 (1969).
14. A. S. Hay, P. Shenian, A. C. Gowan, P. F. Erhardt, W. R. Haaf, and J. E. Theberge, in *Encyclopedia of Polymer Science and Technology*, 1969. Interscience, New York, 1969, Vol. 10, pp. 92–111.
15. A. S. Hay, *Polym. Eng. Sci.*, **16**(1), 1–10 (1976).
16. H. L. Finkbeiner, A. S. Hay, and D. M. White, *High Polym.*, **29** (Polym. Processes). 537–581 (1977).
17. A. S. Hay, *High Perform. Polym.*, *Proc. Symp.*, R. B. Seymour and G. S. Kirshenbaum, Eds., Elsevier, New York, 1986, pp. 209–213.
18. R. N. Johnson, A. G. Farnham, R. A. Clendinning, W. F. Hale, and C. N. Merriam, *J. Polym. Sci., A-1*, **5**, 2375 (1967); A. G. Farnham and R. N. Johnson, U.S. Pat. No. 4,108,837 (1978), to Union Carbide Corporation.
19. R. A. Clendinning, A. G. Farnham, D. C. Priest, and N. L. Zutty, Canadian Pat. 847,963 (1970), to Union Carbide Corporation.
20. R. N. Johnson, in *Encyclopedia of Polymer Science and Technology*, first ed., Wiley, New York, 1967, Vol 11, pp. 447–463.
21. J. E. Harris, "Polysulfone," in *Engineering Thermoplastics: Properties and Applications*, J. M. Margolis, Ed., Dekker, New York, 1985, pp. 177–200.
22. J. B. Rose, in *High Performance Polymers: Their Origin and Development*, R. B. Seymour and G. S. Kirshenbaum, Eds., Elsevier, New York, 1986, pp. 169–185.
23. P. Kovacic and A. Kyriakis, *J. Am. Chem. Soc.*, **85**, 454 (1963).
24. P. Kovacic and R. M. Lange, *J. Org. Chem.*, **28**, 968 (1963).
25. P. Kovacic and F. W. Koch, *J. Org. Chem.*, **28**, 1864 (1963).
26. P. Kovacic and J. Oziomek, *J. Org. Chem.*, **29**, 100 (1964).
27. P. Kovacic, F. W. Koch, and C. E. Stephan, *J. Polym. Sci.*, **2A**, 1193 (1964).
28. P. Kovacic and I. Hsu, *J. Polym. Sci. A-1*, **4**, 5 (1966).
29. P. Kovacic and F. W. Koch, in *Encyclopedia of Polymer Science and Technology*, first ed., Wiley, New York, 1969, Vol. 11, pp. 380–389.
30. J. G. Speight, P. Kovacic, and F. W. Koch, *J. Macromol. Sci. Rev. Macromol. Chem.*, **5**(2),295–386 (1971).
31. P. Kovacic and M. B. Jones, *Chem. Rev.*, **87**(2), 357–379 (1987).
32. W. Kern and R. Gehm, *Angew. Chem.*, **62**, 337 (1950).
33. W. R. Krigbaum and K. J. Krause, *J. Polym. Sci.*, **16**, 3151–3156 (1978).
34. T. Yamamoto and A. Yamamoto, *Chem. Lett.*, 353–356 (1977).
35. T. Yamamoto and Y. Hayashi, and A. Yamamoto, *Bull. Chem. Soc. Jpn.*, **51**(7), 2091 (1978).
36. J. F. Fauvarque, A. Digua, M. A. Petit, and J. Savard, *Macromol. Chem.*, **186**, 2415–2425 (1985).
37. V. Percec and H. Nava, *J. Polym. Sci., Part A: Polymer Chemistry*, **26**, 783–805 (1988).
38. I. Colon and D. R. Kelsey, *J. Org. Chem.*, **51**, 2627 (1986); I. Colon, L. M. Maresca, and G. T. Kwiatkowski, U.S. Pat. 4,263,466 (1981), to Union Carbide Corporation.
39. I. Colon, U.S. Patent 4,400,499 (1983), to Union Carbide Corporation.
40. I. Colon and G. T. Kwiatkowski, *J. Polym. Sci. Part A: Polymer Chemistry*, **28**, 367–383 (1990).
41. J. E. Harris and G. T. Brooks, European Patent Application 376,349 (1990); to Amoco Corporation.
42. J. E. Harris and L. M. Robeson, U.S. Patent 4,713,426 (1987); to Amoco Corporation.
43. A. G. Farnham and R. N. Johnson, U.S. Patent 3,332,909 (1967); to Union Carbide Corporation.
44. J. E. Harris, L. M. Maresca and M. Matzner, U.S. Patent 4,785,072 (1988); to Amoco Corporation.
45. D. A. Barr and J. B. Rose, U.S. Patent 3,634,355 (1972); to ICI, Ltd.

46. G. Darsow, U.S. Patent 3,647,751 (1972); to Farbenfabricken Bayer, A. G.

47. J. E. Harris and M. J. Michno, Jr., U.S. Patent 4,755,556 (1988); to Amoco Corporation.

48 P. A. Staniland, *Bull. Soc. Chim. Belg.*, **98**, 667–676 (1989).

49. P. A. Staniland, U.S. Patent 4,960,851 (1990); to ICI PLC.

50. European Patent Application, 348,717 (1990); to Bayer, A. G.

51. M. Ueda and F. Ichikawa, *Macromolecules*, **23**, 926–930 (1990).

52. H. M. Colquhoun, C. C. Dudman, M. Thomas, C. A. O'Mahoney and D. J. Williams, *J. Chem. Soc., Chem. Commun.*, No. 4, 336–339 (1990); See also: European Patent Application 317,226 (1989); to ICI, PLC.

53. I. Colon, *J. Org. Chem.*, **47**, 2622 (1982); I. Colon, U. S. Patent 4,400,566 (1982); to Union Carbide Corporation.

CELLULOSE-POLYMER COMPOSITES WITH IMPROVED PROPERTIES

P. Gatenholm[1], J. Felix[1], C. Klason[2] and J. Kubát[2]

[1]Department of Polymer Technology
[2]Department of Polymeric Materials
Chalmers University of Technology
S-412 96 Göteborg, Sweden

Summary

In a study of the properties of composites containing wood cellulose and thermoplastics, it was found that the strength and stiffness of these materials were significantly improved by the use of maleic anhydride-modified polypropylene (MAPP). The nature of adhesion was studied with FTIR, ESCA and SEM. The chemical bonding between MAPP and cellulose surfaces was proven to be responsible for promoting interfacial adhesion.

The effect of cellulose fiber surface treatment on the crystallization and interphase morphology of cellulose-polymer composites was studied using polarizing microscopy. Different nucleation patterns of the polypropylene matrix were observed, depending on the molecular structure of the coupling agent.

Prehydrolytic treatment of the cellulose was performed in attempts to improve the dispersibility of cellulose in polymeric matrices. In this way, novel materials with excellent properties were manufactured from municipal waste.

Natural biological composites, with cellulose as reinforcement and bacteria-produced polyhydroxybutyrate (PHB) as a matrix, were manufactured by compounding and injection molding. Such materials are totally biodegradable.

INTRODUCTION

Natural fibers such as wood, cellulose and jute are renewable materials from biological sources with very attractive mechanical properties. For instance, cellulose fibers with moduli of up to 40 GPa can be separated from wood by a chemical pulping process. A growing awareness of environmental problems and the importance of energy conservation make such renewable reinforcing materials of great interest (1). Owing to a lack of compatibility and insufficient interfacial adhesion between hydrophilic cellulose and synthetic polymers such as polyolefins, cellulose is not yet widely used as a reinforcement in composites. Extensive research has been carried out over the past ten years with the aim of improving dispersibility and interfacial adhesion in cellulose-polymer composites. (2-5). It has been shown that the use of coupling agents for the treatment of fibers prior to or added to the compounding step results in improved mechanical properties (6-10).

Paper-contaminated plastics waste collected from the municipal solid waste stream is a potentially inexpensive source of cellulose for bulk composite applications. However, the processing of such waste stream is very difficult because of a high melt viscosity and agglomaration of the cellulose phase. A method for the hydrolytic pretreatment with acid in gas phase of such waste has been recently developed in our laboratories (11).

The aim of this research was to create novel composites for bulk applications using cellulose as reinforcement. We intented to improve properties of polymer composites reinforced with cellulose by pretreating fibers with maleic anhydride modified polypropylene (MAPP). The effect of hydrolytical treatment of cellulose on the dispersibility of fibers in polymers was also investigated. Novel materials based on bacteria-produced polyesters and cellulose were manufactured and their properties evaluated.

EXPERIMENTAL

Materials

The polymers employed for the preparation of composite test bars were injection moulding grade of PP (GY 621 M, ICI Ltd., U.K., M_n=6,500 and M_w=83,600) and PHB (Biopol MBC 100/1277, ICI Biopolymers Ltd., U.K). For the crystallization studies, a commercially available PP film with M_n=50,000 and M_w=268,000 was used (Trespaphan NNA, Hoechst AG, Germany).

Three different coupling agents were used: one alkylsuccinic anhydride and two polypropylenes of different molecular weights but grafted with the same monomer, maleic anhydride. Both polypropylenes contained a 6-weight percent maleic anhydride. The properties of the coupling agents are summarized in Table I:

Table I

Coupling Agent	Chemical Structure	Molecular Weight	Supplier
A	alkylsuccinic anhydride	350	Ethyl Corp.
B	polypropylene grafted with 6 % maleic anhydride	4,500	Epolene 43 Eastman
C	polypropylene grafted with 6 % maleic anhydride	39,000	Hercoprime G Hercules Inc.

The cellulose fiber source was Alpha-pulp (Nymölla AB, Sweden) consisting of >99% cellulose, 0.3% ash, and a neglible amount of lignin.

Methods

Sample bars for mechanical testing were manufactured and tested in the following way: The filler, the additive, and the matrix polymer were mixed and homogenized in a mixing extruder at 180°C. After granulation, the mixes were injection moulded at 180° into tensile test bars whose mechanical properties were measured with an Instron tensile tester. The strain rate was 1.1×10^{-3} s^{-1} and the temperature 23 ± 0.5°C.

An Olympus Vanox Universal Microscope equipped with a Mettler FP82 hot stage was used for the crystallization studies. The samples were prepared by melting a piece of the polymer film onto a microscope slide heated on a hot plate. A few of the fibers under study were placed on the melt, a second layer of polypropylene was laid on the fibers and and a cover glass was then placed on top. The samples were heated at 210°C for 5 min to fully remove the morphological history of the sample. The sample was then cooled at 10°/min to an isothermal crystallization temperature of 130°C.

Hydrolysis of paper-contaminated palstic waste was carried out using formic acid in gas phase temperature of 180°C for 1 hour.

RESULTS AND DISCUSSION

Mechanical properties of cellulose composites

Mechanical properties were evaluated for cellulose-polypropylene composites. Figure 1 shows the effect of fiber loading on the tensile strength with and without the addition of Coupling Agent C.

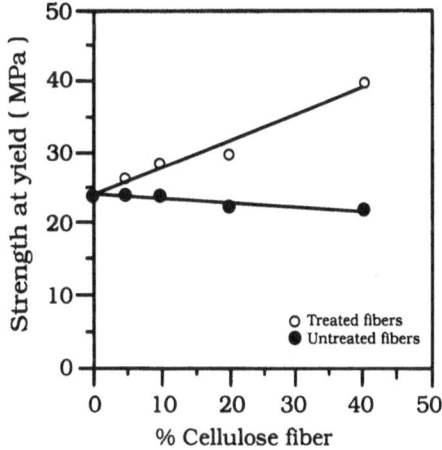

Figure 1. Tensile strength vs. fiber loading with and without the addition of 5% MAPP (Coupling Agent C).

The increased fiber content slightly reduces the tensile strength of the cellulose-polypropylene composite. This changes dramatically with the addition of MAPP coupling agent. Tensile strength is increased proportionally to the fiber content. An improvement in the impact strength is also seen with the addition of MAPP. The increase in ductility produced by MAPP appears to be caused by improved wettability of the fibers by the matrix material. This is confirmed by an investigation of the fracture surfaces (at room temperature) of samples with and without MAPP coupling agent using SEM. Figure 2 provides evidence of the different degrees of adhesion in composites without a coupling agent (Figure 2 a) and with the addition of MAPP (Figure 2 b).

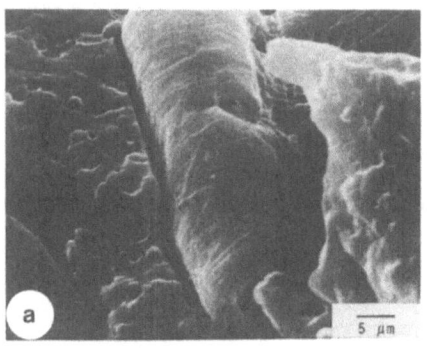

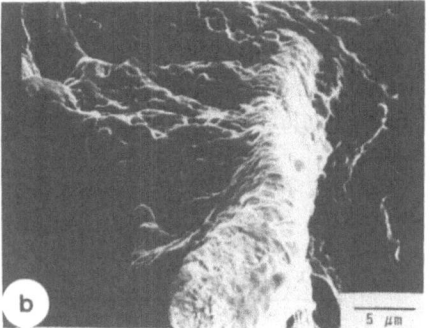

Figure 2. SEM of fracture surfaces
a) without MAPP
b) with MAPP

Coupling reactions

The coupling reactions between cellulose and MAPP were monitored using FTIR. The peak at $1717cm^{-1}$, characteristic of the dimeric form of a dicarboxylic acid, indicated the presence of two carboxyl groups in the as received substance. When heated above 160°C, the two carboxyls are, however, converted into the more reactive anhydride ring, which is able to react with cellulose during compounding at 180°C. Details of the reaction mechanism have recently been reported (9). The results obtained with FTIR explain our observations that the compounding temperature has a crucial effect in adhesion promotion. The mechanical properties were not improved when the precompounding step was carried out at temperatures below 180°C.

Interfacial phenomena

The presence of chemical bonds between MAPP coupling agent and cellulose was confirmed by IR and ESCA analysis. This explains the improvement of adhesion between different phases, as reflected in increased tensile strength levels at higher fiber loadings.

The interfacial morphology and crystallization of the PP matrix onto cellulose fibers were also investigated by polarizing microscopy. The influence of the molecular weight of the MAPP coupling agent on the nucleating ability of the fiber surface was studied, and the results are illustrated by the micrographs in Figures 3a and 3b. It is seen that the higher the molecular weight of the coupling agent used, the larger the number of nuclei that are formed on the fiber surface.

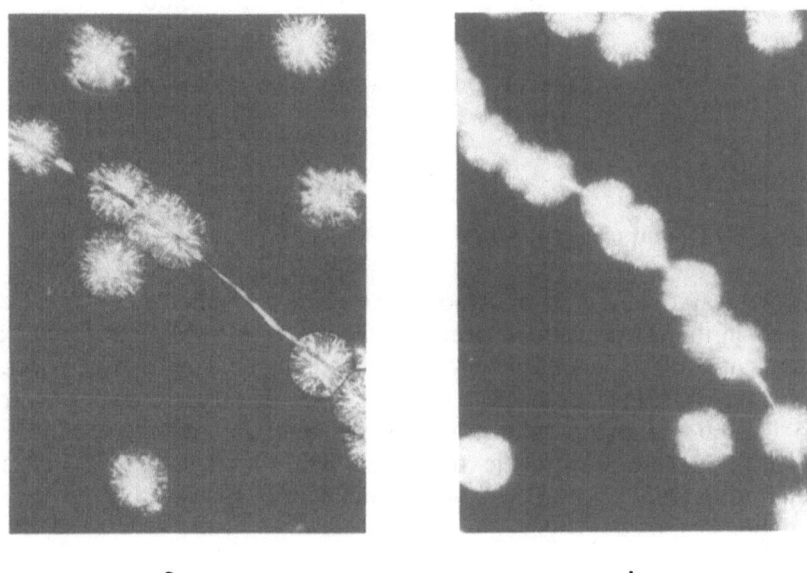

a   b

Figure 3. Crystallization of PP on cellulose fibers
a) treated with low molecular weight MAPP (Coupling Agent B)
b) treated with high molecular weight MAPP (Coupling Agent C)

Effects of interfacial morphology on adhesion

We considered the effect of the structure of MAPP used on interfacial adhesion in composites. Figure 4 shows the tensile strength of the cellulose composite as a function of the molecular weight of the MAPP used. The tensile strength increases with an increased molecular weight of MAPP. The work is ongoing in our laboratories in order to see whether there are any correlations between the crystalization pattern and interfacial adhesion.

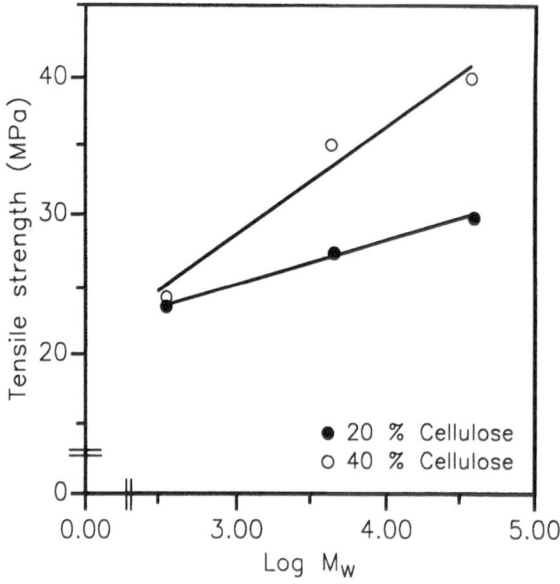

Figure 4. Tensile strength of composites vs. molecular weight of MAPP

Novel materials from waste

An important feature of the cellulose-polymer composites is the degree of dispersion of the fibers in the matrix and the overall homogeneity of the composite structure. Prehydrolytic treatment caused a highly significant improvement in this respect. This has to do primarily with the reduction of particle size of cellulose during hydrolysis. Figure 5 illustrates that the average size of the cellulose content is considerably reduced by 2 hours of hydrolytic treatment with formic acid in gas phase. Processability, measured by melt viscosity, is dramatically improved by hydrolysis.

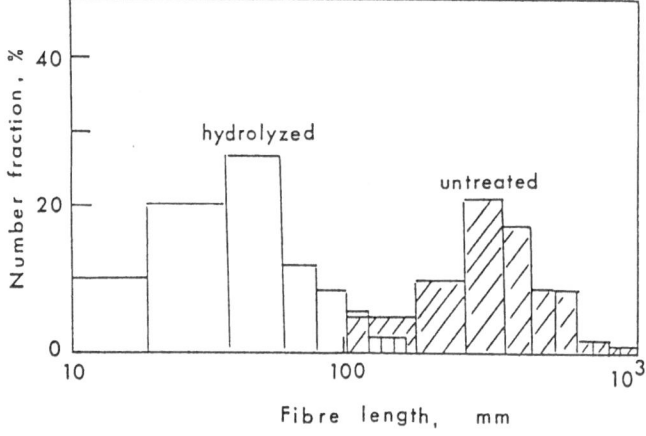

Figure 5. Fiber length distribution of cellulose fibers in injection molded sample bars of prehydrolyzed and untreated samples.

Injection molding of samples containing more than 40% cellulose is limited by the high viscosity of the filled melt. This limit is reached at about 60% by hydrolyzing the fibers. Prehydrolytic treatment has been further optimized in our laboratory and has been applied in a pilot plant for the processing of paper-contaminated plastics waste collected from a municipal solid waste stream. The application of this method to the waste stream not only contributes to a decrease in the amount of of waste but at the same time creates novel composite materials at a very attractive price. Table II demonstrates the differences between the mechanical properties of composites manufactured by treatment of paper-contaminated plastics waste collected from different sources and illustrates the benefits of prehydrolytical treatment.

Table II

Sample*	Tensile Modulus GPa	Tensile Strength MPa	Elongation %
A with hydrolysis	3.9	14.7	1.2
A no hydrolysis	Not processable	----	----
B with hydrolysis	0.5	12	40
B no hydrolysis	0.8	12	4

*Sample A consists of 27% LDPE, 5% HDPE, 2.5% PP, 7% PS, 4.5% PVC, 4.5% PETP, 46% cellulose from packaging and newsprint, and other plastics

*Sample B consists of 58% LDPE, 40% cellulose and other plastics

Natural biodegradable composites

One very important feature of polymers used today is their ability to be either reused or easily biodegradated. To obtain biodagradable materials, bacteria-produced polyhydroxybutyrate (PHB) was used as a matrix in cellulose composites. We found an excellent dispersibility of cellulose material in the PHB matrix and good mechanical properties of this new material.

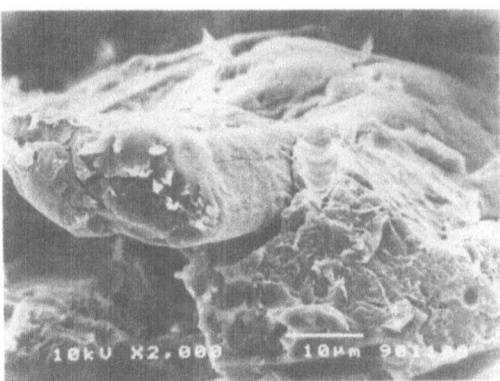

Figure 6. SEM photomicrographs of tensile fracture surfaces of 40% cellulose-polyhydroxybutyrate composite.

Figure 6 shows an SEM micrograph of the tensile fracture surface of a composite containing 40% cellulose and polyhydroxybutyrate as a matrix. Among the observations are good adhesion, no debonding and fractured fibers.

Acknowledgements

The authors wish to express their thanks to the National Swedish Board for Technical Development for financial support of this project. Thanks also go to Mr. A. Mathiasson for his assistance in the experiments.

References

1. P. Zadorecki and A.J. Michell, Polym. Compos., 10 (2) (1989)
2. H. Dalväg, C. Klason, and H.-E. Strömvall, Intern. J. Polymeric Mater., 11, 9 (1985)
3. C. Klason, J. Kubát and P. Gatenholm, in ACS Symposium Series, "Viscoelasticity of biomaterials", Ed. W. Glasser, 1990 in press
4. A.J. Michell, J. E. Vaughan and D. Willis, J. Polymer Sci. Symp., No. 55 (1976) 143
5. D. Maldas, B. B. Kokta and C. Daneault, Intern. J. Polymeric Mater., 12 (1989) 297
6. H. Kishi, M. Yoshioka, A. Yamanoi and N. Shiraishi, Mokuzai Gakkaishi, 34 (1988) 133
7. B. V. Kokta, R. G. Raj and C. Daneault, Polymer-Plastics Tech. Eng., 28 (1989) 247
8. C. Klason, J. Kubát and P. Gatenholm, in Cellulosics Utilization, Eds. H. Inagaki and G. O. Phillips, Elsevier Appl. Sci., London, 1989, p. 87
9. J. Felix and P. Gatenholm, J. Appl. Polym. Sci., 42 (609) (1991)
10. J. Felix and P. Gatenholm, submitted for publication in Polymer
11. C. Klason, J. Kubát and P. Gatenholm, "Processing of Plastic Waste", Proc. European Conf. "Plastic Recycling", Feb 1991, Copenhagen, Denmark

SYNTHESIS OF POLYIMIDES UTILIZING THE DIELS-ALDER REACTION

J.G. Smith Jr.* and R.M. Ottenbrite
Virginia Commonwealth University
Department of Chemistry
Richmond, Virginia 23284

ABSTRACT

New polyimide polymers have been prepared by the Diels-Alder reaction. An exocyclic 1,3-diene system provided monomers and polymers that had very low solubilities and inconsistent composition. These polyimides did form good films, but exhibited no detectable T_g. To enhance the monomer and polymer solubility and to alter the reactivity of the monomer , the 1,3-diene-ring structure was opened. Three new systems were evaluated before a polymer of known composition and with good properties was obtained that provided films with T_g's in the range of 200-260°C.

INTRODUCTION

One class of high performance/thermally stable polymers are the polyimides.[1] They are commonly synthesized by the reaction of an aromatic dianhydride with an aromatic diamine in a polar aprotic solvent (i.e. N,N-dimethylacetamide) under nitrogen to form a soluble poly(amic acid). This amic acid polymer can then be thermally or chemically cyclodehydrated to form the corresponding polyimide.[2-4] Once the imide is formed, the polymer is generally insoluble and infusible. Thus, research has been directed towards developing polyimides that are soluble in common organic solvents, melt processable, and thermally curable without the evolution of volatile byproducts.[3]

One synthetic method of preparing soluble polyimides is by the Diels-Alder reaction between a bis(1,3-diene) and a bismaleimide.[5] Polyimides of low to high molecular weight have been prepared from bisfulvenes[6], bis(butadienyl-2-methyl) carbamates[7], and biscyclopentadienes.[8,9] High molecular weight polyimides have been prepared from the reaction of bismaleim-

---

* Current address: NASA Langley Research Center, Polymeric Materials Branch, Mail Stop 226, Hampton, Virginia 23665-5225

ides and pseudo bisdienes such as substituted thiophene 1,1-dioxides[10], α-pyrones[11], and cyclopentadienones.[12,13] The pseudo bisdienes initially react with the maleimide group to form a cyclic adduct containing a sulfone, lactone, or ketone bridge, respectively. Upon heating, sulfur dioxide, carbon dioxide, or carbon monoxide is expelled to generate a new 1,3-diene which can further react with another maleimide functionality via the cycloaddition process.

The preparation of imide oligomers containing masked dienes in the form of benzocyclobutene[14,15] and biphenylene[16] have been reported. The thermal electrocyclic ring opening of the strained four member ring of benzocyclobutene affords the reactive diene, o-quinodimethane. The oligomers then polymerize by either an intermolecular cycloaddition reaction, giving rise to a linear structure, or by an intramolecular addition route resulting in a highly crosslinked product. In the presence of bismaleimides or bisacetylenes the Diels-Alder reaction is predominant.

In this chapter we report the development of some new polyimide polymers that were prepared by the Diels-Alder reaction. The project involved making changes in the structure of the 1,3-diene monomer in order to influence it's reactivity. We describe four monomer and polymer synthetic phases that we carried out before obtaining a polymer that was soluble in several organic solvents and formed clear colorless films with good T_g values.

PHASE I

Bis (3,4-dimethylenepyrrolidyl) arylenes[18-20]

The monodiene, N-phenyl-3,4-dimethylenepyrrolidine, was observed to readily react with dienophiles (i.e. maleic anhydride, maleimides) by the Diels-Alder reaction.[17] This 1,3-diene system was extended to prepare novel bis(3,4-dimethylenepyrrolidyl) arylenes for polymerization with bismaleimides via the Diels-Alder reaction to afford polyimides without the evolution of volatile byproducts. The bis(3,4-dimethylenepyrrolidyl) arylenes were prepared by the reaction of 4,4'-diaminoarylenes with 2,3-dibromomethyl-1,3-butadiene (Scheme 1).

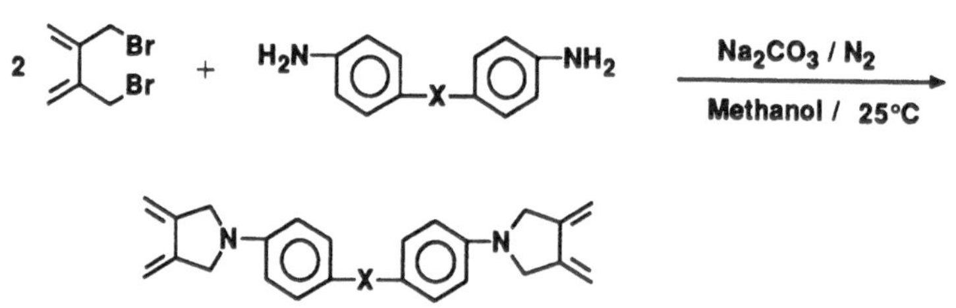

Scheme 1

Purification of the bis(1,3-diene)s proved unsuccessful due to the highly reactive nature of the diene system. This reactivity is due to the 1,3-diene being locked in the reactive

84

cisoid configuration. Self Diels-Alder adducts were observed by ^{1}H NMR for bis[4-(3,4-dimethylenepyrrolidyl)phenyl] methane and was estimated to involve approximately 20-30 % of the monomer.

Polyimides were prepared according to Scheme 2. "Purified" bisdiene monomer was obtained in situ by dissolving the crude monomer in 1,1,2,2-tetrachloroethane (TCE), centrifuging the

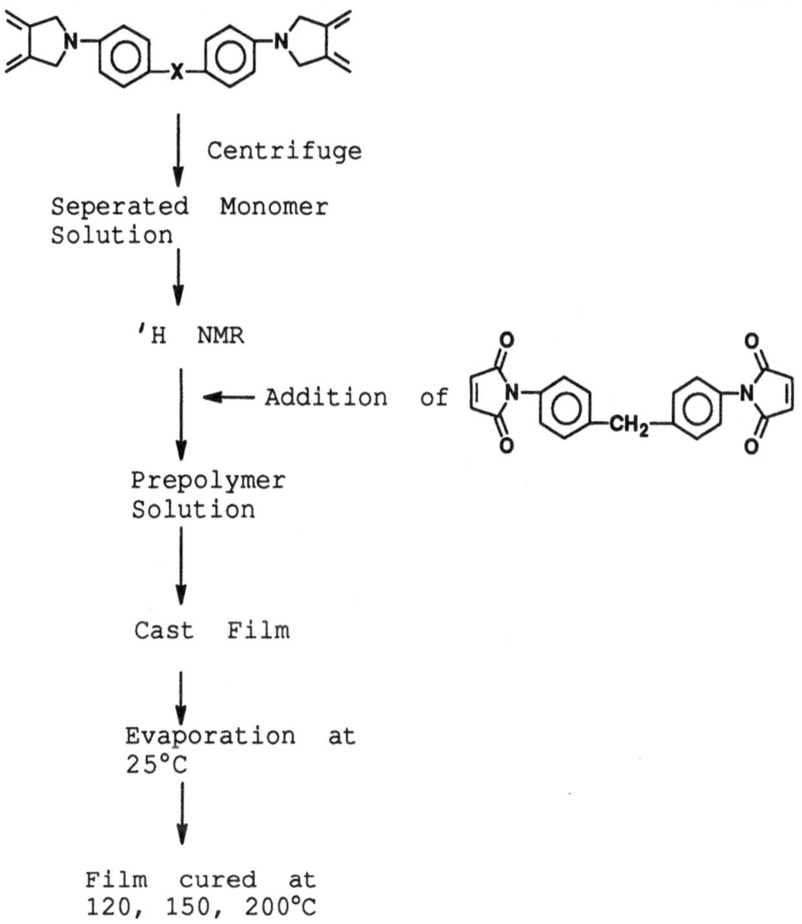

Scheme   2

solution, and subsequently separating the soluble fraction by pipette. The amount of diene monomer in solution was determined by ^{1}H NMR integration of the solvent peak against that of the endo and exocyclic protons of the 1,3-diene moiety. Then an appropriate molar ratio of the bismaleimide was added to the solution and the mixture stirred at room temperature. The prepolymer solution was cast as a film and cured at 120°, 150°, and 200°C for 12 h each. Flexible, creasable films were obtained for molar ratios of 0.7:1.0 - 0.8:1.0 (bismaleimide:bisdiene), while ratios of <0.7:1.0 and >0.8:1.0 afforded brittle material. The polymer films were not soluble in organic solvents and did not exhibit any detectable glass transition temperature (T_g) by differential scanning calorimetry (DSC). Thermogravimetric analysis (TGA) of the films revealed 10 % weight loss at ~380°C in air and ~390°C in nitrogen.

PHASE II

N,N'-Bis(butadienyl-2-methyl),N,N'-diethylarylenes[21,22]

One method of decreasing the reactivity of the bis(3,4-dimethylenepyrrolidyl) arylene system was by modifying the 1,3-diene structure. The 1,3-diene monomer was altered by the removal of one of the ring methylene groups to open the pyrrolidyl ring (Figure 1). This structural change allowed the 1,3-diene group to attain both the unreactive transoid and reactive cisoid configurations, with the former being the predominant species.

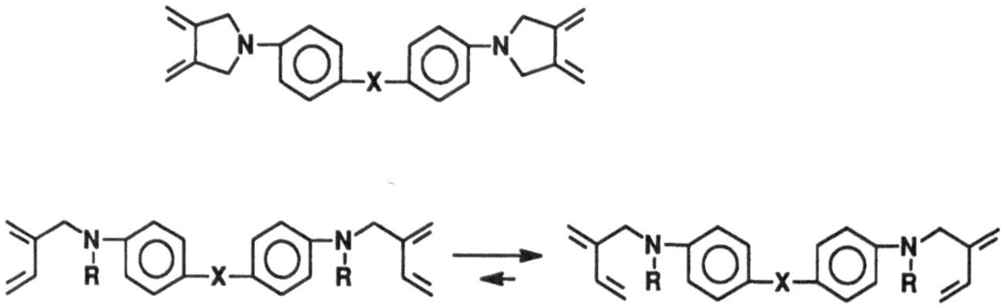

Cisoid   Transoid

Figure 1

The first structurally modified bis(1,3-diene) system was prepared by reacting 2-bromomethyl-1,3-butadiene with secondary aromatic diamines. (Scheme 3) These noncyclic bis(1,3-diene)s were easily purified and soluble in common organic

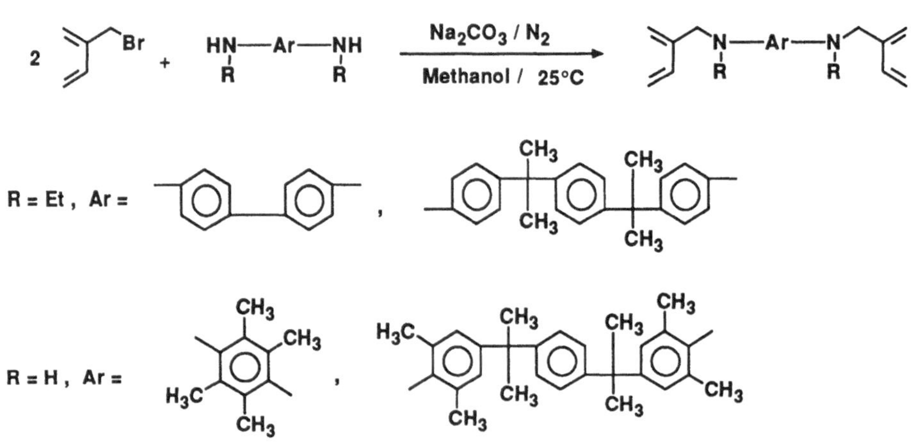

Scheme 3

solvents. No detectable self Diels-Alder adducts were observed by ^{1}H NMR over a 10 day period at 55°C for a 20 % (w/w) deuterated chloroform solution of N,N'-bis(butadienyl-2-methyl), N,N'-diethyl benzidine. This stability was attributed to the 1,3-diene being predominantly in the unreactive transoid configuration.

Polyimides were prepared by the Diels-Alder reaction of stoichiometric quantities of N,N'-bis(butadienyl-2-methyl), N,N'-diethyl arylenes and bis(4-maleimidylphenyl)methane (Scheme 4). These polymers were soluble in m-cresol, DMSO, and chlorinated hydrocarbons. Inherent viscosities (η_{inh}) ranged from 0.08-0.18 dL/g, which implies that these are low molecular weight macromolecules (Table 1). Films cast from chloroform were brittle and flaked off the glass plate. The T_gs ranged from 140-173°C. Moderate thermal stability was observed by TGA with 10 % weight loss occurring at ~360°C in air and ~377°C in nitrogen (Table 1).

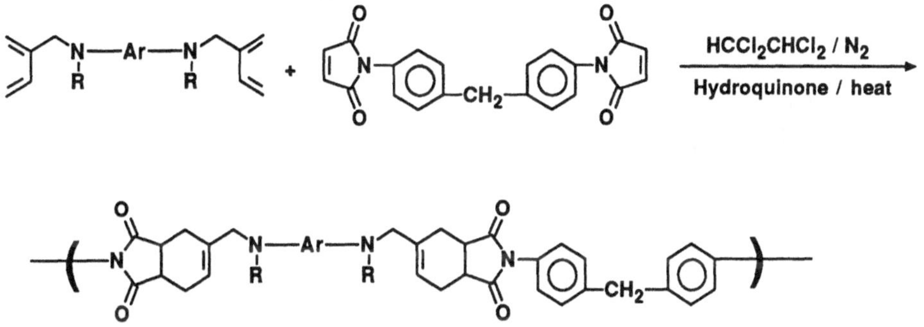

Scheme 4

PHASE III

1,4-N,N'-Bis(butadienyl-2-methyl)-diamino arylenes[21,23]

A second structural modification was made to increase the reactivity of the noncyclic bis(1,3-diene) system in the Diels-Alder reaction by the introduction of steric hindrance at the ortho positions of the aromatic amine functionality (Scheme 3). The presence of the di-ortho methyl substituents also influenced the preparation of only mono-N-alkylated products. The 1,4-N,N'-bis(butadienyl-2-methyl)-diamino arylenes were easy to purify, since they were soluble in many common organic solvents. No detectable self Diels-Alder adducts were observed by ^{1}H NMR over a 10 day period for a 20 % (w/w) deuterated chloroform solution of 1,4-N,N'-bis(butadienyl-2-methyl)-2,3,5,6-tetramethylbenzene heated at 55°C.

Polyimides were prepared by the Diels-Alder reaction of stoichiometric quantities of 1,4-N,N'-bis(butadienyl-2-methyl)-diamino arylenes and bis(4-maleimidylphenyl)methane. These polymers were higher molecular weight macromolecules than

TABLE 1. POLYIMIDE CHARACTERIZATION

Ar	R	T_g[1] °C	η_{inh}[2] dL/g	TGA 10 % wt. Loss[3] air	He or N2
(biphenyl)	Et	173	0.18	346	398
(bisphenol isopropylidene, tetramethyl)	Et	140	0.08	377	355
(pentamethylphenyl)	H	195	0.46	359	343
(substituted bisphenyl isopropylidene, methyl)	H	175	0.25	375	345
(pentamethylphenyl)	$-\overset{O}{\overset{\|}{C}}-C_6H_4-NO_2$	260	0.52	335	359
	$-\overset{O}{\overset{\|}{C}}-$ (naphthyl)	222	0.64	342	365
	$-\overset{O}{\overset{\|}{C}}-CH_3$	219	0.55	330	332
	$-\overset{O}{\overset{\|}{C}}-$ (biphenyl)	218	0.63	333	371
	$-\overset{O}{\overset{\|}{C}}-$ (phenyl)	215	0.81	321	348
	$-\overset{O}{\overset{\|}{C}}-C_6H_4-O-C_6H_5$	202	0.48	318	381

1. Inherent viscosities obtained on 0.3% (w/v) chloroform solutions at 30°C.
2. Glass transition temperatures determined by DSC at a heating rate of 20°C/min.
3. TGA determined on powdered samples at a heating rate of 2.5°C/min.

those obtained in Phase II as indicated by the inherent viscosity data, which ranged from 0.25-0.46 dL/g (Table 1). This increase in the molecular weight was attributed to the enhanced reactivity of the bis(1,3-diene) monomer through steric interactions with the ortho dimethyl groups. These polymers were soluble in m-cresol, DMSO, and chlorinated hydrocarbons. Coherent films could not be cast from chloroform solutions of these polymers. The T_g's ranged from 175-195°C. Thermal stability was determined by TGA with 10 % weight loss occurring at ~367°C in air and ~344°C in nitrogen (Table 1). In air, these macromolecules gained ~5 % weight between 200-300°C. This weight gain was attributed to the oxidation of the secondary amine moiety (NH) in the polymer backbone.

PHASE IV

1,4-N,N'-Bis(butadienyl-2-methyl)-diamido arylenes[21,24]

To protect the secondary amine group (NH) from oxidation, 1,4-N,N'-bis(butadienyl-2-methyl)-2,3,5,6-tetramethylbenzene was modified by reacting it with aromatic acid chlorides to afford the corresponding 1,4-N,N'-bis(butadienyl-2-methyl)-diamido-2,3,5,6-tetramethylbenzene monomers (Scheme 5). The amide pendant group also provides additional steric hinderance surrounding the 1,3-diene system, which would increase the reactivity of the diene in the Diels-Alder reaction by increasing

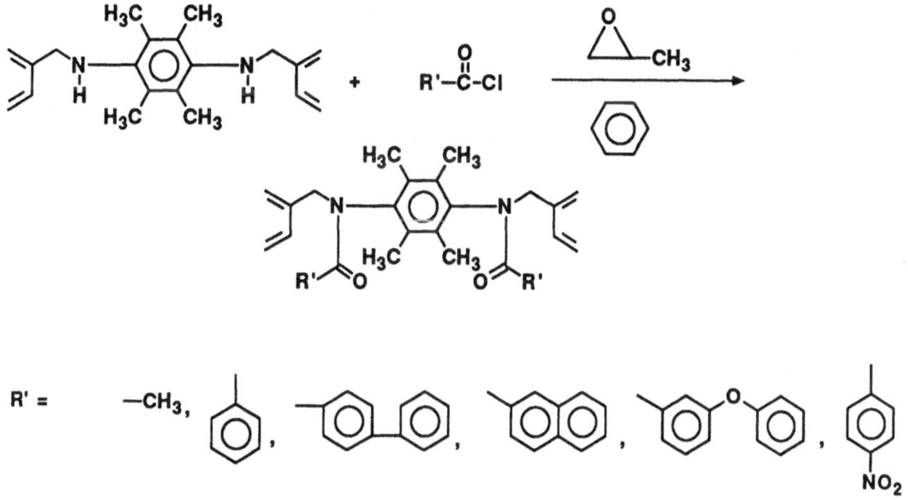

Scheme 5

the relative concentration of the reactive cisoid configuration. These bis(1,3-diene)s were easily purified and soluble in common organic solvents. No detectable self Diels-Alder adducts were observed after 24 h by ^{1}H NMR for a 20 % (w/w) deuterated chloroform solution of 1,4-N,N'-bis(butadienyl-2-methyl)-dibenzamido-2,3,5,6-tetramethylbenzene heated at 55°C. However, this monomer solution did form a gel in the NMR tube after 2 days heating at 55°C, which implies that this system may be more reactive than the previous two noncyclic bis(1,3-diene) systems.

Polyimides were prepared by the Diels-Alder reaction of stoichiometric quantities of 1,4-N,N'-bis(butadienyl-2-methyl)-
diamido-2,3,5,6-tetramethylbenzene monomers and bis(4-maleim-
idylphenyl)methane (Scheme 6). The polymers were soluble in
DMAc and chlorinated hydrocarbons. Inherent viscosity data
ranged from 0.48-0.81 dL/g and suggests that these are moderate to high molecular weight polymers (Table 1). The increase

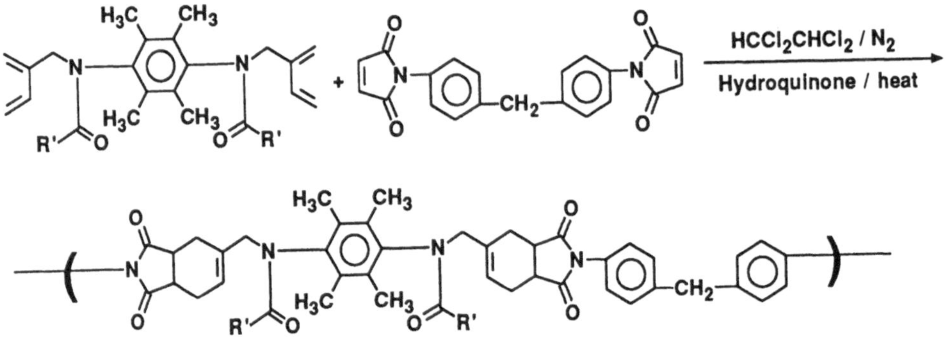

Scheme   6

in the molecular weight over the previous noncyclic bis(1,3-
diene) systems was attributed to the enhanced reactivity of
the bis(amide-1,3-diene) monomer system for reasons previously
mentioned. Several of the polymers afforded flexible, creasable films cast from chloroform. The T_g's ranged from 202-260°C
and was dependent on the amide substituent. Moderate thermal
stability was observed by TGA with 10 % weight loss occurring
at ~330°C in air and ~350°C in helium (Table 1). No weight
gain was observed by TGA in an oxidizing atmosphere (air) for
these polymers.

CONCLUSIONS

The structural modification of the cyclic 1,3-diene system
of Phase I afforded novel noncyclic bis(1,3-diene)s of varying
reactivity in the Diels-Alder reaction with bismaleimides. Low
to high molecular weight polyimides were prepared from the
noncyclic bis(1,3-diene) systems and bis(4-maleimidyl-
phenyl)methane, which exhibited better solubility in organic
solvents than the Phase I system. Moderate thermal stability
was observed by TGA for these polymers.

Experimental

Melting points were determined using either a Thomas melting
point apparatus or Perkin-Elmer DSC-4 differential scanning
calorimeter (heating rate 10°C/min) and are uncorrected. Differential scanning calorimetry (DSC) was performed under a
nitrogen atmosphere at a heating rate of 20°C/min on a Perkin-
Elmer DSC-4 scanning calorimeter, which was controlled by a
Thermal Analysis System 4 microprocessor controller and a TADS
3700 Data station. The reference was a standard DSC cell. The

apparent glass transition temperature (T_g) was taken as the inflection point of the ΔT vs. temperature on the second run after an initial heatup to 300°C followed by quenching to room temperature. Thermogravimetric analysis (TGA) was performed on powdered samples on a Perkin-Elmer TGS-2 Thermogravimetric system in combination with a heater controller and an autobalance AR-2 at a heating rate of 2.5°C/min in air and nitrogen and helium atmospheres at NASA Langley. Inherent viscosities (η_{inh}) were obtained using an Ubeholde viscometer at 30°C on 0.3 % (w/v) solutions in chloroform.

Bis(3,4-dimethylenepyrrolidyl) arylenes (Phase I)

The cyclic bis(1,3-diene)s were prepared by the reaction of 4,4'-diaminoarylenes with 2,3-dibromomethyl-1,3-butadiene in methanol containing sodium carbonate. The mixture was stirred at room temperature under a nitrogen atmosphere for 2 days. The precipitate was recovered by filtration, washed successively with methanol and water and dried under vacuum at room temperature.

Noncyclic bis(amine-1,3-diene)s (Phases II and III)

Noncyclic bis(1,3-diene)s were prepared by the reaction of aromatic diamines with 2-bromomethyl-1,3-butadiene in methanol containing sodium carbonate. The mixture was stirred at room temperature for ~2 days under a nitrogen atmosphere. A precipitate formed, which was recovered by filtration, washed with cold methanol, and dried under vacuum. The bisdiene was dissolved in a methanol:acetone (1:4) mixture and reprecipitated from solution by the dropwise addition of water. The mixture was heated until the precipitate redissolved. The solution was subsequently allowed to slowly cool, during which time the monomer precipitated. The bis(1,3-diene) was isolated by filtration and dried under vacuum.

Noncyclic bis(amide-1,3-diene)s (Phase IV)

1,4-N,N'-Bis(butadienyl-2-methyl)-diamido-2,3,5,6-tetramethylbenzene monomers were prepared by reacting 1,4-N,N'-bis(butadienyl-2-methyl)-2,3,5,6-tetramethylbenzene with aromatic acid chlorides in dry benzene containing propylene oxide. The mixture was stirred for ~2 days at room temperature. The precipitate was filtered, washed with 95 % ethanol, and dried under vacuum. The bis(amide-1,3-diene)s were recrystallized from ethanol or ethanol:chloroform solutions, recovered by filtration, and dried under vacuum.

Polyamineimide Synthesis (Phases II and III)

Into a round bottom flask equipped with nitrogen inlet and mechanical stirrer were added equimolar quantities of the non-cyclic bis(amine-1,3-diene) and bis(4-maleimidylphenyl) methane. Hydroquinone was added as a radical scavenger and 1,1,2,2-tetrachloroethane was added to afford an approximate 30 % (w/w) solution which was stirred at room temperature under a nitrogen atmosphere for 24 h. The reaction temperature was then increased to 60-80°C and maintained for 4 days. The polymer solution was cooled to room temperature and diluted to approximately 10 % solids by the addition of chloroform. The polymer was precipitated by dropwise addition of the reaction mixture into stirred diethyl ether, filtered, and dried under vacuum. Reprecipitation from a chloroform solution into diethyl ether afforded the polyimide.

Polyamideimide Synthesis (Phases IV)

Into a round bottom flask equipped with nitrogen inlet and mechanical stirrer were added equimolar quantities of the noncyclic bis(amide-1,3-diene) and bis(4-maleimidylphenyl) methane. Hydroquinone was added as a radical scavenger and 1,1,2,2-tetrachloroethane was added to afford an approximate 30 % (w/w) solution which was stirred at room temperature under a nitrogen atmosphere for 24 h. The reaction temperature was then increased to 140°C and maintained for 24 h. As the viscosity of the solution increased during these reaction periods, small aliquots of the solvent were added to improve stirring. The polymer solution was cooled to room temperature, precipitated by dropwise addition of the reaction mixture into stirred diethyl ether, filtered, and dried under vacuum. Reprecipitation from a chloroform solution into diethyl ether afforded the polyimide.

ACKNOWLEDGMENTS

The authors wish to thank NASA Langley for their financial support of this research (NAG-1-672) and Shell Chemical Company for their donation of EPON HPT 1061 and 1062 diamines.

REFERENCES

1. P.E. Cassidy, "Thermally Stable Polymers: Synthesis and Properties", Marcel Dekker, New York, 1980.
2. C.E. Sroog, A.L. Eindrey, S.V. Abrams, C.E. Berr, W.M. Edwards, and K.L. Oliver, J. Poly. Sci., Part A: Polym. Chem., 3, 1373 (1965).
3. F.W. Harris, W.A. Feld, and L.H. Lavier, J. Poly. Sci., Polym. Lett. Ed., 13, 283 (1975).
4. P.M. Hergenrother, N.T. Wakelyn, and S.J. Havens, J. Poly. Sci., Part A: Polym. Chem., 25, 1093 (1987).
5. W.J. Bailey in "Diels-Alder Polymerization-Step Growth Polymerization" Vol. 3, D.H. Solomon, Ed., Marcel Dekker, New York, 1972, pp. 279-332.
6. J.E. Reeder (to E.I. du Pont de Nemours and Co.) U.S. Patent 3,344,071, Aug. 1, 1967; Chem. Abstr., 67, 74026t (1967).
7. W.J. Bailey, J. Economy, and M.E. Hermes, J. Org. Chem., 27, 3295 (1962).
8. J.K. Stille and L. Plummer, J. Org. Chem., 26, 4026 (1961).
9. R.W. Upson (to E.I. du Pont de Nemours and Co.) U.S. Patent 2,726, 232, Dec. 6, 1955; Chem. Abstr., 50, 6835f (1956).

10. S.W. Chow and J.M. Whelan Jr. (to Union Carbide Corp.) U.S. Patenet 2,971,944, Feb. 14, 1961 ; _Chem. Abstr._, **55**, 12941e (1961).

11. S.W. Chow (to Union Carbide Corp.) U.S. Patent 3,074,915, Jan. 22, 1963; _Chem. Abstr._, **58**, 10358a (1963).

12. E.A. Kraiman (to Union Carbide Corp.) U.S. Patent 2,890,206, June 9, 1959; _Chem. Abstr._, **53**, 17572e (1959).

13. E.A. Kraiman, _Macromol. Syn._, **2**, 110 (1966).

14. L.S. Tan, _J. Poly. Sci., Part A: Polym. Chem._, **25**, 3159 (1987).

15. L.S. Tan, E.J. Soloski, and F.E. Arnold in ACS Symp.Ser. No. 367, R.A. Dickie, S.S. Labana, and R.S. Bauer, eds., Washington, D.C., 1988, pp. 349-365.

16. S. Stoessel, T. Takeichi, J.K. Stille, and W.B. Alston, _J. Appl. Poly. Sci._, **36**, 1847 (1988).

17. H. K. Chen, Doctoral Dissertation at Virginia Comonwealth University, December 1985.

18. R.M. Ottenbrite, A. Yoshimatsu, and J.G. Smith Jr., _Polym. Prepr._, **28 (2)**, 280 (1987).

19. R.M. Ottenbrite and J.G. Smith Jr., _Polym. Prepr._, **29 (1)**, 263 (1988).

20. R.M. Ottenbrite and J.G. Smith Jr., _Polm. Adv. Tech._, **1**, 117 (1990).

21. J.G. Smith Jr., Doctoral Dissertation at Virginia Commonwealth University, May 1990.

22. R.M. Ottenbrite and J.G. Smith Jr., _Polym. Prepr._, **30 (1)**, 199 (1989).

23. R.M. Ottenbrite and J.G. Smith Jr., _Ibid._, **30 (1)**, 213 (1989).

24. R.M. Ottenbrite and J.G. Smith Jr., _Ibid._, **30 (2)**, 199 (1989).

POLY(IMIDAZOLEAMIDES): A NEW CLASS OF HETEROAROMATIC POLYAMIDES

Ernest L. Thurber, Ramachandran P. Subrayan and
Paul G. Rasmussen*

Department of Chemistry and Macromolecular Research Center
University of Michigan
Ann Arbor, MI 48109

INTRODUCTION

The introduction of heteroaromatic units into polymer backbones has played a major role in the advancement of high performance materials. This strategy has resulted in a number of aromatic heterocyclic polymers and aromatic-heteroaromatic polyamides. For example, Marvel's[1] preparation of polybenzimidazole or Kvenzel's[2] work in which oxadiazole rings were incorporated in aromatic polyamides. We now report the preparation of poly(imidazoleamides), a new class of heteroaromatic polyamides based on cyanoimidazoles.

We have been investigating cyanoimidazoles for the past several years and have shown their use as metal ligands[3] and moderate electron-acceptors.[4] Recently, we have focused our attention on preparing polymers based on cyanoimidazoles.[5] Heteroaromatic polyamides based on a cyanoimidazole backbone (1), as shown below, have a number of attractive properties. The high nitrogen and low hydrogen content of these materials suggests their use in thermally stable and low flammability applications. Substitution of the 1-nitrogen and amido hydrogen of the cyanoimidazole ring permits modification of polymer properties such as solubility. The electron withdrawing effect of nitrile groups may increase thermal and oxidative stability by lowering the LUMO energy level. Belakov and Kosbutskii[6] have shown that electron withdrawing groups increase the thermal stability of aromatic and heteroaromatic polyamides.

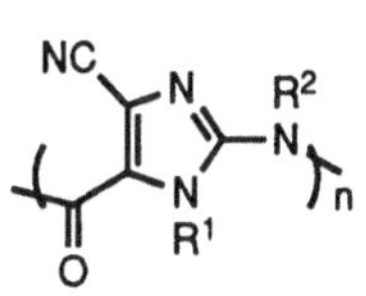

1

EXPERIMENTAL

2-Amino-4-cyano-1-methyl-5-imidazolecaboxylic acid (4a)

A 50-ml flask was equipped with condenser and magnetic stirrer. The flask was charged with 2.25 g of ethyl 2-amino-4-cyano-1-methyl-5-imidazolecarboxylate (3a) (11.2 mol) and 20 ml of water. The reaction mixture was heated to reflux and 4.6 ml of tetraethylammonium hydroxide (40% in water, 11.3 mmol) was added. The reaction was heated to 70-80°C for 30 minutes. The reaction was cooled, filtered and pH adjusted to 1 with HCl(conc.) to give a white precipitate. The white precipitate was filtered, rinsed with acetone and ether to give 1.67 g of a white solid (87%). Mp 212-214°C(dec); IR(KBr) 3396, 3196, 2251, 1656, 1573, 1062, 788 cm^{-1}; ^{1}H NMR(DMSO-d$_6$) δ 6.58(s, 2H), 3.59(s, 3H); ^{13}C NMR(DMSO-d$_6$) δ 159.51, 153.98, 124.99, 116.17, 115.58, 31.08; MS(EI)m/z 166(M+), 122(100%), 121, 86, 44, 42. Anal calcd. for C$_6$H$_6$N$_4$O$_2$: C, 43.38; N, 3.64; N, 33.72. Found: C, 43.20; H, 3.62; N, 33.64.

2-Amino-4-cyano-1-methyl-5-imidazolecarbonyl chloride hydrochloride (5a)

A 100-ml Schlenk flask with a side arm was equipped with condenser, magnetic stirrer and placed under nitrogen atmosphere. The flask was charged with 1.10 g of 2-amino-4-cyano-1-methyl-5-imidazolecarboxylic acid (4a) (6.6 mmol) and 40 ml of thionyl chloride. The reaction was refluxed for 4 hours after which the excess thionyl chloride was removed under vacuum. The addition of 50 ml of HCl(g) solution (1M in ether) afforded a yellow solid. The solid was filtered under nitrogen, rinsed with HCl(g) solution (4 x 10 ml) and anhydrous ether (4 x 10 ml) and collected in a glove box to give 1.20 g of an off-white solid (80%). Mp 250°C(discolors), 300°C(black); IR(KBr) 3075, 2245, 1729, 1682 cm^{-1}; MS(DCI with ammonia -HCl + H)m/z 187(M$^+$+2),187(M$^+$), 166, 149(100%), 123, 121, 77. Anal Calcd. for C$_6$H$_5$N$_4$OCl•HCl: C, 32.60; H, 2.74; N, 25.38. Found: C, 33.00; N, 2.82; N, 24.72.

Ethyl 2-(4-cyano-1-methyl-5-imidazoyl)amino-1-methyl-5-imidazolecarboxylate (6)

A 100-ml three-neck flask was equipped with condenser, magnetic stirrer and placed under nitrogen atmosphere. The flask was charged with 0.26 g of ethyl 2-amino-4-cyano-1-methyl-5-imidazolecarboxylate (3a) (1.3 mmol) and 40 ml of freshly distilled THF. A separate 25-ml flask was charged with 0.39 g of 4-cyano-1-methyl-5-imidazolecarboxylic acid (2.6 mmol) 15 ml of thionyl chloride. The reaction was refluxed for 30 minutes and the excess thionyl chloride was removed under vacuum. The remaining white solid was dissolved with 10 ml of THF and added slowly to the imidazolecarboxlyate solution *via* a cannula. Next, 0.19 ml of pyridine (2.3 mmol) and a catalytic amount of dimethylaminopyridine (DMAP) was added. The reaction was refluxed for 24 hours and the pyridine hydrochloride salt was removed. The THF was evaporated down and the remaining solid was dissolved in EtOAc. The organic layer was washed with 10% HCl(aq) (2 x 25 ml) NaCl(aq) (2 x 25 ml), 10% NH$_4$OH(aq) (1 x 25 ml). A white solid formed upon the final addition of NaCl(aq). The white solid was filtered and saved. The solid was dissolved in hot water and a fluffy white solid formed upon addition of a few drops of HCl(conc). The solid was filtered to give 0.30 g of desired material (70%). The organic layer was dried over MgSO$_4$ and evaporated down to give an additional 0.01 g of solid (72%). TLC(Acetone) R$_f$ 0.54; Mp 185-187°C; IR(KBr) 3220, 3110, 2238, 1727, 1712, 1589, 1390, 1136, 878 cm^{-1}; ^{1}H NMR (DMSO-d$_6$) δ 11.71(bs, 1H), 8.10(s, 1H), 4.40-4.33(q, 2H), 3.85(s, 3H),

3.74(s, 3H), 1.36-1.31(t, 3H); MS(EI)m/z 327(M$^+$), 292, 287, 134(100%), 106, 67, 42. Anal. Calcd. for $C_{14}H_{13}N_7O_3$: C, 51.22; H, 4.30; N, 29.86. Found: C, 50.89; H, 3.96; N, 29.94.

Poly[iminocarbonyl(4-cyano-1-methyl)2,5-imidazole] (7a)

A 25-ml two-neck flask was charged with 0.93 g of 2-amino-4-cyano-1-methyl-5-imidazolecarbonyl chloride hydrochloride (5a) (4.2 mmol) in a glove box. The flask was equipped with a magnetic stirrer and placed under nitrogen atmosphere. The flask was charged with 8 ml of dry hexamethylphosphoramide (HMPA), 0.22 ml of dry pyridine (8.4 mmol) and 0.22 g of 4-dimethylaminopyridine (0.8 mmol). A red-orange precipitate formed 15 minutes after the addition of the base. The reaction was stirred at room temperature for 2 hours and then warmed to 60-80°C for 12 hours. The reaction mixture was poured into 200 ml of 10% HCl(aq) and allowed to stand for 30 minutes. A red-orange solid was filtered, washed several times with water and hot methanol, collected and dried to yield 0.25 g of the desired material (41%). η_{inh} 0.18 dl/g at 30°C for a 0.5 g/dl solution in H_2SO_4; Mp 310°C(turns black); DSC 100°C(endotherm, loss of H_2O), > 350°C (endothermic decomposition), scan rate of 10°C/min; IR(KBr) 3420, 3260, 1683, 1617, 1584, 1540, 1402, 1388, 1360, 1145 cm^{-1}; Anal. Calcd. for $C_6H_4N_4O \cdot H_2O$: C, 43.38; H, 3.64; N, 33.72. Found: C, 43.34; H, 3.63; N, 32.35.

Poly[iminocarbonyl-(5-cyano-1-methyl)2,4-imidazole] (7b)

A 10-ml two-neck flask was equipped with condenser, magnetic stirrer and placed under a nitrogen atmosphere. The flask was treated with a trimethylchlorosilane solution (1/9 trimethylchlorosilane/CHCl3). The flask was charged with 0.31 g of 2-amino-5-cyano-1-methyl-4-imidazolecarboxlyic acid (4b) (1.9 mmol) and 5 ml of dry pyridine. The reaction was cooled to 0°C and 0.16 ml of silicon tetrachloride was added (1.3 mmol) to the reaction. A white solid formed soon after the addition of silicon tetrachloride. The white solid turned a tan color as the reaction was heated. The reaction was refluxed for 24 hours and poured into 100 ml of 10% HCl(aq). The tan solid was filtered, washed with acetone and SiO_2 was removed after dissolving the solid in DMAc. The polymer was reprecipitated by pouring DMAc solution into a water/methanol mixture. The tan solid was filtered, washed several times with water, hot methanol, acetone, collected and dried to give 0.12 g of the desired material (44%). A brittle film was cast from DMAc η_{inh} = 0.12 dl/g at 30°C for a 0.5 g/dl solution in H_2SO_4; Mp 300°C(turns black), 310°C(dec); DSC 100°C(endotherm, loss of H_2O), 325°C(endotherm, loss of CO_2), > 350°C (endothermic decomposition), scan rate of 10°C/min; IR(KBr) 3444, 3426, 2234, 1696, 1624, 1575, 1540, 1507 cm^{-1}. ^{1}H NMR (DMSO-d$_6$) δ 11.14(bs, 0.2H),10.58(bs, 0.1H), 6.66(bs, 0.2H), 3.69(bs, 3H); Anal. Calcd. for $C_6H_4N_4O \cdot H_2O$: C, 43.38; H, 3.64; N, 33.72. Found: C, 43.98; H, 3.08; N, 33.23.

RESULTS AND DISCUSSION

AB monomers, based on cyanoimidazoles, were prepared by selective alcoholoysis of one of the nitrile groups and alkylation of the 1-nitrogen. This was accomplished in the following manner. Ethyl 2-amino-4(5)-cyano-5(4)-imidazolecarboxylate (2), prepared by Japanese workers,[7] was alkylated by deprotonation of the 1-hydrogen, pKa = 7.2, with sodium hydride and followed by the addition of dimethylsulfate. Fortunately, the resulting regioisomeric mixture could be separated by fractional crystallization. Next, the desired amino acids (4a) and (4b) were prepared by selective hydrolysis of the ethylester group with

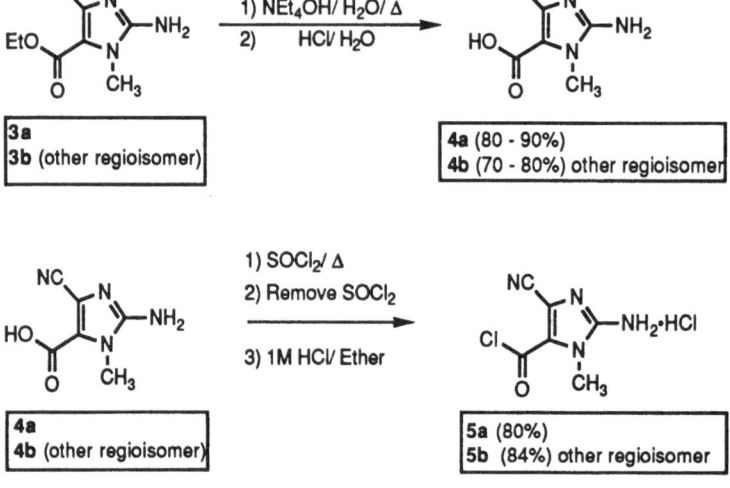

tetraethylammonium hydroxide. Nuclear Overhauser Effect (NOE) studies were carried out on the decarboxylated amino acids to determine the correct assignment of the regioisomers.[8] In addition, the amine hydrochloride salts of the acid chlorides (5a & 5b) were isolated under nitrogen atmosphere.

AB type aromatic polyamides, such as poly(benzamide) are prepared from acid chlorides or by the activation of the amino acids. These activation methods include triphenylphosphite/LiCl,[9] triphenylphosphine/hexachloroethane,[10] silicon tetrachloride[11] and dimethyldichlorosilane.[12] Preliminary investigation of these polymerization methods were conducted on model systems. For example, the dimer model (6) was prepared in good yield by reacting compound (3a) with a cyanoimidazolecarbonyl chloride. The low nucleophilicity of the amine group, due to the electron withdrawing effect of the nitrile, was partially overcome by the use of dimethylaminopyridine (DMAP).

Poly(imidazoleamides) (7a & 7b) were prepared in moderate yields by reacting the acid chloride (5a) or (5b) in HMPA with pyridine and catalytic amount of DMAP. The precipitation of polymer was observed early on in the reaction. Unfortunately, attempts to increase polymer solubility with the addition of lithium salts did not have any effect. In addition, poly(imidazoleamides) (7a & 7b) were prepared by dissolving the amino acid (4a) or (4b) in pyridine followed by the addition of silicon tetrachloride.

Poly(imidazoleamides) are yellow to dark red in appearance. Luminescence studies are being conducted to determine if this color is due to conjugation along the polymer backbone. These polymers are soluble in sulfuric acid and some polar aprotic solvents. For example, brittle films have been cast from DMAc and NMP solutions. Inherent viscosity (η_{inh}) measurements range from 0.1 to 0.2 dl/g in sulfuric acid at 30°C. Combustion analyses, as well as differential scanning calorimetry (figure 1), indicate these materials retain water even after rigorous drying procedures. This phenomenon is common for imidazoles[13] and polyamides.[14] Thermogravimetric analyses (figure 2) indicate these polymers are thermally stable in excess of 300°C, under nitrogen atmosphere.

SUMMARY AND CONCLUSIONS

Poly(imidazole)amides, a new class of heteroaromatic polyamides, were prepared from the polymerization of amino acids and acid chlorides to yield low to moderate molecular weight polymers. These materials are thermally stable in excess of 300°C, under nitrogen atmosphere Presently, we are investigating methods to increase molecular weights of these polymers as well as preparing new AB monomers, as shown below.

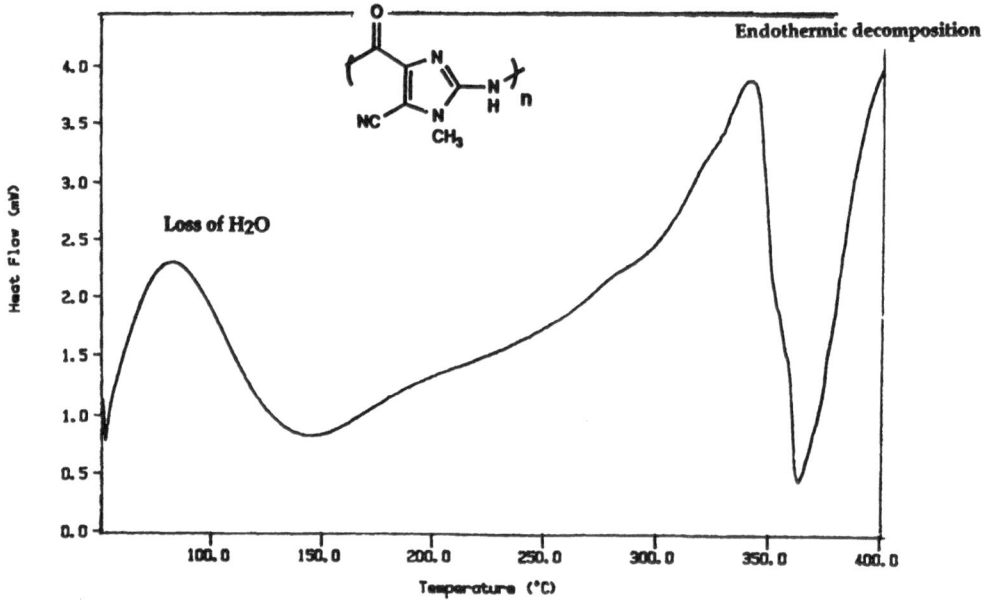

ACKNOWLEDGEMENT

We gratefully acknowledge support of this work from the Amoco Chemical Corporation.

Figure 1. Differential Scanning Calorigram of Poly(imidazoleamide)

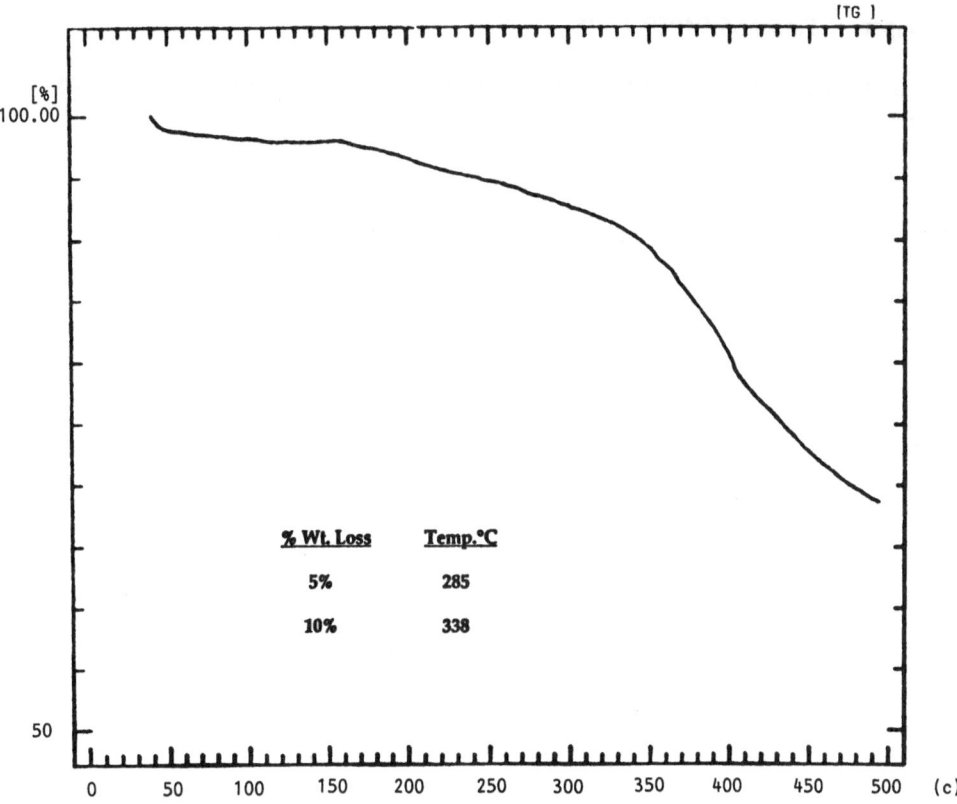

Figure 2 Thermogravimetric Analysis of Poly(imidazoleamide)

REFERENCES

1. Vogel, H.A.; Marvel, C.S. *J. Polym. Sci.* **1963**, *A3*, 1665.

2. Kvenzel, H.E.; Wolf, G.D.; Bentz, F.; Blankenstein, G.; Nischk, G.E. *Makromol. Chem.* **1969**, *130*, 103

3. For examples see: (a) Rasmussen, P.G; Kolowich, J.B.; Bayón, J.C. *J. Amer. Chem. Soc.* **1988**, *110*, 7042; (b) Rasmussen, P.G.; Anderson, J.G.; Bayón, J.C. *Inorg. Chem. Acta* **1984**, *87*, 159; (c) Rasmussen, P.G.; Bayón, J.C. *Inorg. Chem. Acta* **1984**, *81*, 115.

4. Allan, D.S.; Bergstrom, D.F.; Rasmussen, P.G. *Synthetic Metals* **1988**, *25*, 139.

5. (a) Allan, D.S.; Thurber, E.L.; Rasmussen, P.G. *J. Polym. Sci., Polym. Chem. Ed.* **1990**, *28*, 2475; (b) Allan, D.S.; Thurber, E.L.; Apen, P.G.; Kim, Y.K.; Subrayan, R.P.; Francis, A.H.; Rasmussen, P.G. *Proceedings of the ACS Division of Polymeric Materials: Science and Engineering* **1989**, *61*, 335; (c) Rasmussen, P.G.; Allan, D.S.; Apen, P.G.; Thurber, E.L. *Polymer Prepr.* **1988**, *29*, 325.

6. Belyakov, V.K.; Ksobutskii, V.A. *Vysokomol. Soedin., Ser. A* **1976**, *18*, 2452.

7. Japanese Patent 55-94959 **1979**.

8. Subrayan, R.P.; Thurber, E.L.; Rasmussen, P.G. *Tetrahedron Letters* **1991**, *submitted*.

9. Higashi, I.; Gota, M.; Kakinok, H. *J. Polym. Sci., Polym. Chem. Ed.* **1980**, *18*, 1711.

10. Wu, G.C.; Tanaka, H.L ; Sanuik, K.; Ogata, N. *Polymer J.* **1982**, *14*, 571.

11. Strohriegel, P.; Heitz, W. *Makromol. Chem., Rapid Commun.* **1985**, *6*, 111.

12. Akar, A.; Galioglu, O. *Makromol. Chem., Rapid Commun.* **1988**, *9*, 19.

13. MacDonald, R.N.; Cairncross, A.; Sieja, J.B.; Sharkey, W.H. *J. Polym. Sci., Polym. Chem. Ed.* **1974**, *12*, 664.

14. Keinath, S.E.; Morgan, R.J. *Thermochimica Acta* **1990**, *166*, 17.

CURE KINETICS OF EPOXY CRESOL NOVOLAC ENCAPSULANT FOR

MICROELECTRONICS PACKAGING

Rolf W. Biernath† and David S. Soane*

Department of Chemical Engineering
University of California
Berkeley, CA 94720

INTRODUCTION

A finished semiconductor device typically comprises diverse materials, including metal conductors, a patterned semiconductor substrate, and various interlayer or encapsulating dielectrics. Pronounced stresses can therefore be generated within the device as a result of mismatched thermal expansion coefficients. Both polyimides and epoxy molding compounds cured at elevated temperatures develop appreciable thermoelastic stresses upon cooling. During normal operation, temperature transients repeatedly strain wire bonds, which can lead to metallurgical fatigue. Large chip sizes for high levels of integration have resulted in new failure modes, including aluminum pattern deformation, passivation cracks, and chip fracture. The development of rubber-filled, low-stress molding compounds is largely motivated by the need to reduce these encapsulant-induced stresses.

Stress evolution in polymers stems from a number of factors in addition to mismatched coefficients of thermal expansion. Curing-induced shrinkage, contraction due to the evaporation of solvents and volatile by-products, and relaxation accompanying physical aging are three examples. Stress concentration near topographically sharp features, viscoelasticity, competition between curing kinetics and thermal processing history, as well as ambient environment further complicate the issue.

-- As part of an ongoing research program at Berkeley to elucidate the various factors contributing to stress evolution of packaged semiconductor devices and to design process ing conditions to minimize stress we have conducted a chemical kinetics study to determine the rates of epoxy cresol novolac curing. This paper presents our findings, together with a brief literature survey and a few comments on the network structure formation accompanying cure.

PART 1. REVIEW OF LITERATURE

Historically, the reaction of epoxy with phenol has been employed only in a few commercial processes, foremost amongst which was the formation of high molecular weight α,ω-terminated epoxy resins. This involved the reaction of epoxy and the phenolic hydroxyl of bisphenol A (BPA) in order to achieve a linear chain extension of the epoxy, in which BPA was the linkage. The reactions were generally performed at fairly low temperatures, under dilute conditions in solvent in which excess phenol or epoxy was present. Catalysts were found to improve reaction selectivity and rate.

---

† Current address: 3M Corporation, St. Paul, MN.
* To whom correspondence should be addressed.

Contemporary Topics in Polymer Science, Vol. 7., Edited by
J.C. Salamone and J. Riffle, Plenum Press, New York, 1992

Until recently, the epoxy-phenol reaction was considered "unimportant" economically.[1] The fraction of epoxy on the commercial market which is reacted with a phenolic reactive group is in fact quite small. This lack of use, however, belies the high quality and high precision required by applications, namely microelectronics packaging and encapsulation in which these reactants are used. Epoxy resins cured with phenolic novolacs are used extensively as a base material for microelectronics packaging. Newer "high-tech" resins which have been proposed for IC packaging applications also rely on the epoxy-phenolic reaction to achieve crosslinking.[2]

Encapsulants formulated out of epoxy novolac resins are transfer molded to form the protectant shell of electronics packages. Molecular weight increase and crosslink network formation due to chemical reaction govern the rheological and mechanical properties of the encapsulants. It is particularly important to understand cure kinetics because of its effect on stress distribution in the IC package during processing through the changing material properties as the encapsulant is cured. Preliminary experiments using bending beam experiments and specially designed stress chips confirm the significant effects of cure chemistry, stoichiometry, and temperature regimen on the stress state.[3]

The mathematical model for stress prediction under development at Berkeley hinges on structure-property relations for thermosetting polymers. Network structure must be accurately correlated with the degree and rate of epoxy conversion. Epoxy-hardener stoichiometry and chemistry, as well as catalyst loading, influence the kinetics and final network structure. Hence, these influences must be understood before successful stress predictions can be made.

The objective of this chapter is to establish correlations between structure and reaction conditions for a standard commercial formulation consisting of epoxidized cresol novolac (ECN) and phenolic novolac (PN) as they are catalyzed by triphenyl-phosphine (TPP). These compounds are shown in Figure 1. The reactive groups are the epoxide on the ECN and the phenolic hydroxyl on the PN.

Two types of reaction are possible in the overall reaction of phenol and epoxy. The *primary* reaction is that of phenol with epoxide:[4]

(1)

This reaction generates secondary hydroxyls, which also exhibit reactivity with epoxide, and results in the *secondary*, or *branching*, reaction.

(2)

Reaction (Rxn.) 1 creates a second-order crosslink; Rxn. 2 creates a third-order crosslink (or higher if further hydroxyls react). The selectivity between these two reaction mechanisms (and resulting material structure) depends on the catalyst, the active hydrogen compound being added (phenolic or aliphatic hydroxyl), catalyst concentration, temperature, and stoichiometry.[1] Another important factor in thermosets is crosslink density and the resulting mobility restrictions, which would favor reaction by the more mobile reactants in the later stages of cure. Finally, diffusion limitations may quench or hinder the reaction if the cure temperature is less than the ultimate Tg of the fully cured resin.

The various reaction products and their distribution will determine the evolving network structure and its mechanical properties. One goal, then, for property

Epoxidized Cresol Novolac (ECN)

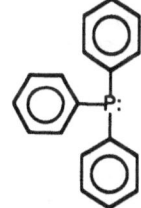

$n = 1 - 2$

Phenolic Novolac (PN)

$n = 1 - 2$

Triphenyl Phosphine (TPP)

Figure 1.   Chemical structures of the reactants used in this study.

optimization is to ascertain which propagation mechanism produces the most desirable material properties, and under what conditions that mechanism is favored.

Overview of Mechanisms

Competing ideas drawn from the literature on the catalyzed epoxy-phenolic curing mechanism are reviewed here and will be evaluated in subsequent chapters with regards to the results of experiments carried out in this study; the most common catalysts employed are Lewis bases. Many reaction mechanisms have been suggested for the Lewis-base catalysis of the epoxy-phenolic reaction.[4-20] The basic reactions these authors attempted to describe are shown in Rxns 1 and 2. These models differ significantly in their description of which elementary molecular reactions occur to achieve the net reaction. In some of these models, the Lewis-base acts as a true catalyst; in others, it first forms a complex with one or two of the reactants before the catalytic nature of the complex is activated.

Reaction quenching due to diffusion limitations or steric limitations are also significant issues, as is catalyst degradation. For example, a commonly used catalyst is triphenyl phosphine, TPP. In the case of TPP, it is important to consider the significance both of a possible oxidation reaction, which may degrade TPP into inert triphenylphosphine oxide, and of the subsequent effect of such loss on the epoxy reaction rate.

The existing literature will be reviewed here in several subsections. The first is a review of the uncatalyzed epoxy-phenol reaction. Catalytic systems are then discussed. The literature of catalyzed model systems will then be reviewed, followed by a review of catalyzed bulk reactant systems.

Uncatalyzed Epoxy-Phenol Reaction

Shechter and Wynstra were the first to examine the uncatalyzed epoxy-phenol reaction.[4] They found that no reaction occurred when equimolar quantities of phenol and phenyl glycidyl ether (PGE, a low molecular weight epoxy) were held at 100°C. When held at 200°C, however, epoxide disappeared at a much faster rate than the phenol, with the net result that about 60% of the reaction was epoxide with phenol (Rxn. 1) and 40% was epoxide with secondary hydroxyl (Rxn. 2). The extent of the second reaction is particularly significant when one considers that the secondary hydroxyl concentration was originally zero and became finite only when some epoxide had reacted with phenol.

Shechter demonstrated that phenol serves as a catalyst for the secondary hydroxyl-epoxide reaction. This was shown by reacting a mole of PGE with a mixture of one mole of phenol and a half mole of dipropylene glycol at 200°C. The results of this reaction indicated that in 8 hours 63% of the epoxy reacted with the alcohol and 32% with the phenol. In contrast, only 40% of the epoxy reacted with the alcohol in 16 hours when there was no phenol present, thus indicating that the phenol is the active catalyst in the epoxy-secondary hydroxyl reaction.

Catalysis Considerations

The uncatalyzed epoxy reaction occurs far too slowly for high-speed integrated circuit packaging procedures, which require reaction to be essentially complete in a few minutes at temperatures around 170°C. Tighter control over the reaction selectivity leads to greater product consistency and reliability. An appropiate catalyst or combination of catalysts can speed up the reaction and provide tighter control over the reaction selectivity. Many types of catalysts have been used; these include acids and bases, group 5a compounds (tertiary amines, phosphines), and quaternary ammonium complexes.[13-16]

Triphenylphosphine

Banthia & McGrath summarized the arguments for using TPP as a catalyst most convincingly.[13] In general, triaryl nucleophiles of group 5a elements, such as TPP and tertiary amines, are nucleophiles of very low basicity and have sterically hindered central atoms. Both of these properties have been shown to reduce the side reaction of the epoxy with the secondary hydroxyl group under dilute conditions in solvents such as dioxane and tetrahydrofuran in which there is excess phenolic and low temperatures (80°C-100°C). These factors also reduce the probability of an undesirable epoxide-epoxide side reaction.

In a study comparing TPP to several other catalysts, TPP was found both to accelerate the epoxy-phenol reaction most rapidly and to show the highest selectivity of the phenolic hydroxyl of bis-phenol A toward epoxy reaction under mild temperatures and dilute concentrations. Banthia and McGrath concluded that little or no side reaction of epoxy with secondary hydroxyls took place. The reasoning was that a necessary catalytic intermediate for the side reaction, the formation of an ion pair between the secondary hydroxyl group and the phosphorous atom, was hindered by the low basicity of the triphenylphosphine and also because the phosphorous was sterically encumbered. However, no concrete data is presented to support this interpretation.[13]

It is unclear, *a priori*, whether the conclusions reached for mild reaction conditions in dilute solutions of DGEBA and bis-phenol A apply to the bulk epoxy novolac-phenolic novolac crosslinking reaction at high temperatures. The present work seeks to contrast fast curing reactions in this bulk state (20-60 minutes) with those carried out in a dilute solvent solution over long reaction times (48-72 hours). This comparative work is partially motivated by the findings of Lunsford which claim that milder reaction conditions, such as those used by Banthia and McGrath, are expected to decrease the rate of epoxy-secondary hydroxyl reaction.[12] Conversely, Lunsford's findings imply that significant secondary hydroxyl reaction with epoxide may occur under the harsher reaction conditions seen in the novolac reaction.

Chemical kinetics studies performed on the epoxy-phenolic reaction which use tertiary amine catalysts are helpful in understanding TPP catalysis because these catalysts share similar chemical properties. Phosphorous, for example, falls in the same elemental group as nitrogen, group 5a. The most significant differences in chemical properties would lie in the basicity of the phosphine as compared to the amine, steric effects, catalytic strength, and possible oxidatative pathways.

A potential disadvantage of TPP is that it can oxidize at high temperature into TPP=O (triphenyl phosphine oxide). Triphenylphosphine oxide is an unreactive material whose presence does not degrade the properties of the resin.[6, 15] If such deactivation of the TPP occurs through oxidation, the kinetics would be made substantially more complicated.

TPP, ECN and PN are all friable solids, and each enables stable molding powders to be formulated. An added advantage of TPP is that it does not catalyze the reaction at room temperature; consequently, very stable resin mixtures can be prepared.[17]

Reaction Conditions

Most of the kinetics experiments reported in the literature for the epoxy-phenol reaction have been performed under dilute conditions in solvents such as dioxane and tetrahydrofuran. Many were performed with excess phenolic, and low temperatures

(80°C-100°C) were another common factor. These factors in and of themselves serve to reduce the probability of the undesirable side reactions between epoxy-epoxy and epoxy-secondary hydroxyl.[12] Studies on monofunctional model compounds, such as PGE and phenol, will not exhibit the steric and diffusional limitations which are important for crosslinking resins.

The subsequent discussion of the reaction mechanisms will center on catalyzed reactions on the model systems - those which do not crosslink - and on catalyzed reactions in bulk systems - those which do crosslink. Finally, the catalyst deactivation issue will be examined.

Catalyzed Reactions in Model Systems

Shechter and Wynstra (1956)

Shechter and Wynstra explored phenyl glycidyl ether (PGE) reactions with a variety of reactants, including alcohols and phenols.[4] Base catalysts, such as tertiary amines and potassium hydroxide, were found to improve the epoxy-phenol reaction selectivity and to lower substantially the activation energy. Shechter and Wynstra titrated for unreacted epoxy directly using pyridinium chloride to measure the epoxide and phenol concentrations over time as the reaction progressed.

Shechter and Wynstra deduced that tertiary amines initiate epoxide polymerization by adding to the epoxide, which is then followed by a reaction with alcohol:

$$(3)$$

$$(4)$$

This first step had been proposed earlier by Narracot.[21] The necessity of the second step was deduced from the effect of isopropyl alcohol concentration on the rate of epoxide consumption. The reaction rate was found to depend sharply on the alcohol concentration: no reaction occurred when there was no alcohol present, and a more rapid reaction occurred as the alcohol concentration was increased. Shechter believed these results indicate that the quaternary base formed in Rxn. 3 is not a catalyst, since the proximity of the base's positive and negative charges may diminish its catalytic ability.

Further support for this interpretation was found in the observation that steric factors are of considerable importance in determining the catalytic power of the tertiary amine on the alcohol-glycidyl ether reaction: the larger the pendant groups on the tertiary amine, the less effective the catalyst. The most effective catalysts were those with methyl groups, the least effective were those with 2-ethylhexyl groups. For the largest pendant groups, the rate of reaction was no greater than the noncatalyzed reaction.[4] The larger pendant groups make it impossible for the catalytic center to get in proper position relative to the reactants in order to catalyze the reaction.

Shechter further studied Rxns. 3 and 4 with aromatic compounds. The phenoxide ion, ϕO^- (where ϕ represents the aromatic benzyl), formed in Rxn. 4 proceeds to attack the epoxide as follows:

$$\text{ww}\langle\bigcirc\rangle\text{-O}^- + \underset{O}{CH_2\text{-}CH}\text{wwww} \longrightarrow \text{ww}\langle\bigcirc\rangle\text{-O-CH}_2\text{-}\underset{O^-}{CH}\text{wwww} \tag{5}$$

This alkoxide ion reacts immediately with phenol present in excess because of its high order of basicity in order to regenerate the phenoxide ion. This cycle is repeated and the possibility of side reactions taking place is diminished:

$$\text{ww}\langle\bigcirc\rangle\text{-O-CH}_2\text{-}\underset{O^-}{CH}\text{wwww} + \text{ww}\langle\bigcirc\rangle\text{-OH} \tag{6}$$

$$\longrightarrow \text{ww}\langle\bigcirc\rangle\text{-O-CH}_2\text{-}\underset{OH}{CH}\text{wwww} + \text{ww}\langle\bigcirc\rangle\text{-O}^-$$

In their initial studies, Shechter and Wynstra had demonstrated that these catalysts could achieve high selectivity of the epoxy-phenol reaction in a solution of excess phenol. They later reversed the situation and studied the reaction selectivity in a solution of excess glycidyl ether in which the phenol was used as the limiting reagent. The excess of epoxide over phenol was measured until phenol had practically disappeared, and the results in all cases indicated high selectivity towards the epoxy-phenol reaction. The tertiary amine, benzyldimethylamine, was somewhat more effective than potassium hydroxide; benzyltrimethylammonium hydroxide was even more powerful. First-order kinetics were observed for all reactions. Since it was postulated that the phenoxide ion was common to all these reactions, the observed differences in reaction rates were linked to the cation. In was not determined, however, whether the cation effect is one of different degrees of dissociation of the phenol salts or of some other phenonomenon.

Sorokin and Shode (1966)

Sorokin and Shode investigated the reaction of epoxy with phenol in the presence of tertiary amines.[9, 10] The epoxy studied was 1,2-epoxy-3-phenoxypropane (EPP), which is a short-chain monofunctional α-epoxide. The experiments were performed in solvent at 50°C.

The first observation made was of a chemical reaction between the epoxy and the tertiary amine. Sorokin and Shode observed the formation of a quaternary ammonium phenoxide in stoichiometric concentrations of epoxy and amine, which was due to the reaction of the epoxy with the tertiary amine. No product of the epoxy-phenol reaction was noted:

$$\underset{O}{\text{ww}CH\text{-}CH_2} + HO\text{-}\phi + R_3N \tag{7}$$

$$\updownarrow$$

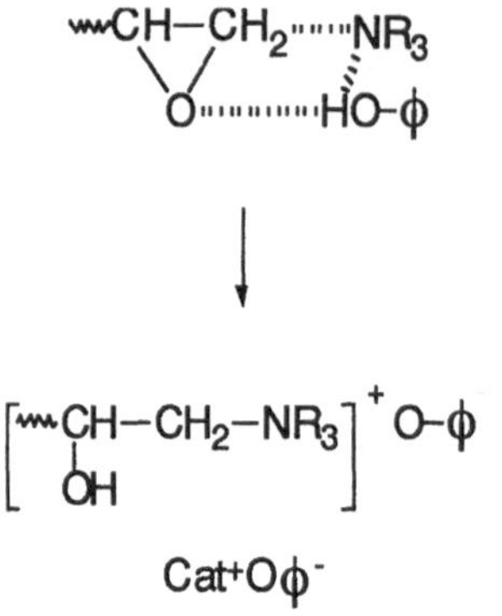

$$
\begin{array}{c}
\text{wwCH-CH}_2\text{''''''NR}_3 \\
\diagdown\!\diagup \quad \vdots \\
\text{O''''''''HO-}\phi
\end{array}
$$

$$\Big\downarrow \tag{8}$$

$$
\left[
\begin{array}{c}
\text{wwCH-CH}_2\text{-NR}_3 \\
|\\
\text{OH}
\end{array}
\right]^{+} \text{O-}\phi
$$

$$\text{Cat}^+\text{O}\phi^-$$

Elemental analysis of extraction residue was used to identify reaction products. Nearly all of the tertiary amine was converted into the quaternary ammonium phenoxide in 90-95% yield (based on the amine) at amine concentrations lower than or equal to the epoxy concentration. The product of the reaction between epoxy and phenol, 1,3-diphenoxy-2-propanol, was formed only when the initial amine concentration was less than the epoxy concentration. The yield of 1,3-diphenoxy-2-propanol was dictated by the difference between the initial epoxy and amine concentrations. Thus, their results indicate that the reaction of epoxy with phenol in the presence of tertiary amines must initially go through the stage in which the quaternary ammonium phenoxide, $\text{Cat}^+\text{O}\phi^-$, is formed.

Sorokin and Shode deduced that the formation of 1,3-diphenoxy-2-propanol, the reaction product of EPP-phenol, is catalyzed by the strong base, $\text{Cat}^+\text{O}\phi^-$:

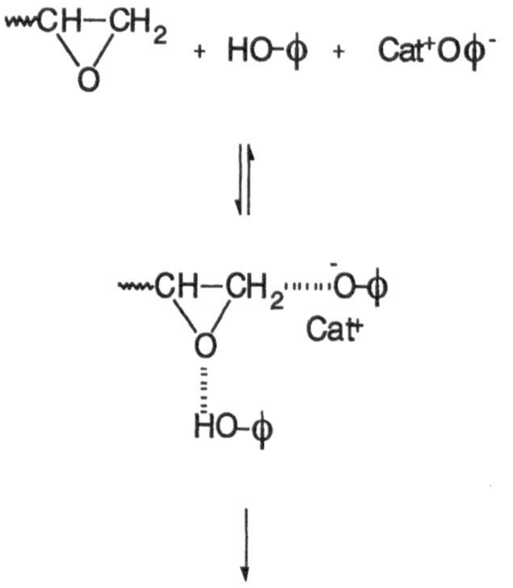

$$
\begin{array}{c}
\text{wwCH-CH}_2 \\
\diagdown\!\diagup \\
\text{O}
\end{array}
+ \text{HO-}\phi + \text{Cat}^+\text{O}\phi^-
$$

$$\Big\Updownarrow$$

$$
\begin{array}{c}
\text{wwCH-CH}_2\text{''''''O-}\phi \\
\diagdown\!\diagup \qquad \text{Cat}^+ \\
\text{O} \\
\vdots \\
\text{HO-}\phi
\end{array}
\tag{9}
$$

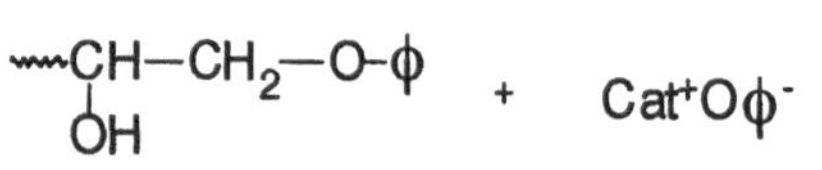

$$\Big\downarrow$$

$$
\begin{array}{c}
\text{wwCH-CH}_2\text{-O-}\phi \\
|\\
\text{OH}
\end{array}
+ \quad \text{Cat}^+\text{O}\phi^-
\tag{10}
$$

Analysis of the reaction products showed that the reaction of epoxy with the tertiary amine, Rxns. 7 & 8, is preferred over formation of 1,3-diphenoxy-2-propanol, Rxns. 9 & 10. Moreover, only the reaction of epoxy with the tertiary amine occurs at equimolar proportions of all reactants. Thus, the tertiary amine reacts with excess epoxy and phenol to form the catalytic complex until all the amine is consumed. The catalyzed epoxy-phenol reaction *then* occurs until the limiting reagent, either epoxy or phenol, is depleted.

The formation of the trimolecular transition state, Rxn. 9, is more complicated for phenols which have more acidic hydroxyl groups, such as p-nitrophenol and 2,4-dinitrophenol with pK_as of 7.15 and 4.11, respectively (that of phenol is defined as 10.00). These highly acidic phenols form strong hydrogen bonds with the tertiary amine, resulting in strongly associated phenol-amine complexes so that there exists almost no free amine to participate in the reaction. (It is also anticipated that a more basic amine or phosphine catalyst may similarly form strong hydrogen bonds with a less acidic phenol.)

$$R_3N \;+\; HO\text{-}\phi \;\overset{K_{assoc}}{\rightleftarrows}\; R_3N''''HO\text{-}\phi$$

$$\Big\Updownarrow K_{ion}$$

$$\overset{+}{R_3NH} \;+\; \overset{-}{O}\text{-}\phi \;\overset{K_{dissoc}}{\rightleftarrows}\; \overset{+}{R_3NH}''''\overset{-}{O}\text{-}\phi \tag{11}$$

The equilibrium constants for the association of the amine with the phenol and the degree of ionization of the binary complex are directly proportional to the acidity of the phenol and the basicity of the amine.

The highest nucleophilic activity (hence, the highest catalytic activity) is possessed by the free amine.[9, 10] The polar complex $R_3N\cdots HO\phi$ is very much less active as a result of the donor-acceptor interaction between the tertiary amine and the proton of the phenol. The polarity of its molecular compound with the base rises and the nucleophilic activity falls with increasing acidity of the phenol. Also, the polarization of the epoxide ring in the trimolecular complex increases and the nucleophilic activity of the tertiary amine bound in the complex diminishes considerably with increasing proton-donor properties of the phenol.[9, 10]

The magnitudes of the association constants for the $R_3N\cdots HO\phi$ complex are very high: $K_{assoc} = 600$ for p-nitrophenol.[22] For such highly acidic phenols, $Cat^+O\phi^-$, which is believed to be the catalyst for weaker phenols, is not formed because of interference by the stong polar complex; this practically excludes the trimolecular reaction altogether. Sorokin and Shode concluded that there must be a different catalyst and, accordingly, they proposed a new mechanism for such cases:

$$\text{wwCH}-\text{CH}_2 \;+\; \overset{+}{R_3NH}''''\overset{-}{O}\text{-}\phi$$

$$\Big\Updownarrow$$

$$\tag{12}$$

$$\text{wwCH}-\text{CH}_2$$

$$\overset{+}{R_3NH}''''\overset{-}{O}\text{-}\phi$$

$$(13)$$

$$
\begin{array}{c}
\overset{\displaystyle \|}{\text{O}} \\
\end{array}
$$

$$
\text{\textasciitilde CH}-\overset{\delta+}{\text{CH}_2}\cdots\overset{\delta-}{\text{O}}-\phi
$$
$$
\underset{\delta-}{\text{O}}\cdots\underset{\delta+}{\text{H}}\text{NR}_3
$$

$$(14)$$

$$
\text{\textasciitilde}\underset{\overset{|}{\text{OH}}}{\text{CH}}-\text{CH}_2-\text{O}-\phi \quad + \quad \text{R}_3\text{N}
$$

In this reaction, the catalytic role of the tertiary amine resides in the intensification of the nucleophilic activity of the proton-donor molecule, i.e., the phenol, by the withdrawal of a proton. The oxygen of the epoxide ring can form a hydrogen bond with the - proton of the ammonium in which there is polarization of the bond in the ring. The reaction proceeds through a cyclic transition state in which there is an alternation of charges.

Some additional points may be made here in regards to how Sorokin and Shode's findings apply to tertiary phosphine catalysis of the epoxy-phenol reaction. An excess of phenolic hydroxyl drives the equilibrium toward the dissociated state (Rxn. 11). An increase in temperature also drives the equilibrium toward dissociation. Thus, the phenoxide ion formed by the equilibrium may enable the ionic mechanism of Rxns 5 and 6 to propagate the epoxy cure. The electrons in the outer shell of the phosphorous in TPP are more loosely held than those of the nitrogen in a corresponding tertiary amine. The nitrogen has 3 electrons in its 2p outer shell, whereas the phosphorous has 3 electrons in its 3p outer shell. This may draw the TPP into forming ionic complexes with less acidic phenols. Therefore, two conceivable means of creating an ionic species potentially capable of propagating the reaction for TPP-based systems have been demonstrated.

Banthia and McGrath (1979)

Banthia and McGrath proposed a mechanism for the TPP catalysis of the reactions between epoxy resins and various curing agents, such as carboxylic acids, anhydrides, thiols and phenols.[13] The reactive pathway they proposed for the epoxy-phenol reactions is:

$$
\phi_3\text{P} \quad + \quad \text{HO}-\text{\textcircled{O}}\text{\textasciitilde} \longrightarrow \phi_3\overset{+}{\text{P}}\text{H}\ \overset{-}{\text{O}}-\text{\textcircled{O}}\text{\textasciitilde}
$$

$$(15)$$

$$
\phi_3\overset{+}{\text{P}}\text{H}\ \overset{-}{\text{O}}-\text{\textcircled{O}}\text{\textasciitilde} \quad + \quad \underset{\text{O}}{\overset{\displaystyle \text{CH}_2-\text{CH}\text{\textasciitilde}}{\triangle}}
$$

$$\longrightarrow \quad \phi_3 \overset{+}{P} \; H \overset{-}{O} \atop \text{www} \overset{|}{C}H - CH_2 - O - \langle O \rangle \text{www}$$

$$(16)$$

$$\longrightarrow \quad \underset{\text{www}}{\overset{OH}{\underset{|}{C}H}} - CH_2 - O - \langle O \rangle \text{www} \;\; + \;\; \phi_3 P$$

$$(17)$$

In this mechanism, the TPP first forms an activated ion-pair complex with the phenol (Rxn. 15), which then attacks the epoxide group to generate another ion-pair complex (Rxn. 16). This complex then degenerates, and this results in an epoxy-phenol linkage which allows for the release of TPP (Rxn. 17). This mechanism is similar to the one proposed by Sorokin and Shode for acidic phenols.

Catalyzed Reactions in Bulk Systems

More recently, the epoxy-phenol reaction has been studied in the more concentrated bulk state.[6, 7, 11, 23] In these reactions, no solvent was present and temperatures were higher (150°C to above 200°C) than in the previous studies. Steric and diffusional limitations may also affect the reaction.

Romanchick (1981)

Romanchick used TPP as the catalytic agent to study the reaction of an elastomer-modified epoxy with bisphenol A.[6, 7] The goal was to achieve a linear chain extension of the epoxy with bisphenol A by linking a pair of the elastomer modified epoxies. Based on his experimental results, he postulated that the reaction proceeds in the following manner:

$$\phi_3 P \;\; + \;\; \underset{O}{CH_2 - CH\text{www}} \;\; \longrightarrow \;\; \underset{\overset{-}{O} - \overset{|}{C}H}{\overset{\phi_3 \overset{+}{P} - CH_2}{}}$$

$$(18)$$

$$\underset{\overset{-}{O} - \overset{|}{C}H}{\overset{\phi_3 \overset{+}{P} - CH_2}{}} \;\; + \;\; HO - \langle O \rangle \text{www}$$

$$\longrightarrow \;\; \underset{HO - \overset{|}{C}H}{\overset{\phi_3 \overset{+}{P} - CH_2}{}} \;\; + \;\; \overset{-}{O} - \langle O \rangle \text{www}$$

$$(19)$$

$$\xrightarrow{\hspace{1cm}} \underset{\overset{|}{OH}}{\sim\sim CH} - CH_2 - O - \bigcirc - \sim\sim \quad + \quad \phi_3 P \tag{20}$$

Here, the nucleophilic attack by TPP opens the epoxide and then produces a zwitterion (Rxn. 18). Proton abstraction from phenol yields the phenoxide ion (Rxn. 19), which subsequently reacts with the electrophilic carbon attached to the positive phosphonium ion and, thereby, regenerates the catalyst (Rxn. 20).

Romanchick proposed that two side reactions may occur during isothermal aging of the rubber-modified epoxies at 175°C for up to 8 hours. It was believed that these occurred via non-catalytic routes. The first is an epoxy homopolymerization reaction:

$$\tag{21}$$

The experimental results indicated that this homopolymerization reaction (Rxn. 21) is an unlikely explanation of how the epoxy chemically ages. The second aging reaction is that of epoxy with the secondary hydroxyl generated from the epoxy-phenolic reaction:

$$\tag{22}$$

$$\tag{23}$$

The uncatalyzed secondary hydroxyl-epoxy reaction seems to explain the long-time aging data best; however, this reaction occurs very slowly compared to the catalyzed epoxy-phenol reaction.

Gagnebien (1985)

Gagnebien, et al., have performed the most thorough study to date of the epoxy-phenol reaction in the melt using a series of tertiary amine catalysts.[11] NMR and gel permeation chromatography results indicate the presence of two types of side reactions in addition to the main epoxy-phenol reaction: a branching reaction of epoxy onto the secondary hydroxyl formed by the main reaction, and an epoxy homopolymerization which is initiated by the catalyst. The corresponding mechanisms for these reactions are described by the following schemes.

The *main* epoxy-phenol reaction proceeds through the following steps:

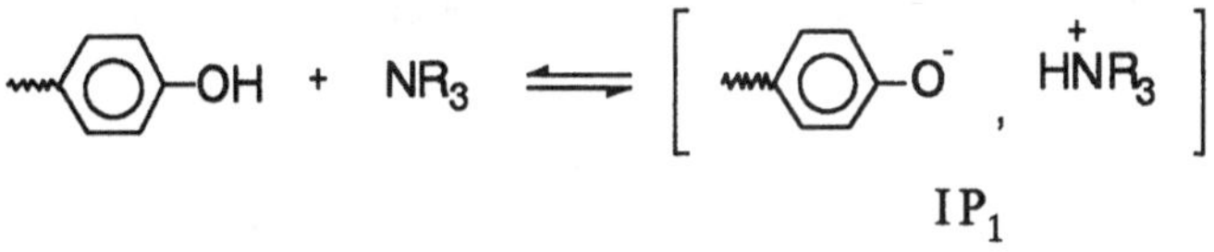

$$\text{~~}\langle\bigcirc\rangle\text{—OH} + NR_3 \rightleftharpoons \left[\text{~~}\langle\bigcirc\rangle\text{—O}^- , \; H\overset{+}{N}R_3 \right]$$

$$IP_1 \tag{24}$$

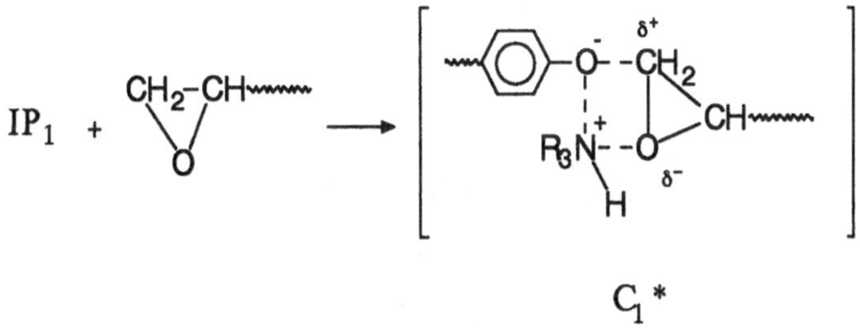

$$IP_1 + \; {}^{CH_2-CH\text{~~}}_{\diagdown O \diagup} \longrightarrow \left[\text{~~}\langle\bigcirc\rangle\text{—}\overset{\delta^-}{O}\text{-}\text{-}\overset{\delta^+}{CH_2} \;\; {}_{CH\text{~~}} \right.$$

$$\left. R_3\overset{+}{N}\text{-}\text{-}\underset{H}{\overset{\delta^-}{O}} \right]$$

$$C_1^* \tag{25}$$

$$C_1^* \xrightarrow{3} \text{~~}\langle\bigcirc\rangle\text{—O-CH}_2\text{—}\underset{OH}{CH}\text{~~} + NR_3$$

$$P \tag{26}$$

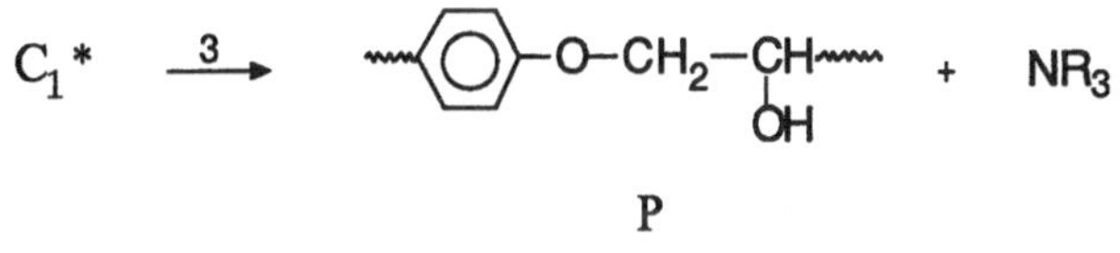

This mechanism is similar to the Sorokin and Shode acidic phenol mechanism. However, a somewhat different transition state, C_1^*, is encountered. In the *main* reaction, the phenol and the amine associate instantaneously and form an ion pair, IP_1, which is probably tightly associated, as Sorokin and Shode observed.[9, 10] IP_1 leads to the complex, C_1^*, through a quadripolar interaction with the epoxy ring (Rxn. 25). C_1^* then dissociates rapidly to form the β-hydroxyether, P, and the amine. Hence, the tertiary amine, NR_3, is here postulated to be a true catalyst for the main reaction.

Gagnebien proposes three possible side reactions and mechanisms other than the *main* reaction. The first is *epoxy homopolymerization* as initiated by the free tertiary amine. The initial step is the formation of a zwitterion, Z, in Rxn 27. The zwitterion then attacks another epoxy ring, which forms a larger zwitterion (Rxn 28). This process could repeat *ad infinitum* until all the epoxy is incorporated into the zwitterion:

$$NR_3 \; + \; \underset{O}{\overset{CH_2-CH\text{\textbf{\textasciitilde\textasciitilde\textasciitilde}}}{\triangle}} \;\longrightarrow\; \underset{O^-}{\overset{+}{R_3}N-CH_2-CH\text{\textbf{\textasciitilde\textasciitilde}}}$$

$$Z \tag{27}$$

$$Z \; + \; \underset{O}{\overset{CH_2-CH\text{\textbf{\textasciitilde\textasciitilde\textasciitilde}}}{\triangle}} \;\longrightarrow\; \overset{+}{R_3}N-CH_2-\underset{\underset{O^-}{|}}{\overset{|}{CH}\text{\textbf{\textasciitilde\textasciitilde}}}O-CH_2-CH\text{\textbf{\textasciitilde\textasciitilde}}$$

$$\tag{28}$$

Shecter and Wynstra found that this reaction requires the presence of alcohol to proceed, and Gagnebien did not find this reaction to be a significant means of epoxy consumption.

The second side reaction proposed is a *branching* reaction, and is more complex than the epoxy homopolymerization:

$$\text{P} \qquad\qquad\qquad\qquad\qquad \text{M} \tag{29}$$

$$C_2{}^* \tag{30}$$

$$C_2{}^* \;\longrightarrow\; NR_3 \; + \; \cdots \tag{31}$$

Free tertiary amine associates with the secondary hydroxyl groups and leads to an amino alcohol, M (Rxn. 29). The amino alcohol forms an activated complex, $C_2{}^*$, with the epoxy dipole (Rxn. 30). $C_2{}^*$ dissociates to form the branched product (Rxn. 31). This mechanism accounts for the observation that branching takes place only after a certain extent of epoxy conversion, i.e., when the free amine concentration increases as it is released from the phenol ion pair IP_1 in Rxn. 24. Free amine is a true catalyst in this *branching* reaction since it is regenerated.

The third side reaction is *zwitterion catalyzed branching*. This can take place since the zwitterion, Z (formed in Rxn. 27), can react with phenol (Rxn. 32) to form an associated ion pair, IP$_2$, which is able to catalyze epoxy-phenol reaction (Rxn. 33), thereby leading to an alcoholate, IP$_3$.

$$R_3\overset{+}{N}-CH_2-\underset{\underset{O^-}{|}}{C}H\sim\sim \;+\; \sim\sim\!\!\bigcirc\!\!-OH \;\longrightarrow\; \left[\sim\sim\!\!\bigcirc\!\!-O^-\;,\;Q^+ \right]$$

Z   IP$_2$

(32)

$$IP_2 \;+\; \overset{CH_2-CH\sim\sim}{\underset{O}{\bigtriangleup}} \;\longrightarrow\; \left[\sim\sim\!\!\bigcirc\!\!-O-CH_2-\underset{\underset{O^-}{|}}{C}H\sim\sim\;,\;Q^+ \right]$$

IP$_3$

(33)

IP$_3$ can behave in one of two ways: either it can be protonated by free phenol, which results in the formation of the main reaction product, β-hydroxyether (Rxn. 34), or it can react with an epoxy to yield a branched product (Rxn. 35). Gagnebien concludes that epoxy homopolymerization is much less important than branching from the secondary hydroxyl group:

$$\sim\!\!\bigcirc\!\!-OH \;\longrightarrow\; \sim\sim\!\!\bigcirc\!\!-O-CH_2-\underset{\underset{OH}{|}}{C}H\sim\sim \;+\; IP_2$$

P

(34)

IP$_3$

$$\overset{CH_2-CH\sim\sim}{\underset{O}{\bigtriangleup}} \;\longrightarrow\; \left[\begin{array}{c} \sim\sim CH_2-CH\sim\sim \\ \underset{O-CH_2-\underset{\underset{O^-}{|}}{C}H\sim\sim}{|} \end{array} \quad Q^+ \right]$$

(35)

IP$_3{'}$

Hale (1988)

A recent study of the reaction of epoxide with phenol in the presence of tertiary amines was performed by Hale.[23] Hale formulated his analysis according to the the mechanism of Sorokin and Shode. The reaction mechanism is here described in a short-hand notation (compare to Rxns. 7 - 10):

$$E + P + T \dashrightarrow C \qquad \text{(rate } k_1) \tag{36}$$

$$E + P + C \dashrightarrow R + C \qquad \text{(rate } k_2) \tag{37}$$

where E represents the epoxide, P the phenol, and T the tertiary amine or phosphine, C is an intermediate catalytic complex (the quaternary ammonium phenoxide) which is the true catalyst, and R is the product of the addition of phenol to epoxide. Sorokin and Shode found Equation 36 to be first-order with respect to epoxide and to the amine; however, they observed that the apparent order with respect to phenol to changes as the concentration is changed. Equation 37 is also first-order with respect to epoxide and with respect to the catalyst, C; again, no definite order with respect to phenol was observed. The order of the reaction with respect to phenol is used as an adjustable parameter in the development of a phenomenological kinetic equation.

The analysis proceeds by assuming a stoichiometrically balanced epoxy-phenol system. By noting that the concentration of catalyst is proportional to the conversion of accelerator, x_T, the kinetic equations can be written in terms of fractional conversion as:

$$dx_E/dt = k_1 (1-x_E)(1-x_P)^r (1-x_T) + k_2 (1-x_E)(1-x_P)^r x_T \tag{38}$$

$$dx_T/dt = k_1 (1-x_E)(1-x_P)^r (1-x_T) \tag{39}$$

where x_E is the fractional conversion of epoxide groups, x_P is the fractional conversion of phenol, x_T is the fractional conversion of the accelerator, and r is an adjustable reaction order. Several assumptions have been made here. The order of reaction with respect to phenol, r, is assumed to be the same in both Rxns. 36 and 37. Also, epoxy side reactions are not considered. The desired range of applicability for this analysis was limited to the pregel region, corresponding to $x_E < 0.25$. In this region there are always plenty of phenol groups available, and this favors a clean epoxy-phenol reaction. For this stoichiometrically balanced system, the terms $1-x_E$ and $1-x_P$ are equal, and Equations 38 and 39 become:

$$dx_E/dt = k_1 (1-x_E)^n (1-x_T) + k_2 (1-x_E)^n x_T \tag{40}$$

$$dx_T/dt = k_1 (1-x_E)^n (1-x_T) \tag{41}$$

This set of equations validly describes the Sorokin and Shode mechanism. In an attempt to simplify the equations into one simple phenomenological equation, it is assumed that the reaction is autoaccelerating, such that x_T scales with x_E:

$$x_T = x_E^m \tag{42}$$

This assumption, however, is questionable and will be discussed later. When Equation 42 is inserted into Equation 40, the equation becomes:

$$dx_E/dt = (k_1 + k_2' x_E^m)(1-x_E)^n \tag{43}$$

where $k_2' = k_2 - k_1$.

Equation 43 has the same form as the expressions used to fit kinetic data for amine-cured epoxies, which are known to be *autoaccelerating*. The amine-cured epoxy reaction is autoaccelerating because *more catalyst is produced* as the reaction proceeds. The catalyst for this reaction is the set of all hydroxyl groups (including water and the secondary hydroxyls formed by the reaction).

There is a net increase in the number of hydroxyl groups due to the opening of the epoxide ring as the epoxy-amine reaction progresses. Since hydroxyl groups serve to catalyze the reaction, the amount of available catalyst and, hence, the reaction rate increases in proportion to the extent of reaction of the epoxy. This phenomenon is called *autoacceleration*, and is characteristic of chemistries which generate more catalyst in the course of curing.

The situation is different for the epoxy-phenol reaction. No catalyst is generated as the epoxy-phenol reaction proceeds. An initial charge of catalyst is placed in the system, and reaction proceeds until the limiting reagent is consumed. The maximum concentration of active catalyst is limited to the initial concentration of catalyst. Secondary hydroxyl groups are generated during the reaction but do not catalyze the reaction according to any of the reaction mechanisms that have been proposed. Therefore, the reaction is not *autoaccellerating*.

An accelerating effect will be observed, however, according to Sorokin and Shode's mechanism, due to the necessary formation of the active catalytic complex, C. Initially no active complex is available and epoxy-phenol reaction does not occur. As Rxn 36 generates the active complex, Rxn. 37 consumes epoxy and phenol more rapidly in proportion to this active complex becoming available.

This accelerating behavior appears similar to the autoacceleration phenonomen observed for the S_N2 addition due to the initiation reaction (Rxn. 36), yet there are substantial and fundamental differences between the two phenomena. First, in the autoaccelerating reaction the increasing active catalyst concentration is directly proportional to the extent of epoxy reaction. This is not the case for the epoxy-phenol reaction. The active complex formation reaction, Rxn. 36, is *independent* of the propagation reaction, Rxn. 37.; thus, any observed proportionality is simply coincidental. Second, the rate of production of active catalytic complex will decrease as the tertiary amine or phosphine is consumed in Rxn. 36. The maximum concentration of active catalytic complex is limited by the initial charge of catalyst. These considerations are not accounted for in the autoacceleration model.

These weaknesses become apparent upon examination of the modeling results. The parameters for Equation 43 were obtained by non-linear regression from DSC rate-versus-conversion data. The reaction orders m and n were found to be 3.33 and 7.88, respectively. This empirical fit is valid only to about 35% conversion of the epoxy for temperatures above 150°C. This empirical fit was adequate since only a description of the kinetics through the gel point of the system was needed (x_{gel} = 0.25). However, this fit is inadequate for describing the reaction kinetics and resultant network formation through the *full course* of the reaction, and it misrepresents the known physics of the reaction.

Catalyst Deactivation

Possible deactivation of TPP through oxidation is a complicating feature of this chemistry. Two major pathways for catalyst loss are through oxidation and volatilization:

$$\phi_3P \xrightarrow{O_2} \phi_3P=O \tag{44}$$

$$\xrightarrow{\Delta} \phi_3P \text{ (vapor)} \tag{45}$$

Volatilization of the catalyst is in all probability negligible, given the low diffusivity of the bulky TPP and its high propensity for association with phenol. Direct oxidation of TPP is also likely to be slow, given the dilute concentration of dissolved oxygen in the resin.

Wittig, Romanchick, and Uejima have described some potential reaction paths by which oxidation may occur. [6, 18, 24] Wittig reported a mechanism by which TPP may be oxidized into TPP=O through reaction with an epoxide:

$$\phi_3P + \underset{O}{\overset{CH_2-CH\text{\large\textasciitilde}}{\triangle}} \longrightarrow \underset{O-CH}{\overset{\phi_3\overset{+}{P}-CH_2}{\mid}} \longrightarrow \underset{CH_2=CH\text{\large\textasciitilde}}{\overset{\phi_3P=O}{+}} \tag{46}$$

Romanchick reported that this final step, wherein the zwitterion decomposes into a terminal olefin and triphenylphosphine oxide, can only occur when the phenol is exhausted. Therefore, deactivation of the TPP is not a significant factor in the epoxy-phenol kinetics. Romachick's data also indicate that TPP oxidation is helpful in reducing unwanted branching at the secondary hydroxyls when the primary epoxy-phenol reaction has run its course.

It is clear from the above discussions that in different mechanisms for the reaction of epoxy with phenol have been reported in the research literature . Some explain the role of the tertiary amine or phosphine as being that of a catalyst. Others deem its role to be part of a larger activated catalytic complex. Still others explain its role as being that of an ionic reaction initiator. Next,these proposed mechanisms will be evaluated with regards to their ability to explain the experimental observations for an epoxy cresol novolac-phenolic novolac resin.

PART II. KINETICS EXPERIMENTS

Materials and Experimental Methods

System Studied

The resin system studied consists of epoxidized cresol novolac (ECN, Shell DPS-164) which was cured with a phenolic novolac (PN, Borden 173 1) and catalyzed by triphenyl phosphine (TPP, Aldrich). The epoxy equivalent weight of the ECN is 194 gm/mole epoxide; the phenolic equivalent weight of the PN is 135.7 gm/mole phenolic hydroxyl. Density measurements were performed using a displacement method on degassed samples. This yielded a value of 1.22 ± 0.03 gm/cm^3 for the ECN, PN and mixed formulations, and this value was found not to vary significantly during cure. All concentrations were calculated on the basis of 1.22 gm/cm^3 density, assuming ideal mixing.

A range of stoichiometries and catalyst concentrations were studied. The stoichiometric ratio, R, is defined as the ratio of moles of epoxy reactive groups to phenolic reactive groups. The unit of phr refers to parts per hundred epoxy resin; for example, 0.5 phr TPP means 0.5 grams of TPP per 100 grams epoxy cresol novolac. A variety of sample compositions were used in the study covering a range of stoichiometric ratio, epoxy. phenolic hydroxyl and TPP concentrations

Sample Preparation

All the ingredients are solids at room temperature. Samples were prepared by both solvent mixing and by melt mixing since it is impossible to achieve uniform mixing on the microscopic scale required for accurate DSC, FTIR and microdielect-rometry experiments by simply mixing the finely ground powders,

For solvent mixing, the solid ingredients were individually weighed to an accuracy of 4 decimal places so as to achieve the desired stoichiometric ratio of degassed ECN and PN, as well as TPP. The solids were then dissolved (~10 wt% solids) in spectroscopic grade acetone under agitation with a stir bar until visual observation indicated that all ingredients were dissolved (about 10-15 minutes). The acetone was then evaporated under high vaccuum at room temperature for 8-24 hours, until the sample had dried completely. The test for dryness was to determine whether the foam, which had formed during evaporation of the solvent, was brittle throughout and exhibited neither stickiness nor any scent of acetone. The very fine powder resulting from this step was densified by melting at 60°C for 10-15 seconds, and resulting samples were stored at room temperature in tightly sealed glass bottles. There is no indication that the epoxy reaction progressed in the solid state at room temperature.

Most of the samples in this kinetics study were prepared by the solvent-mixing process; a few, however, were prepared by the melt-mixing process. Melt mixing was done as follows. ECN was weighed into a metal can, while PN was weighed into a separate can. Each was heated to 150°C, melted, and degassed in a high vacuum oven until the molten liquids ceased emitting gas bubbles. Weight loss due to the degassing process was determined to be negligible. TPP was then added to the PN and was mixed vigorously. The TPP-PN mixture was then added to the the ECN and also was mixed vigorously for about 20 seconds until uniformly mixed. The mixture was poured rapidly into 5x9 inch aluminum pans suspended in a bath of dry ice & isopropyl alcohol. The formulations were solidified throughout after 5-25 seconds. These samples were stored in the same way as the others. No differences were observed in the cure behavior of samples prepared by either the solvent- or melt-mixing methods.

Tools Used

Differential Scanning Calorimetry

DSC measurements were carried out using a Perkin-Elmer differential scanning calorimeter, model DSC-7. The instrument was calibrated using high purity indium and tin standards, and the calorimeter block was cooled using an ice water bath. Samples of about 8-12 mg of the formulations were placed in hermetically sealed aluminum pans and introduced into the DSC at room temperature. Ramp-cures were performed at 20°C per minute from 10°C to 340°C. The reaction exotherm was quantified by integrating the peak area. Glass transition temperatures were determined from heating runs at 20°C/minute, and the mid-point Tg was used throughout this study.

For isothermal cures, the samples were heated rapidly (200°C/min) to the reaction temperature from room temperature; cure temperatures from 130°C to 200°C were utilized. The samples were cured isothermally until no further sign of reaction was observed. The isothermal baseline was established for each temperature by running dummy samples through the identical cure cycle. The data files were transferred to a PC, and the heat flow was integrated over time after subtraction of the baselines.

In order to determine the fraction of epoxy reacted from the reaction exotherm, the heat from the exotherm was scaled to the ideal heat of reaction expected from complete conversion of the epoxy. It is assumed throughout this paper that the epoxy-phenol reaction and epoxy-secondary hydroxyl reaction generate the same amount of heat per reacted epoxy. The ideal heat of reaction for epoxy with hydroxyl groups was found by Hale to be 85.8 kJ/mole epoxy.[23] This value was tested and confirmed by running ramp-cures of ECN-PN formulations consisting of 1 part epoxide to 2 parts phenolic hydroxyl (0.50 stoichiometry). Ample excess phenolic novolac ensured complete reaction of the epoxy with the phenolic. The results from these tests were found to be consistent with the hypothesis that epoxy reacted completely with the phenol; accordingly, the value found by Hale was confirmed.

Titration

The procedure entitled "Determination of α-Epoxy Group in EPON Resins (WPE)" in the Shell Chemical Company EPON Resin Structural Reference Manual was employed to determine the fraction of epoxy reacted during mixing of the formulations. (A similar method is ASTM D1652-73, which uses a hydrogen bromide acetic acid reagent.) The Shell procedure consists of reacting the epoxy with excess tetraethyl-ammonium bromide and then titrating for a pH change with 0.100 normal perchloric acid ($HClO_4$) in glacial acetic acid. The molecular weight per unreacted epoxide (WPE) was thereby determined to be

$$WPE = \frac{(\text{grams of epoxy})(1000 \text{ ml/l})}{(\text{ml HCLO}_4)(\text{N of HCLO}_4)} \tag{47}$$

If Q is defined as $1/WPE$, then Q represents the number of available epoxy functional groups in a given mass of the epoxy resin. Q_0 (= $1/WPE_0$) represents the number of epoxy groups initially available in an unreacted sample and is easily obtained from a titration of the pure ECN. Consequently, the fraction of epoxy unreacted is

$$x_e = 1 - Q/Q_0 \tag{48}$$

The manual states that this method is accurate to within 2%. Several trials were performed and found to yield x_e repeatedly to within a variation of about 0.01.

A deviation from the standard titration method was employed in cases in which the sample had gelled. The sample was ground very finely with mortar and pestle, dissolved into the titrating solution and titrated with perchloric acid until endpoint was achieved. The sample was then agitated for another hour to make sure all the epoxide within the gel had opportunity to react with the titrating solution. This further reaction was noted when the solution changed back from its green endpoint to blue or purple. Additional perchloric acid was added to achieve the green endpoint again. This wait/titrate process was repeated as necessary for all samples until no hints of color change were noted within 2 hours after the last titration. In these cases, the sum total of the titrant was used to determine WPE and x_{epoxy}. It is estimated that the accuracy for these numbers was diminished to 3 or 4% because of the accumulation of volume measurement errors.

Microdielectrometry

Microdielectrometry enables ionic conductivity measurements to be made during cure. This experimental technique and apparatus have been described elsewhere.[25,26] In this study, low conductivity sensors were employed with a Micromet Eumetric System II microdielectrometer. A thermal diode is on the sensor surface for temperature measurements. The dielectric sensors are comb-type electrodes on the surface of an integrated circuit chip; the individual combs are spaced 16 microns apart. On-chip transistors amplify the output signal. Since the electrode spacing stays constant during an experiment, the permittivity and loss factor can be determined directly from current loss and phase lag measurements across the electrodes through the use of tables contained in the microdielectrometer firm-ware. This yields more useful and accurate information than more traditional parallel plate electrodes in which it is difficult to keep the electrode spacing constant and repeatable, especially during heating or cooling.

The microdielectrometer is capable of measuring the dielectric permittivity (dielectric constant) and the dielectric loss factor over a frequency range of 0.005-10,000 Hz. This broad range of frequency span enables the ionic conductivity to be tracked over the full course of the cure. As a rule, the lower the frequency level, the more the ions dominate the dielectric loss and permittivity measurements.

The permittivity, E', and the loss factor, E", measured by the dielectric spectro-meter for a given resin system are influenced by dipole motion, ionic conduction, and electrode polarization. The means by which ionic conductivity is calculated from these measurements can be found in literature.[25,26]

Static charge problem

In order to use the microdielectrometry technique to study the ECN-PN resin, very thick coatings of the resin had to be avoided because of errors induced in the loss factor measurements when such coatings were used. An example of the noise, which is

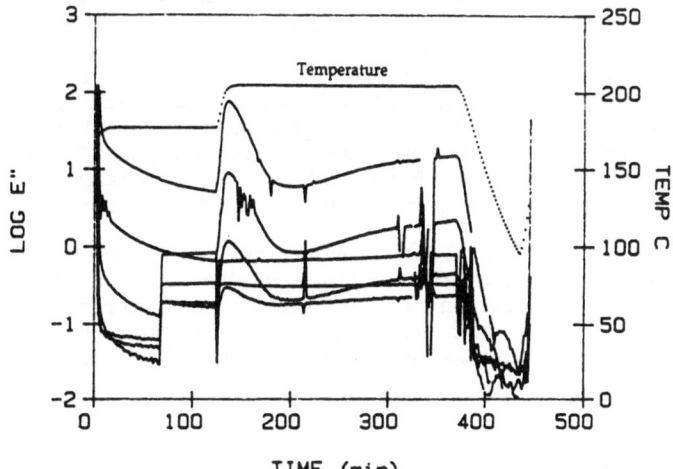

Figure 2.   Noisy loss lactor, E", data resulting from electrical interference
during cure of a 1.18 stoichiometry ECN-PN coating with 1.0 phr TPP.
The coating was approximately 100 mil thick. The temperature is the
top curve: initially 170°C, then 200°C after 120 minutes. The other
curves from top to bottom are E" at 0.1-10000 Hz, increasing by orders
of magnitude.

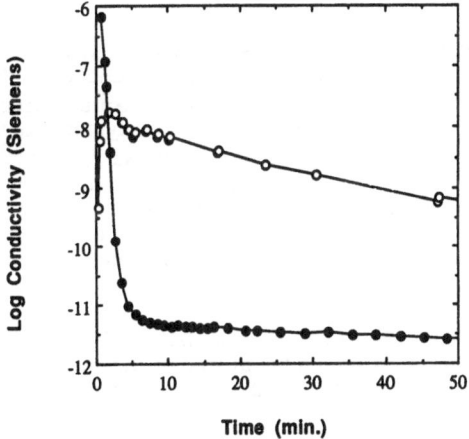

Figure 3.   Ionic conductivity curves for two 1.00 stoichiometry ECN-PN resins
cured at 200°C. The hollow circles represent cure with no TPP. The
solid circles represent cure with 0.5 phr TPP. The catalyzed reaction
proceeds much more rapidly than the uncatalyzed one.

123

attributed to electrical charging, is shown in Figure 2. Difficulties due to static charge when using the low conductivity sensor have also been observed by others.[27] Charge buildup in a bulk resin sample (a casting) was measured using a Monroe Electronics model 230B static electricity detector; it was found to be greater than 1000 Volts surface potential, corresponding to an approximate charge of 10^{-6} Coul/m^2. This charge buildup may be due to a number of causes: mechanical work done in crushing and mixing the ECN, PN, and TPP; fan-forced hot air in the curing oven; and possible ionization during reaction due to an ion-based reaction mechanism. Regardless of which of these causes is responsible, the noise problem is eliminated by using thin coatings of the resin on the sensor. This is because the lower mass of a thin coating stores less total charge than does a thicker coating. The lower charge in the vicinity of the on-sensor circuitry enables the sensor circuitry to perform correctly.

Coatings of resin which are five to ten mils thick were melted onto the sensor surface at 70-100°C. DSC experiments indicate negligible reaction during the few seconds needed to melt the resin onto the sensor in this temperature range. The coated sensors were then placed in a preheated oven for isothermal cures. Time lag for the sensors to equilibrate thermally ranged from 30 seconds to 1 minute, depending on the cure temperature.

Exploration of Key Issues

Our brief literature review discussed in Part I showed that epoxy-phenolic chemistry is complex. In this section, criteria are developed in order to differentiate these various reaction mechanisms.

Reaction Selectivity

Epoxy Homopolymerization

Figure 3 shows the decrease in ionic conductivity resulting from the cure of two different formulations at 200°C: the first is ECN with an equimolar amount of PN but no TPP; the second is an equimolar stoichiometry of ECN and PN with 0.5 phr TPP.

The uncatalyzed ECN-PN reaction is very slow compared to the catalyzed epoxy-phenol reaction. The uncatalyzed equimolar epoxy-phenolic reaction was also found to be sluggish by Shechter.[4] A cure carried out at 200°C required 12 hours to consume 90% of the epoxy. In addition, the catalyzed epoxy homopolymerization reaction (Rxns. 27 and 28) was found to be very slow by Shechter and by Narracot.[4,29] It was actually necessary to add alcohol in order to trigger any reaction, and then reaction was observed between only epoxy and the alcohol. It was concluded that the zwitterion formed by the ring opening reaction between the tertiary amine and the epoxy (Rxn. 27) is unable to react with epoxy. This was believed to be due to the proximity of the zwitterion's positive and negative charges.[4]

These cases indicate that neither epoxy homopolymerization nor uncatalyzed epoxy-phenol reaction can play a significant role in the catalyzed epoxy-phenol reaction in the short timescales (i.e., 20-40 minutes) during which the catalyzed reaction occurs.

Secondary Reaction

A series of DSC experiments was performed in order to determine whether the secondary reaction between epoxy and secondary hydroxyl is of any significance in the epoxy novolac formulation. In these experiments, formulations consisting of 2 parts epoxy to 1 part phenol were cured with increasing amounts of triphenyl phosphine. The cures were done by ramping at 20°C per minute from room temperature to 300°C. The reaction exotherm was integrated and scaled to the ideal heat of 85.8 kJ/mole epoxy. An example DSC trace resulting from a ramp cure is shown in Figure 4; determination

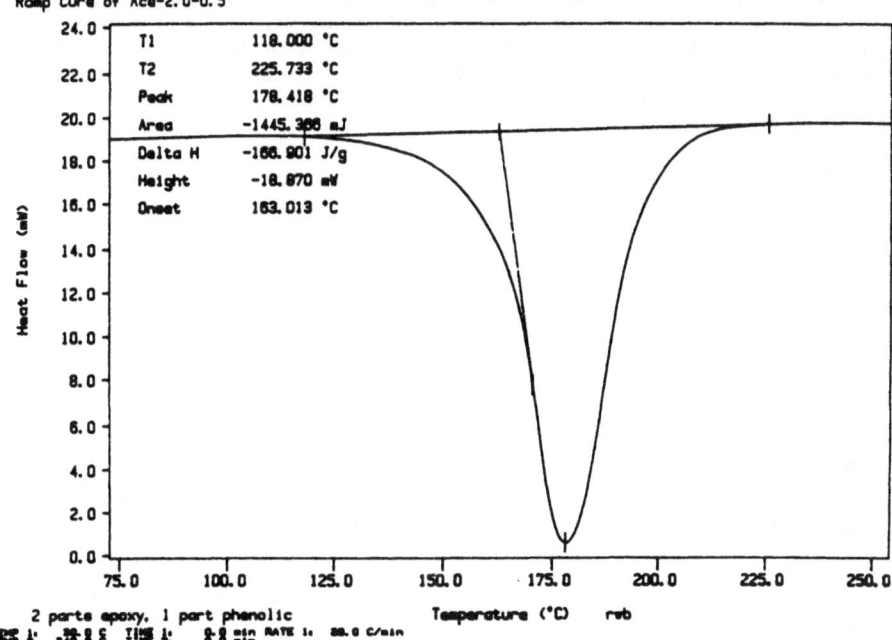

DSC Data File: ni008
Sample Weight: 8.660 mg
Tue Jan 10 17:46:50 1989
Ramp Cure of Ace-2.0-0.5

PERKIN-ELMER
7 Series Thermal Analysis System

T1   118.000 °C
T2   225.733 °C
Peak   178.418 °C
Area   -1445.366 mJ
Delta H   -166.901 J/g
Height   -18.870 mW
Onset   163.013 °C

2 parts epoxy, 1 part phenolic

Figure 4. Heat flow measured by DSC during 20°C/min. ramp cure of 2.00 stoichiometry ECN-PN formulation as catalyzed by 0.5 phr TPP. Also shown is the a peak area calculation as used to determine the fraction of epoxy reacted. The exotherm onset and peak temperatures are also marked.

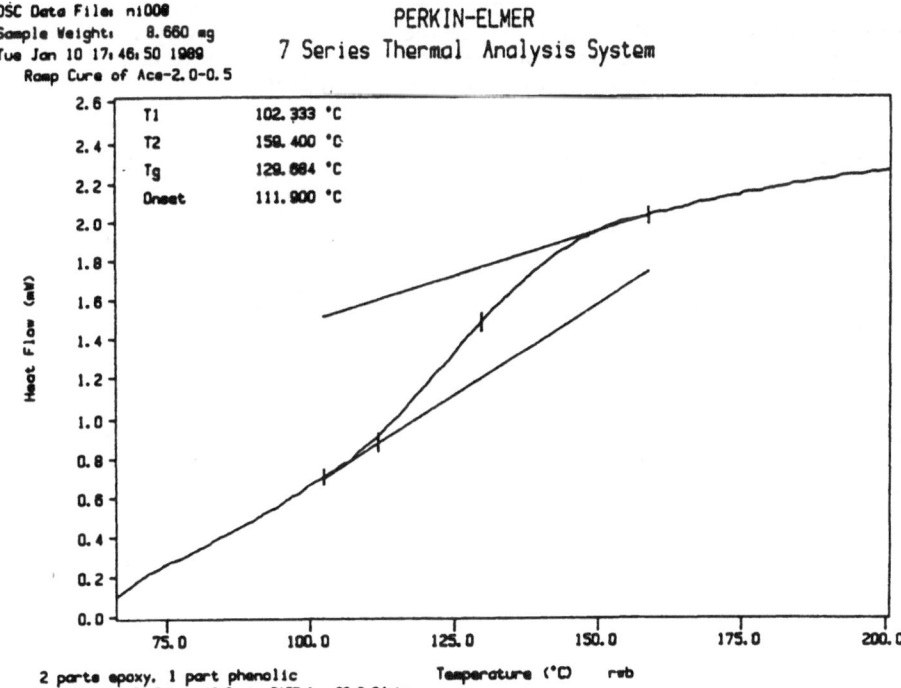

DSC Data File: ni008
Sample Weight: 8.660 mg
Tue Jan 10 17:46:50 1989
Ramp Cure of Ace-2.0-0.5

PERKIN-ELMER
7 Series Thermal Analysis System

T1   102.333 °C
T2   159.400 °C
Tg   129.684 °C
Onset   111.900 °C

2 parts epoxy, 1 part phenolic

Figure 5. Determination of Tg by DSC subsequent to ramp cure of a 2.00 stoichiometry formulation catalyzed by 0.5 phr TPP. Heating rate was 20°C/min.

of the Tg is shown in Figure 5, and the results are summarized in Table 1. The first two columns give the concentration of TPP in parts per hundred resin (phr) and millimolar (mM) units. The next column gives the initial fractional epoxy conversion resulting from the mixing process as it was obtained by titration. Column 4 gives the fraction epoxy reacted during the thermal ramp as it was obtained from the reaction exotherm. The sum total of these two conversions is given in Column 5. Columns 6 and 7 show the temperatures of the onset of the reaction exotherm and its peak. The last column gives the glass transition temperature obtained immediately after the thermal ramp.

Table 1. Results from Ramp Cure Study of 2.00 Stoichiometry.

TPP (phr)	TPP (mM)	X_{epoxy} initial	X_{epoxy} exotherm	X_{epoxy} total	Exotherm Onset (°C)	Exotherm Peak (°C)	Tg (°C)
0.5	17.2	0.04	0.48	0.52	163.0	178.4	129.7
1.0	34.3	0.04	0.48	0.52	130.2	156.3	133.4
2.0	68.7	0.12	0.53	0.65	128.2	153.6	139.3
4.0	137.	0.33	0.33	0.66	113.7	139.1	142.5
8.0	274.	0.40	0.29	0.69	106.0	131.7	147.6

The epoxy conversion for the 2.00 stoichiometry should be 50% if only the primary epoxy-phenolic reaction occurs, due to exhaustion of the phenolic limiting reagent. For TPP concentrations up to 1.0 phr, the total conversion is very near to that which corresponds to 50% conversion of the epoxy. Thus, for ramp cures in this concentration range, the reaction appears to consist solely of the epoxy-phenolic addition.

As the catalyst concentration is further increased, the initial conversions obtained by titration reveal that these 2.00 stoichiometry formulations have become so reactive as to undergo significant reaction during the room-temperature solvent-mixing process. In fact, the 4.0 and 8.0 phr samples were very difficult to grind into a powder, since they had reacted well beyond gelation. The lowering of the exotherm onset and peak temperatures provides additional support for the conclusion that high catalyst concentrations significantly increase the reactivity of the formulations (see Figure 6b).

For the cases in which the catalyst concentration is 2.0 phr or higher, the epoxy reacts beyond the theoretical maximum conversion of 50% for epoxy reacting only with phenol. This is evidence that some secondary reaction is catalyzed by TPP during the ramp cure of these formulations. Figure 6a shows the dependency of the initial epoxy conversion and total epoxy conversion on the TPP concentration. There appears to be some threshold concentration between 1.0 and 2.0 phr TPP in which the secondary reaction becomes significant. Hale reported a threshold concentration in the imidazole catalysis of the epoxy-phenolic reaction as well.[23]

Figure 6b shows the dependency of Tg on the TPP concentration. The final Tgs range between 131°C and 147°C for the range of catalyst concentrations, with a slight increasing trend as the catalyst concentration is increased. This rise in Tg is further evidence both of increased epoxy conversion via the branching reaction and, hence, also of a higher crosslink density.

These results contrast with the findings of Banthia and McGrath.[13] The model they postulated assumed that the TPP is in fact the active catalyst for the reaction. It had been suggested that the inability of TPP to catalyze the secondary reaction is due to steric effects. The bulky TPP cannot maneuver into an appropriate position to facilitate reaction between the secondary hydroxyl (which is closely attached to the main chain backbone) and the epoxy. In contrast, the phenolic hydroxyl has less steric hindrances nearby; it has more rotational degrees of freedom as well, thereby enabling TPP to facilitate reaction with epoxy. However, as discussed previously , other ionic-type reaction mechanisms have been proposed (Shecter and Gagnebien), as have reactions which require formation of an activated complex which is then, in turn, the active catalyst to the reaction (Sorokin). Either of these mechanisms could conceivably allow the secondary reaction to occur.

It will be shown later that the 2.00 stoichiometry also exhibits the secondary branching reaction during isothermal cures throughout the range of catalyst concentrations. No clear evidence of secondary reaction was seen in these experiments for formulations of stoichimetric ratio, R, of 1.30 or less (R is defined as the ratio of epoxy reactive groups to phenolic reactive groups). Thus, the only significant reaction evidenced is the epoxy-phenol reaction for both isothermal cures and ramp cures. The 2.00 stoichiometry will be treated as a special case throughout this dissertation, since it is made more complex by the competition between the primary and secondary reactions.

Reaction Initiation

Microdielectrometry Study

Figure 7 shows the effects of catalyst concentration on isothermal cure of the resin at 150°C. Catalyst concentrations spanning the range of 0.05 phr to 1.0 phr TPP were studied in a 1.18 stoichiometry ECN-PN formulation. After the microdielectric sensors were coated with the formulations, they were placed into an oven onto a half-inch thick aluminum slab which also had been preheated. Temperatures were monitored during the cure by an on-chip thermal sensor. Since the sensor indicated temperature variations of less than 2°C during the cures, there was neglibible sample heating due to the reaction exotherm in these experiments.

When the ionic conductivity is graphed on a log-log plot with time as one axis, a reaction initiation period is clearly seen for all the catalyst concentrations. One measurement of the initiation period is the time to reach the maximum rate of change in ionic conductivity. Figure 8a shows that the initiation time measured in this way depends inversely on the catalyst concentration.

For the purpose of comparison, a phenomenological "reaction rate" is defined as the time rate of change of the log conductivity. The maximum reaction rate calculated for each catalyst concentration gives insight into how the actual reaction depends on the catalyst concentration. Plotting the maximum reaction rates against catalyst concentration reveals a square (quadratic) dependency on catalyst concentration, as shown in Figure 8b.

When the temperature dependency is examined using the same epoxy formulation, initiation periods are again noted (see Figure 9). The length of the initiation period increases with decreasing cure temperature, thereby signaling a retarded complex or ionic species formation, which is due either to activation energy or mass-transfer limitations. In addition, the long-time plateaus vary with the curing temperature. The higher conductivity at elevated temperatures indicates higher free-volume fractions, a higher thermal energy, and a correspondingly higher ion mobility. These results assume that the active ion concentration does not depend much on temperature. The trapped free-volume may be due to the fact that reaction has progressed rapidly and the equilibrium free-volume state has not been established.

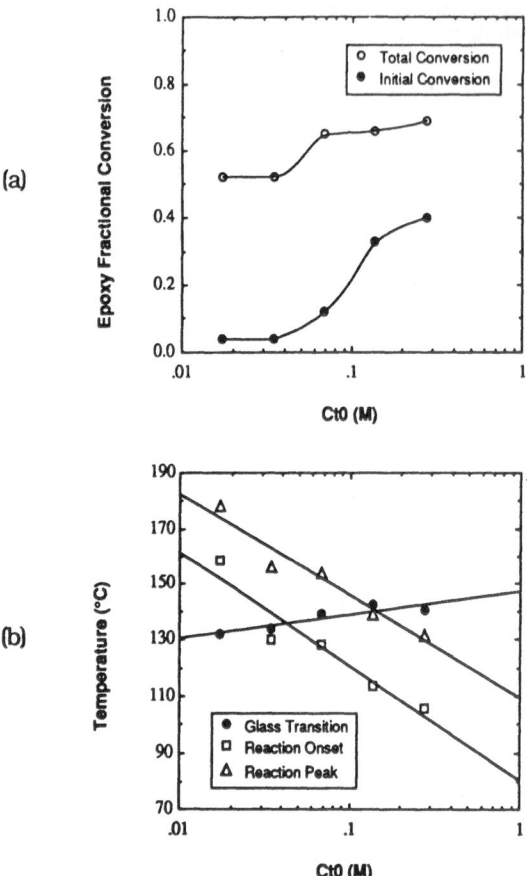

(a)

(b)

Figure 6.   Figure (a) shows dependence of the initial epoxy conversion and total
epoxy conversion on the TPP concentration for the 2.00
stoichiometry. Figure (b) shows dependence of the glass transition
temperature, exotherm onset temperature, and exotherm peak
temperature on the TPP concentration.

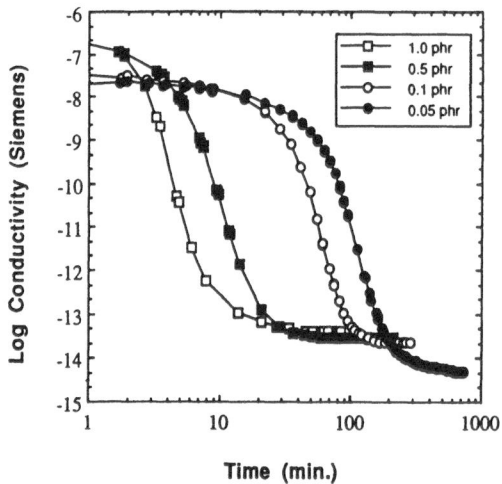

Figure 7.   Ionic conductivity variation during 150°C isothermal cures of a 1.18
stoichiometry ECN-PN catalyzed by TPP. The TPP concentration
(phr) varies according to the insert.

128

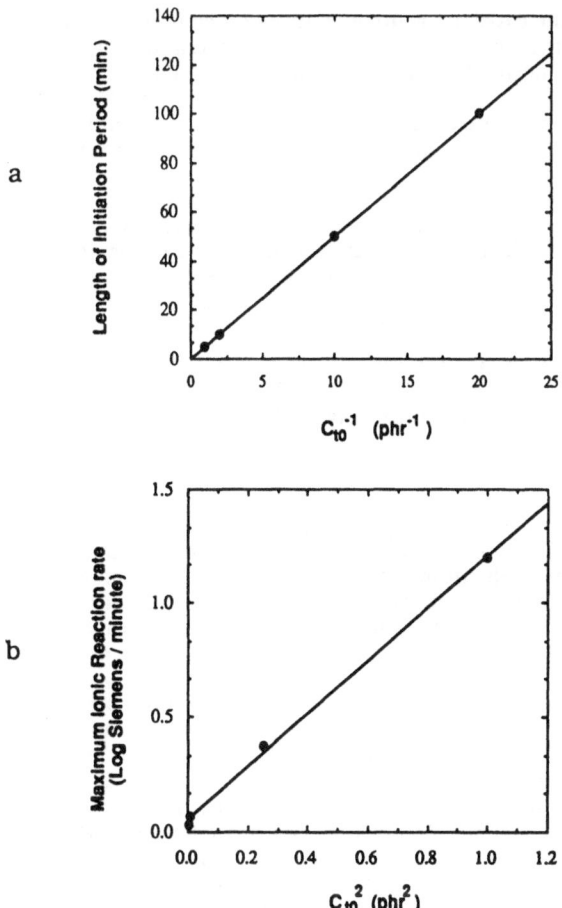

Figure 8.   Figure (a) shows how the length of the initiation period varies with the inverse of the TPP concentration, C_{t0}^{-1} (phr^{-1}). Figure (b) shows maximum ionic "reaction rate" plotted against the square of the TPP concentration, C_{t0}^2 (phr^2).

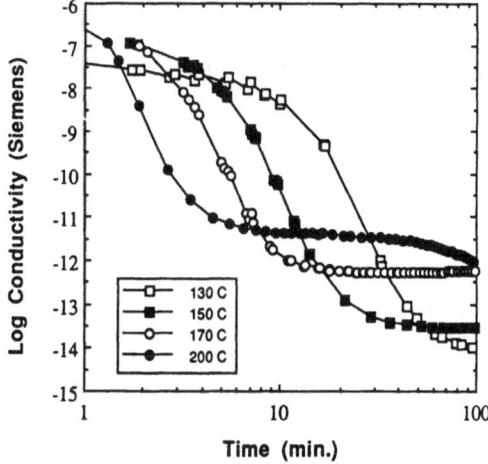

Figure 9.   Ionic conductivity as a function of time during isothermal cures of a 1.18 stoichiometry ECN-PN formulation catalyzed with 0.50 phr TPP.

Correspondingly, the expected volume contraction cannot keep pace with the rapid reaction. Eventually, when reaction slows down, the free volume decreases appreciably so the plateau level begins to exhibit a downward shift (200°C curve).

Figure 10a correlates the initiation time in an Arrhenius diagram and shows an energy-activated dependence on temperature. (Note that the initiation time increases with decreasing temperature, so the apparent activation energy from this plot would be negative.) Similarly, the maximum reaction rate is described by an Arrhenius dependency (Figure 10b). The fit is excellent, which indicates that the maximum reaction rate is controlled by energetic considerations rather than mobility. Absolute magnitudes of activation energies obtained from these data are 44 kJ/mole for the initiation time and 54 kJ/mole for the maximum reaction rate. These activation energies are lower than those reported in the literature for similar catalyzed reactions: 68 to 92 kJ/mole for epoxy with phenolic catalyzed by imidazoles[30] and 57 to 110 for epoxy with anhydrides as catalyzed by amines.[31] The reduced activation energies suggest that the TPP may promote epoxy ring opening.[32]

DSC Study

Microdielectrometry has provided some insights into the manner in which the reaction initiates. However, direct correlations between ionic conductivity and the extent of reaction are difficult to obtain because of the dependency of the ionic conductivity on both temperature and catalyst concentration. The results of the ionic conductivity studies have been useful for guiding other, more quantitative experiments, such as DSC.

A grid study, which varies both stoichiometric ratio and catalyst concentration, was performed isothermally at 150°C using differential scanning calorimetry. In one column of the grid, the concentration ratio of TPP to phenolic reactive groups (C_{t0}/C_{p0}) was kept constant; in the second column, the concentration of TPP was kept constant; and in the third column, the concentration ratio of TPP to epoxy reactive groups (C_{t0}/C_{e0}) was kept constant. An example DSC trace for the isothermal cure of the 1.00 stoichiometry formulation is shown in Figure 11; this formulation served as the pivot point for the grid study. Figures 12-14 show how the rate of epoxy reaction as a function of time changes with stoichiometry. Figure 12 shows this rate for a constant initial C_{t0}/C_{p0} ratio, Figure 13 for a constant TPP concentration of 13.42 mM, and Figure 14 for a constant initial C_{t0}/C_{e0} ratio. This study clearly reveals initiation periods in which the epoxy reaction rate accelerates. In all these cases, the initiation period grows more apparent as the stoichiometric ratio, R, increases.

The initiation period indicates that TPP by itself is probably not the active catalyst. Rather, the TPP must first complex or react with some other constituent to produce an activated catalytic complex. Substantial initiation periods are observed for low phenol concentrations (high R, high epoxy content), yet almost no initiation period for high phenol concentrations (low R, low epoxy content) are observed. This suggests that a rate-limiting step in the formation of the activated catalyst complex occurs as some type of interaction between the TPP and phenol. Such an interaction would occur more rapidly for higher PN concentrations and more slowly for lower PN concentrations.

Reaction Propagation

First-Order Reaction Propagation

Figures 15-17 show the same rate-of-reaction data plotted against fractional conversion of the epoxy. The fractional conversion of the epoxy was obtained by integration of the rate-of-reaction data. The linear relation of reaction rate to conversion as the maximum conversion is approached is particularly impressive. This linear relationship is seen for all stoichiometries and variations of TPP concentration. Some deviation is noted in the 2.00 stoichiometry cases and appears to be a result of a secondary reaction. This anomolous behavior in the isothermal cure of the 2.00

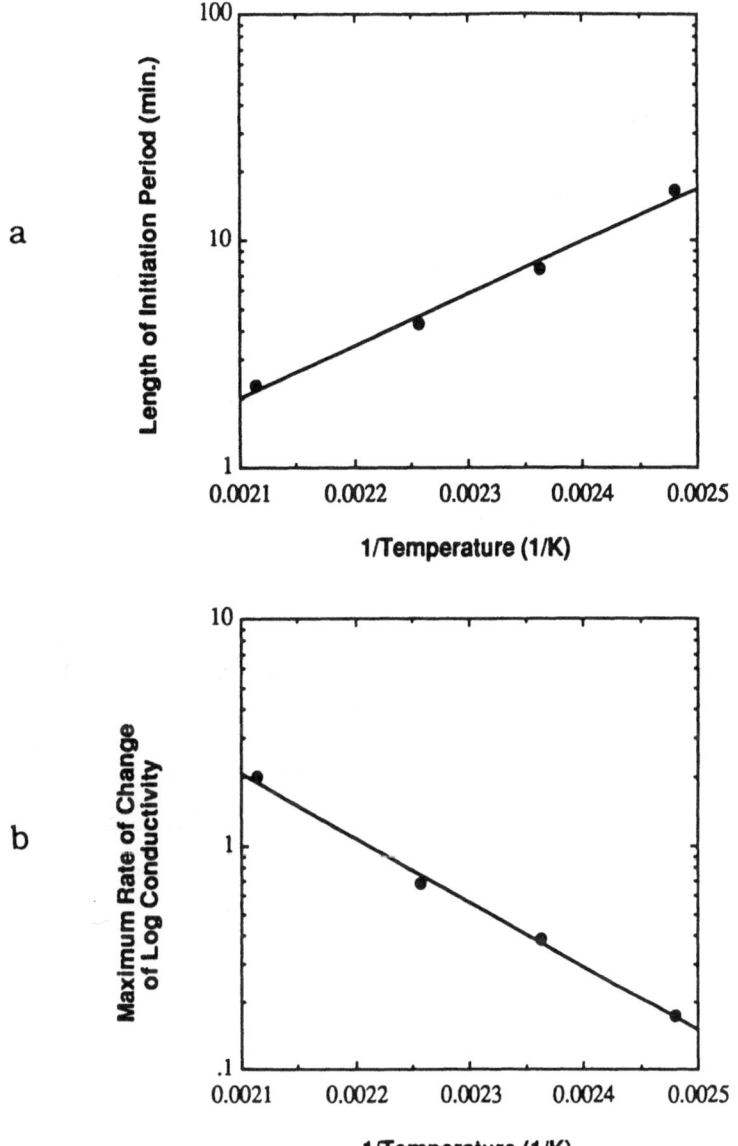

Figure 10.   Figure (a) shows the Arrhenius diagram for the inition time obtained from ionic conductivity measurements for the epoxy formulation cured at different temperatures. Figure (b) shows the Arrhenius diagram for the maximum rate of change of log ionic conductivity for the epoxy formulation cured at different temperatures.

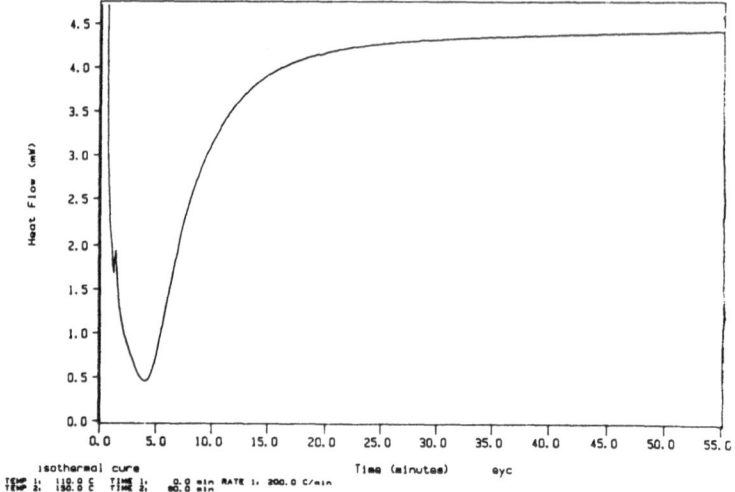

Figure 11. Heat flow measured by DSC during isothermal 150°C cure of 1.00 stoichiometry ECN-PN formulation as catalyzed by 0.50 phr TPP.

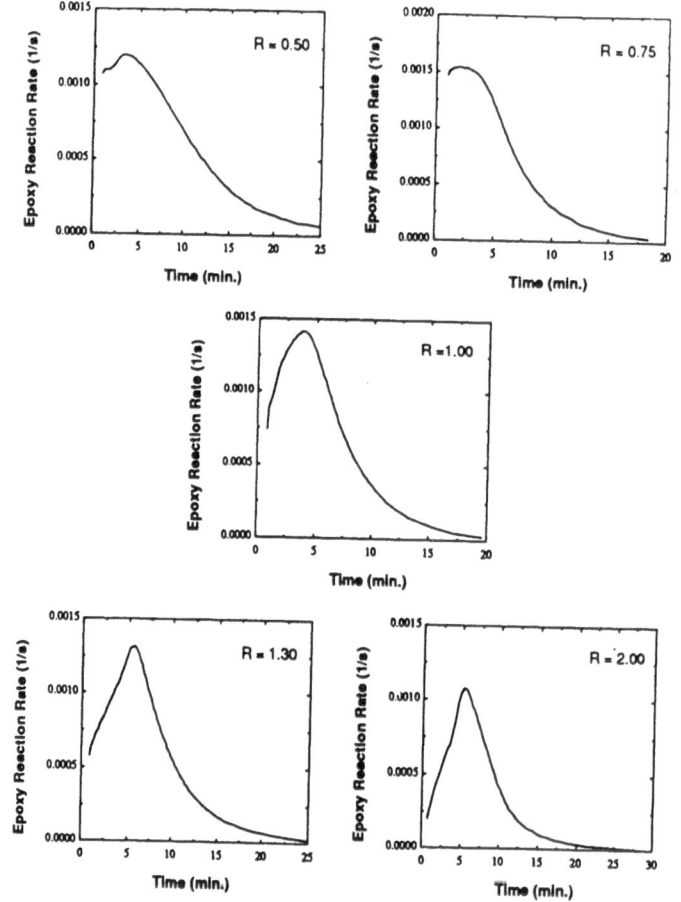

Figure 12. Rate of reaction as a function of time for various stoichiometries at 150°C. Concentration ratio of TPP to phenolic hydroxyl was held constant.

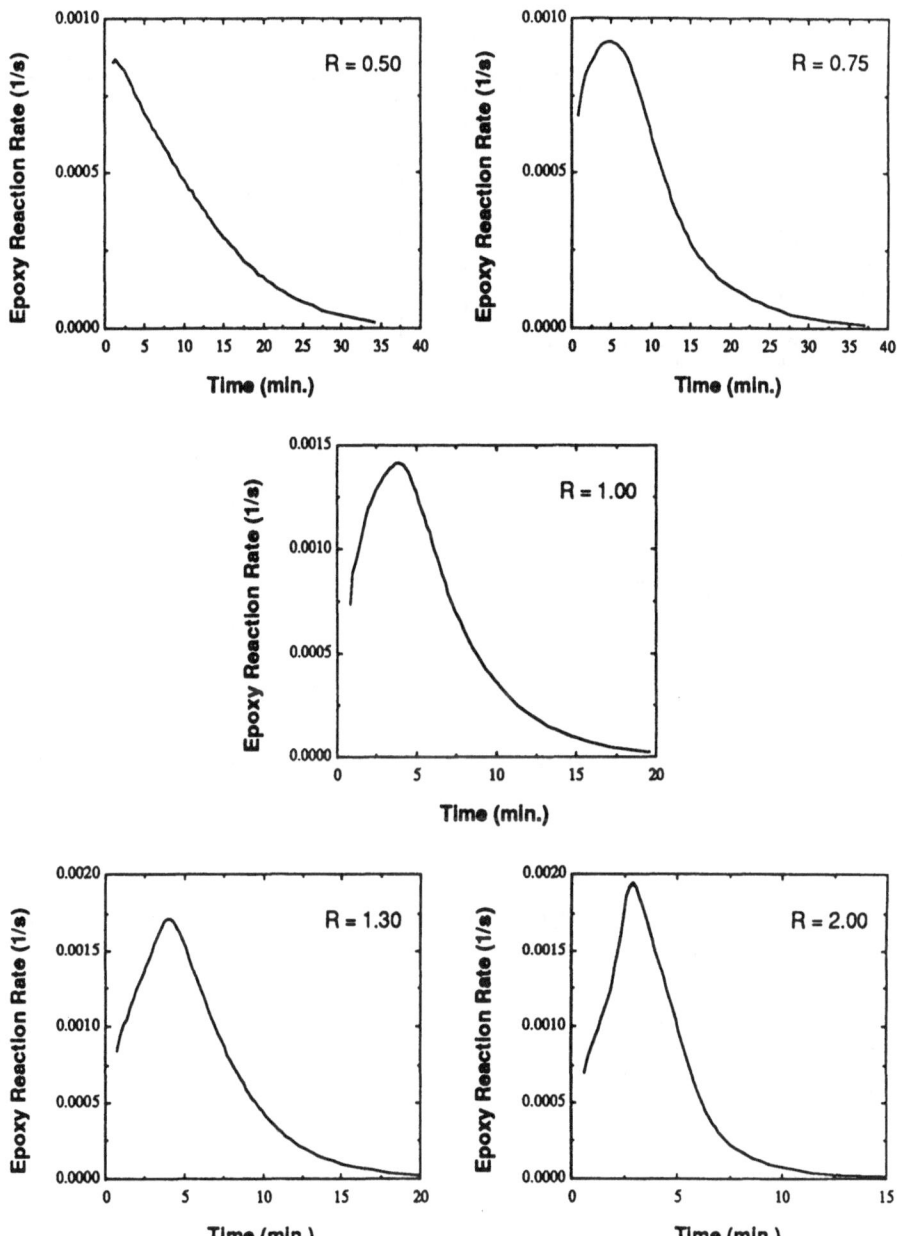

Figure 13.   Rate of reaction as a function of time for various stoichiometries at 150°C. Concentration of TPP was held constant.

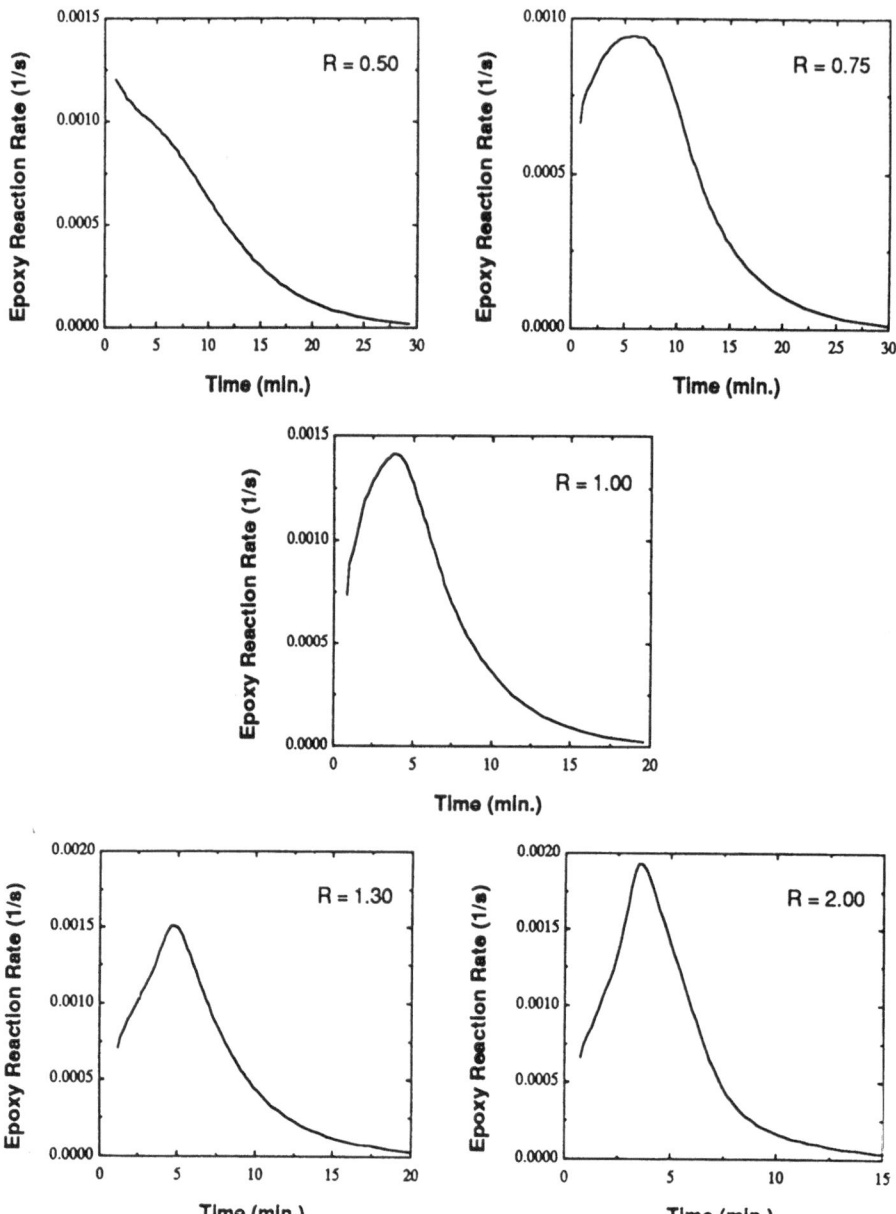

Figure 14.   Rate of reaction as a function of time for various stoichiometries at 150°C. Concentration ratio of TPP to epoxide was held constant.

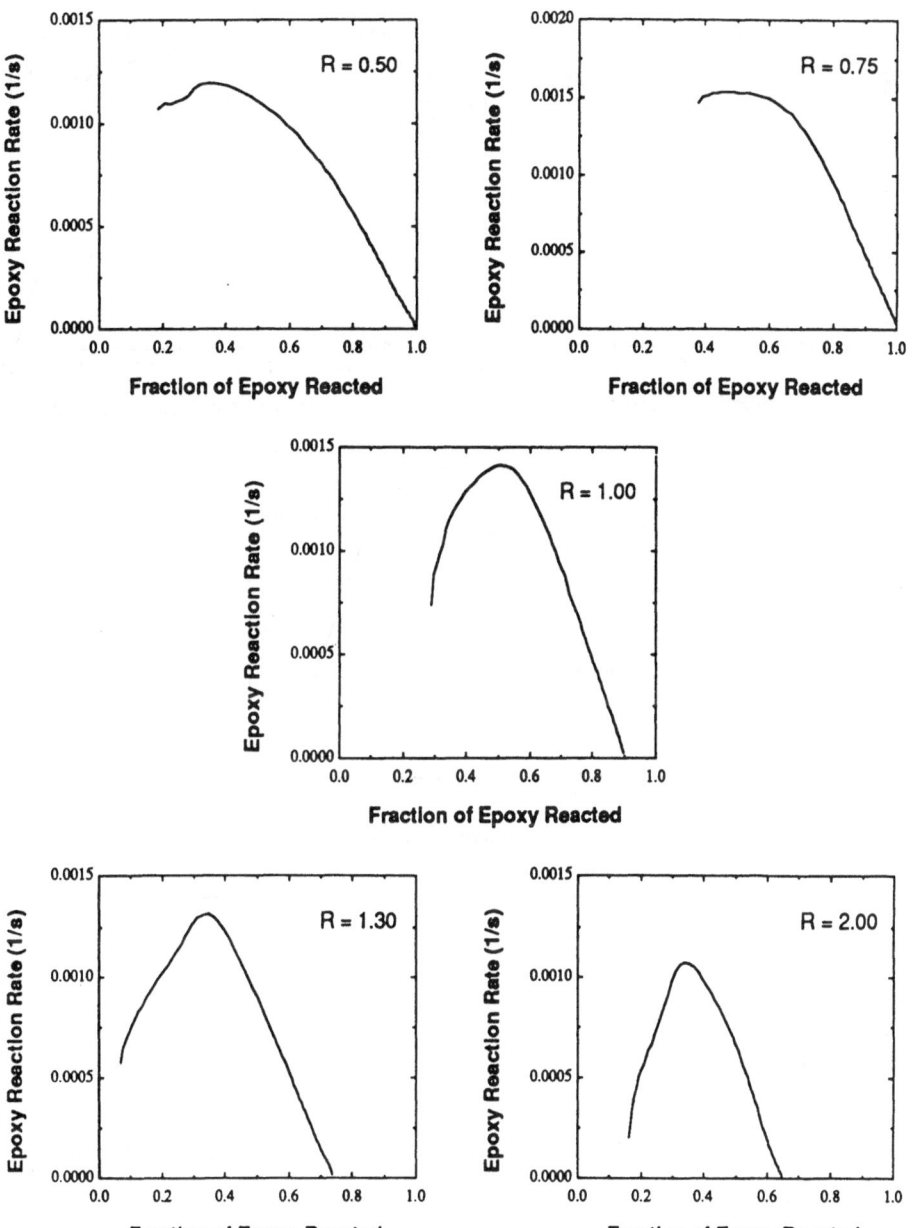

Figure 15.   Rate of reaction as a function of fractional conversion for various
stoichiometries at 150°C. Concentration ratio of TPP to phenolic
hydroxyl was held constant.

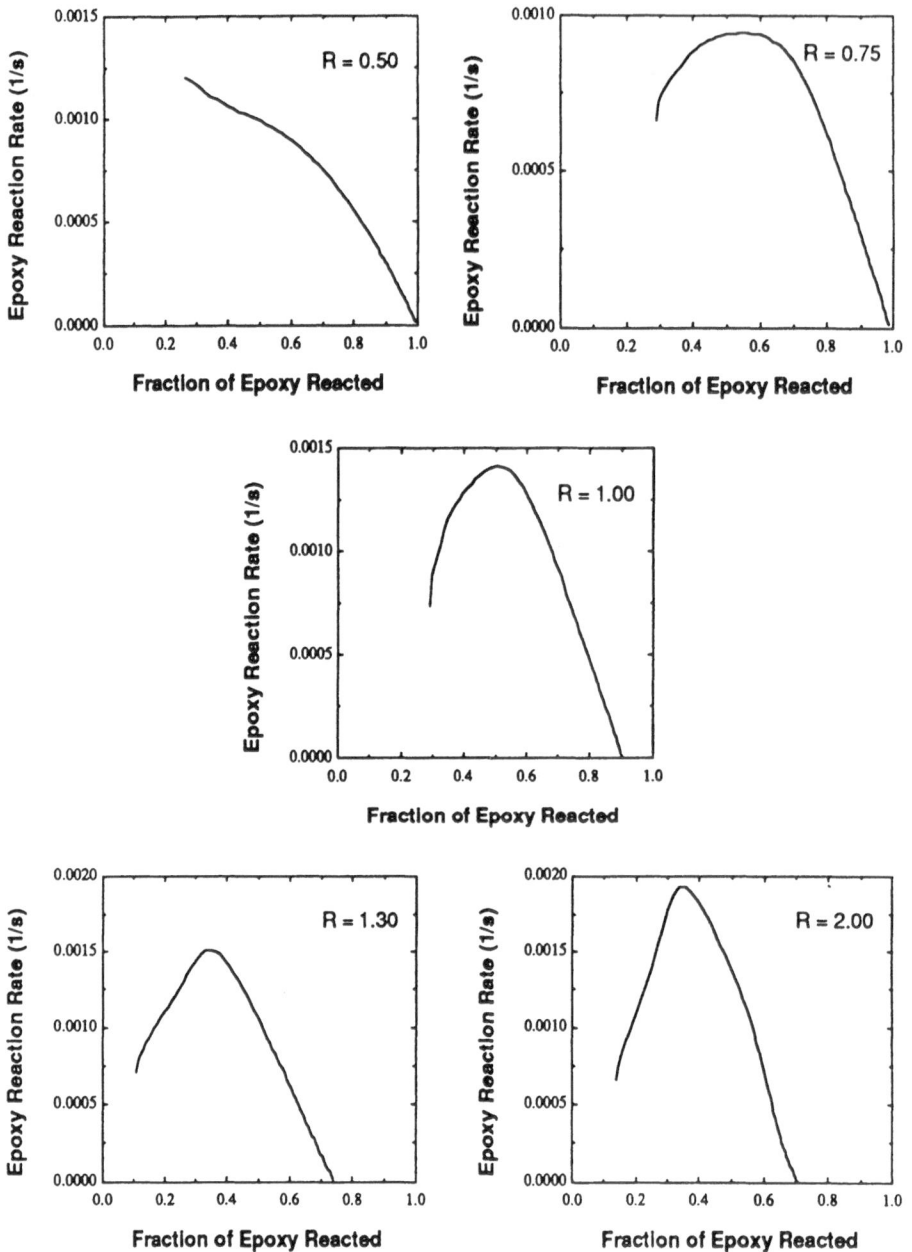

Figure 16. Rate of reaction as a function of fractional conversion for various stoichiometries at 150°C. Concentration of TPP was held constant.

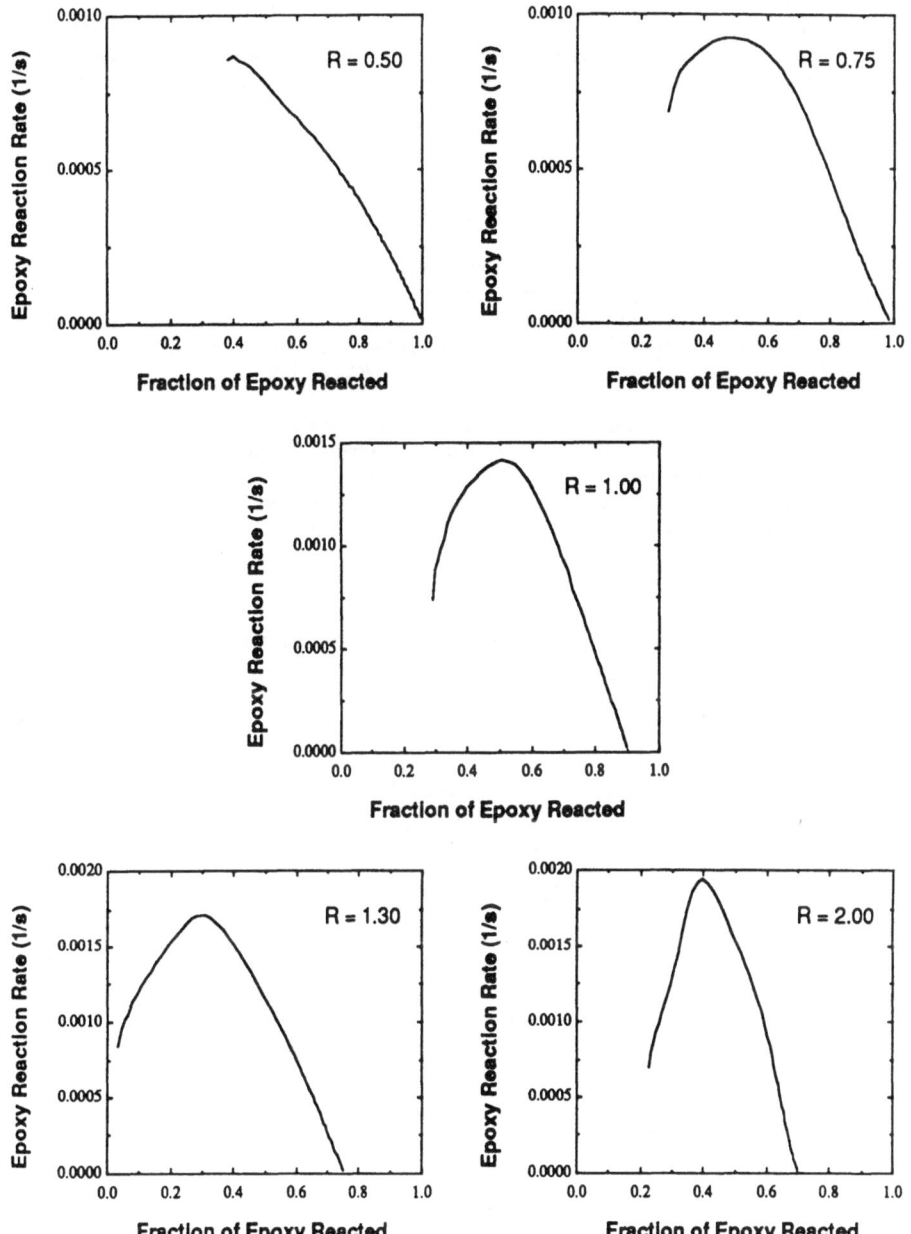

Figure 17. Rate of reaction as a function of fractional conversion for various stoichiometries at 150°C. Concentration ratio of TPP to epoxide was held constant.

stoichiometry is observed for all the catalyst concentrations studied and is explored further in the next part.

Figure 18 illustrates the linear decay in reaction rate by focussing on the equimolar case. The slope of the linear region yields a first-order rate constant consistent with a rate law of $dx/dt \sim k(x_{max}-x)$ in the latter stage of the reaction. Figure 19 plots the logarithm of the reaction rate versus conversion for the reaction at 150°C. Later in this chapter, it is shown that the 1.00 stoichiometry exhibits a sterically hindered maximum epoxy conversion of 0.94. Reaction quenching before this maximum conversion is probably due to diffusional limitations. The narrow region over which this quenching occurs demonstrates that the diffusion-limited reaction regime occurs over only a small span of conversion. Thus, diffusion effects in this reaction system are minimal; they serve only to quench the reaction rapidly at some maximum epoxy conversion, $x_{e,max}$.

Activation Energy and Diffusion Limitations

The temperature dependence of the rate constant was also investigated. First order rate constants were determined from the linear regions of many isothermal cures over a range of temperature, stoichiometry, and catalyst concentration. The cures were performed at 130, 150, 170, and 200°C. The stoichiometric ratios investigated were 0.50, 0.75, 1.00, 1.18, 1.30, and 2.00 (at a constant C_{t0}/C_{e0} ratio).

An Arrhenius plot of the first order rate constant for the range of stoichiometries is presented in Figure 20. The 2.00 stoichiometry is left out because it exhibits considerable secondary reaction, whereas the other stoichiometies do not. A linear correlation is observed throughout the range of temperature and stoichiometry. The activations energies at various stoichiometries are as follows:

Table 2. Activation Energies Determined from Reaction Rate Data of Figure 20.

R	E_k
(-)	(kJ/mole)
0.50	68.1
0.75	63.9
1.00	65.3
1.18	62.8
1.30	62.1
Average	64.4

The activation energies show a tight distribution around the average of 64.7 kJ/mole, with a standard deviation of 2.4 kJ/mole. This activation energy approximates those reported in the literature for similar catalyzed epoxy reactions.[30,31] It is higher than those obtained from the microdielectrometry experiments. This may be because these experiments measure only the aggregate property of ionic conductivity, which is

138

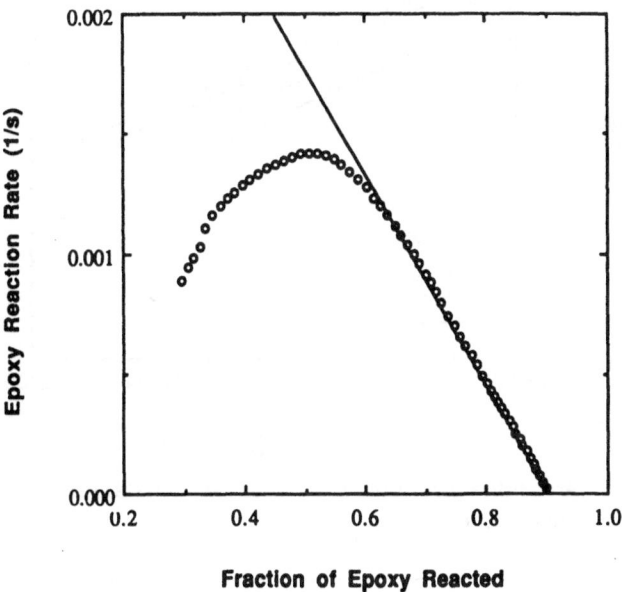

Figure 18.   Rate of reaction as a function of fractional conversion for R=1.00 at 150°C. The slope of the linear region yields a first-order rate constant.

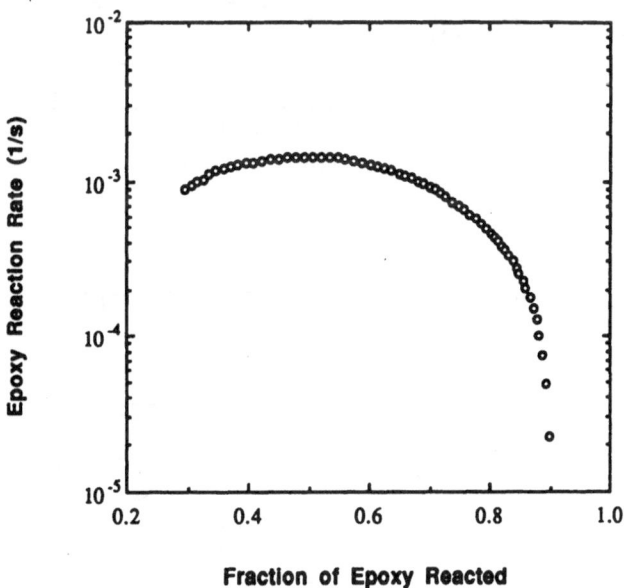

Figure 19.   Logarithm of the DSC reaction rate versus conversion for 1.00 stoichiometry ECN-PN formulation with 0.50 phr TPP.

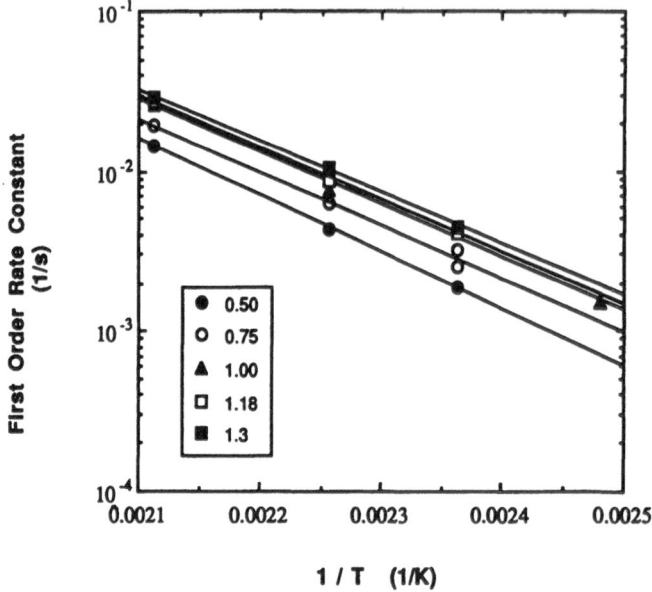

Figure 20.   Arrhenius plot of the first-order rate constant, K_1, for various stoichiometries.

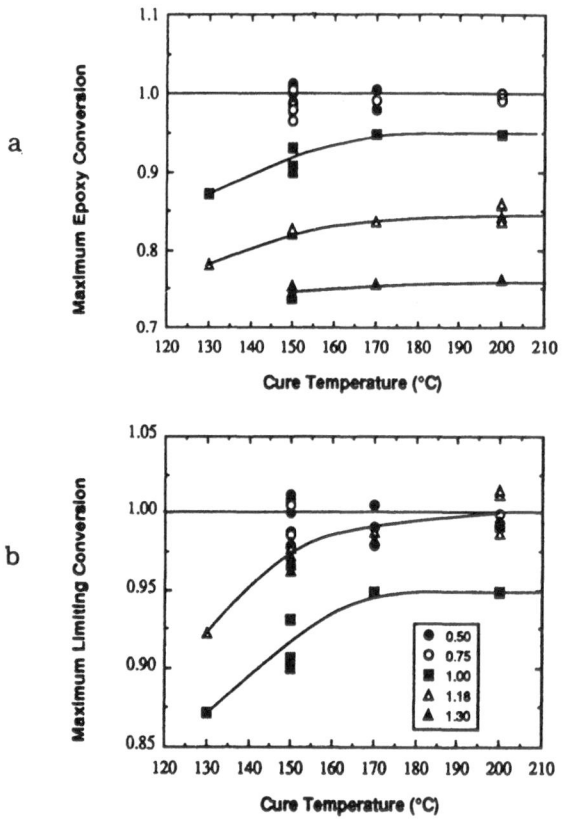

Figure 21.   Figure (a) shows maximum epoxy conversion for various stoichiometries. Figure (b) shows maximum limiting reagent conversion for the same stoichiometries.

affected by the state of cure, viscosity, and concentration of ions. The data from the DSC experiments provide a more direct measure of the rate of reaction, especially in the latter part of the reaction; hence, it provides a more accurate measure of the activation energy.

Several conclusions can be drawn from the Arrhenius plots. The most important deduction is that the reaction in the first-order kinetics regime is clearly driven by energetic considerations: diffusion-limitations appear inconsequential throughout the major part of the reaction. The linearity of the Arrhenius plot is a significant issue and implies one of two things. Either it means that the reaction mechanism is the same over the entire temperature range 130 to 200°C; experimentally there is no observed competition between mechanisms in the stoichiometric range 0.50 to 1.30. Or it means that any competing mechanism must have the same activation energy so that the competition is indiscernable in the Arrhenius diagram.

Reaction Termination

Reaction termination refers to the cessation of reaction due to the depletion of the limiting reagent, diffusional limitations, or steric hindrance. The most common cause of reaction termination is depletion of the limiting reagent. Diffusional limitations become operative as the mobility of molecules in the reacting mixture diminishes due to molecular weight increase and crosslink density increase. For very high crosslink densities or very bulky reactive groups, steric hindrances may prevent the reactants either from coming close enough to each other to react or from coming into the orientations necessary for reaction to occur.

Depletion of Limiting Reagent

Figure 21a plots maximum epoxy conversion versus cure temperature. This shows that the maximum epoxy conversion is dependent upon the stoichiometric ratio and the cure temperature. The maximum conversion reported here has a minimum random error equal to that of the initial conversion obtained by titration, $\Delta x_e \approx 0.02$. Additionally, a small amount of error must be allowed due to the inherent thermal equilibration period in DSC isothermal cures, $\Delta x_e \approx 0.02$. Thus, the y-axis error bars around these data points are estimated to be ±0.02.

Most of the stoichimetric dependence results from the fact that epoxy is not always the limiting reagent; the phenolic hydroxyls are the theoretical limiting reagent for R > 1. If the conversion of the limiting reagent is plotted, as in Figure 21b, it becomes clear that the reaction progresses in most cases until the limiting reagent is depleted. This is also true for the 0.50 and 0.75 stoichiometries; for those closer to one-to-one stoichiometry and epoxy-rich (R = 1.00, 1.18, and 1.30), a combination of diffusional and steric limitations are observed.

Steric Limitations

The 1.00 stoichiometry betrays steric limitations quite clearly at high temperature cures. In spite of being cured at temperatures well above its ultimate glass transition temperature of 170°C, incomplete cure is observed and reaches a maximum of 94%.

The concept of an ultimate conversion lower than 100% is reasonable in view of steric restrictions in a highly crosslinked network. Oleinik has reported an ultimate conversion of 92% in a diepoxide-diamine system.[33] This ultimate conversion value is attributed to topological limitations and is consistent with computer simulations [33]. Hale reports an experimental ultimate conversion of less than 85% for a 1:1 stoichiometry ECN-phenolic cresol novolac (PCN) system which is similar to the one in this study.[23] The lower ultimate conversion seen in the ECN-PCN system is in agreement with greater steric hindrances arising from the cresol pendant group. The topological constraints on the ultimate conversion will increase as functionality of the reactive molecules also increases.

Diffusional Limitations

Diffusional limitations are seen for the 1.00 and 1.18 stoichiometries during the low temperature cures of 130°C and 150°C. Diffusional effects on reaction quenching have been studied by many researchers recently, and the resins in these studies usually exhibit substantial diffusional limitations over the temperature range of interest.[34-36] The maximum conversion data in Figure 21a demonstrates that the ECN-PN reaction catalyzed by TPP shows only a small such influence over a narrow range of stoichiometry at temperatures 150°C and under.

As a short summary of the experimental portion of this work, we note that DSC and microdielectrometry provide useful information concerning the ECN-PN reaction catalyzed by TPP over a range of stoichiometries, catalyst concentrations, and temperatures. Complex curing behavior is observed, which suggests the involvement of phenolic as well as secondary hydroxyls. There are indications that the catalyst first forms an activated complex before it is able to catalyze the epoxy-phenol reaction. It is assumed that this activated complex is ionic in character because of the electrical noise observed in ionic conductivity measurements. TPP promotes epoxy reaction selectivity toward reaction phenol rather than the branching reaction with secondary hydroxyls at elevated curing temperatures in concentrated resin systems. However, the selectivity is not perfect as it is under milder and more dilute conditions. The 2.00 stoichiometry exhibits the branching reaction at all catalyst concentrations during isothermal cures and at catalyst concentrations above 1.0 phr for ramp cures at 20°C/min. Diffusional limitations do not play a dominant role in the reaction over the range of interest.

PART III. DETERMINATION OF RATE LAW

Based on the DSC and microdielectrometry studies, the kinetics of the TPP catalyzed epoxy-phenol reaction can be described through an *initiation* regime and a "steady state" *propagation* regime. Diffusion limitations do not appear to have substantial influence within the temperature range of interest.

DSC experiments indicate that the epoxy reaction rate is first-order in the latter "steady state" regime of the reaction:

$$\frac{d\,x_e}{d\,t} \;=\; k_1 \left(x_{e,max} - x_e \right) \tag{49}$$

Reaction Initiation
where x_e is the fraction of epoxy reacted, $x_{e,max}$ is the maximum fraction of epoxy reacted at the given stoichiometry and temperature, and k_1 is the first-order kinetic constant.

The goal of this section is to examine how this simple rate law may be extended to describe the influences of stoichiometry, catalyst concentration, and temperature in both the linear, first-order regime and the nonlinear initiation regime.

Reaction Propagation

Determination of First-Order Rate Constant

First-order reaction rate constants were determined over a wide variety of temperatures, stoichiometries, and TPP concentrations by using the isothermal DSC experiments as described previously. For the sake of comparison, the first-order rate constants at 150°C are plotted with respect to the same scaling parameters as were used in the grid study in Figures 22 and 23. The data exclude the 2.00 stoichiometry because it shows indications of secondary reaction.

No correlation between the rate constant and TPP concentration, C_{t0}, is seen in Fig. 22a, nor is there a correlation between the rate constant and the TPP to ECN (epoxide) ratio, C_{t0}/C_{e0}, in Fig. 22b. However, the first-order rate constant *does* scale linearly with the initial ratio of TPP to PN (phenolic hydroxyl), C_{t0}/C_{p0}, in Fig. 23. A least-squares fit of the data has a correlation coefficient of 0.903 and passes through the origin. The linear correlation between the rate constant and initial TPP to PN ratio implies some type of interaction between TPP and PN and/or between ECN and TPP. Plots of the reaction rate constant with respect to other variables do not improve over the fit seen with respect to C_{t0}/C_{p0}.

The reaction-rate constant vanishes as the TPP to PN ratio approaches zero, whereas an uncatalyzed reaction would be evidenced by a nonzero reaction rate constant when C_{t0}/C_{p0} equals zero. The intersection at the origin implies that *no significant* uncatalyzed reaction occurs. The linearity of the plot indicates that the only catalytic mechanisms which occur are those which are fundamentally dependent on the same types of interactions between TPP and the other reactants in the range of stoichiometries examined (R = 0.50 to 1.30).

Table 3. Temperature Dependency of First-Order Rate Constant.

Temperature (°C)	q_{k1} (s^{-1})	r^2
130	.2953	0.838
150	1.044	0.903
170	2.066	0.932
200	6.361	0.984

The functional dependency of the rate constant on temperature is examined in Figure 24. The first-order rate constant is plotted as a function of C_{t0}/C_{p0} at 130, 150, 170 and 200°C, and linear fits to the data are shown. The best-fit lines were found to pass very close to the origin; accordingly, fits were then performed which "forced" each line to pass through the origin. In all of these cases, excellent correlation with the data was maintained, and the results are summarized in Table 3. Here q_{k1} is the slope of the least squares line and r^2 is its correlation coefficient.

Figure 25, which is an Arrhenius plot of these data, yields a good exponential fit to the data with a correlation coefficient of 0.990. Therefore, the steady state first-order kinetic constant, $k_{1,ss}$, may be described as:

$$k_{1,ss} = m_{k1} \left(\frac{C_{t0}}{C_{p0}} \right) e^{-E_{k1}/R_g T}$$

(50)

where $m_{k1} = 2.14 \times 10^8 \text{ s}^{-1}$, $E_{k1} = 68.0$ kJ/mole, R_g is the ideal gas constant (8.314 J/mole °K), and T is absolute temperature. The activation energy obtained here is similar in magnitude to that obtained by DSC.

Determination of the Maximum Conversion

Maximum conversion is determined based on the limitations discussed in Part II. In summary, the epoxy conversion corresponding to complete conversion of the limiting reagent is the maximum conversion for stoichiometries 0.75 and lower, or 1.30 and higher. Diffusional and steric limitations lower the maximum conversion for those stoichiometries near 1.00. However, a model for the 2.00 stoichiometry is not employed because of inadequate data with which to quantify the secondary reaction.

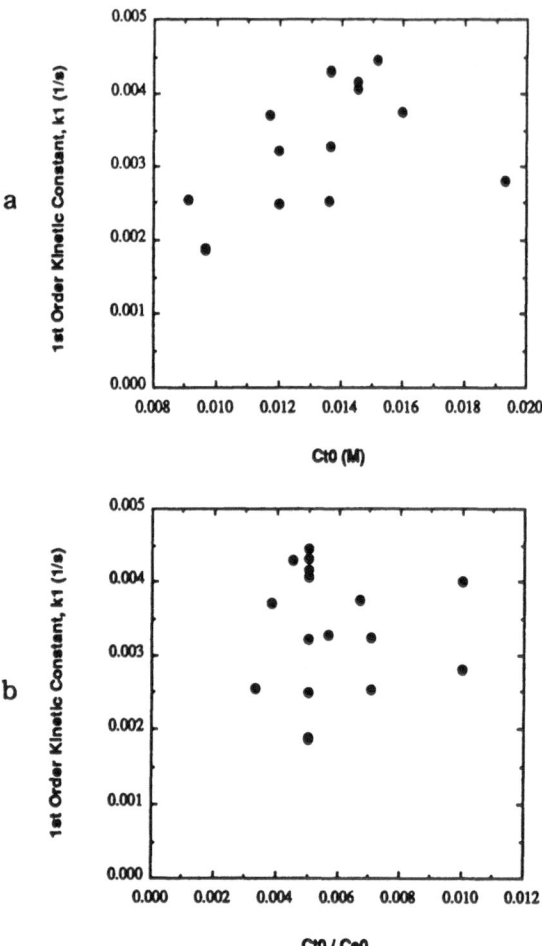

Figure 22.   Figure (a) shows the scaling of the first-order rate constant, k_1, with C_{t0} at 150°C. Figure (b) shows the scaling of the first-order rate constant, k_1, with C_{t0}/C_{e0} at 150°C. No correlation is observed in either case.

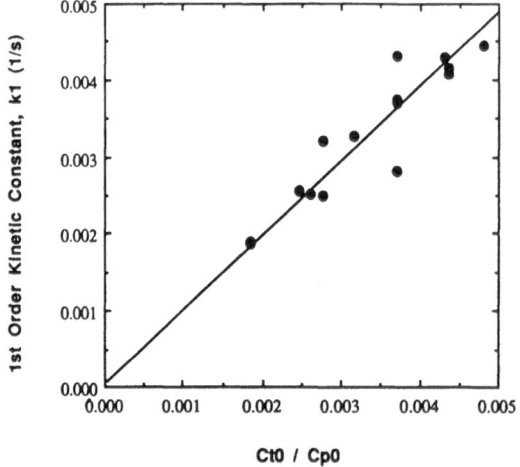

Figure 23.   Scaling of the first-order rate constant, k_1, with C_{t0}/C_{p0} at 150°C. A linear correlation is indicated.

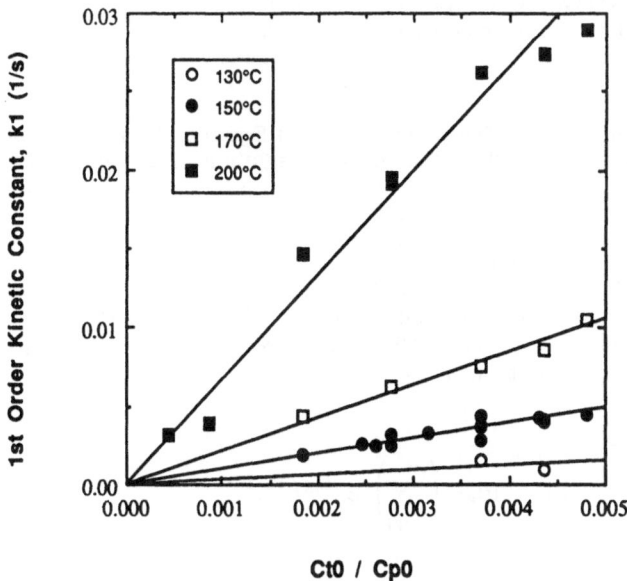

Figure 24. First-order kinetic constant, k_1, linear correlation with C_{t0}/C_{p0} for the range of temperatures 130-200°C.

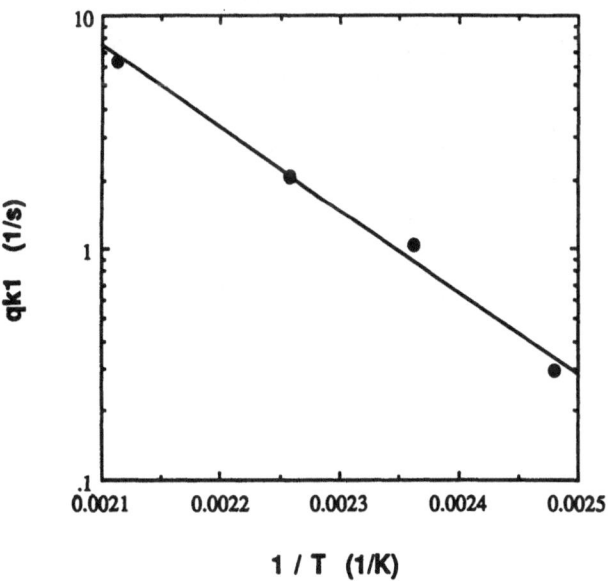

Figure 25. Arrhenius plog of q_{k1}, used in calculation of the temperature dependency of the first-order rate constant, k_1.

Reaction Initiation

The rate law describing reaction initiation is more complicated than the law for the "steady state" region. Several of the reaction mechanisms discussed in Part I involve steps in which the catalyst is somehow activated. Sorokin and Shode found that the catalyst eventually reacts completely to form the active catalytic species.[9] A "steady state" is reached when the TPP has completely reacted to form the active catalytic species.

The dependency on TPP concentration in Equation 50 for the first-order constant, $k_{1,ss}$, provides a clue as to how the magnitude of the rate constant depends on the concentration of the active catalytic species. All the mechanisms which possess an activation step require the creation of one catalytic or reactive species per TPP molecule: the Sorokin model predicts a quaternary phosphonium phenoxide complex, the Shechter and Gagnebien model predicts a zwitterion. Thus, the maximum concentration of activated complexes, $C_{ac,max}$, in a given formulation is limited by the concentration of TPP. If this activated complex is denoted generically by C_{ac}, then $C_{ac,max} = C_{t0}$ and the fraction of active complex formed is, $x_{ac} = C_{ac}/C_{t0}$. Equation 50 for k_1 can be rewritten in terms of C_{ac}, as

$$k_1 = m_{k1} \left(\frac{C_{ac}}{C_{p0}}\right) e^{-E_{k1}/R_g T}$$

(51)

k_1 at steady state is achieved when $x_{ac} = 1$, i.e. when $C_{ac} = C_{ac,max} = C_{t0}$.

Since C_{ac} affects the epoxy reaction rate through k_1, it is assumed that any reduction in reaction rate from the "steady state" first-order reaction rate is due to an incomplete formation of the active catalytic species. This assumption is supported by the fact that $k_1 \rightarrow$ zero as C_{ac} (or C_{t0}) $\rightarrow$ zero, as was seen in Figure 24. Thus, the fraction of active complex formed can be determined from the scaling of the the *actual* reaction rate with the *expected* first-order reaction rate.

This scaling is illustrated in Figure 26. A first-order plot of epoxy reaction rate versus the fraction epoxy reacted is fit with a linear least-squares line in the steady state region. This first-order line is extrapolated to the vertical axis. The difference between the rate from the first-order line and the actual reaction rate is ratioed to the rate expected at zero epoxy conversion. This ratio scales with the fraction of active catalytic complex formed, x_{ac}. This can be stated mathematically as

$$x_{ac} = 1 - \frac{\left(\frac{d x_e}{d t}\right)_{extrapolation} - \left(\frac{d x_e}{d t}\right)_{actual}}{\left(\frac{d x_e}{d t}\right)_{x_e = 0}}$$

(52)

This equation give us x_{ac} as a function of x_{epoxy} from the first-order plot. Parametrically mapping x_{ac} onto the time associated with the corresponding x_{epoxy} point gives x_{ac} as a function of time.

The results of this analysis are illustrated in Figure 27. Several interesting observations may be made about these data. First, the rate of increase in x_{ac} is approximately constant as a function of time, thereby implying a pseudo-zeroth-order reaction rate law for the formation of the active complex. Secondly, $x_{ac,0}$ is shown to be a function of the stoichiometric ratio, R, based on a linear extrapolation of x_{ac} to

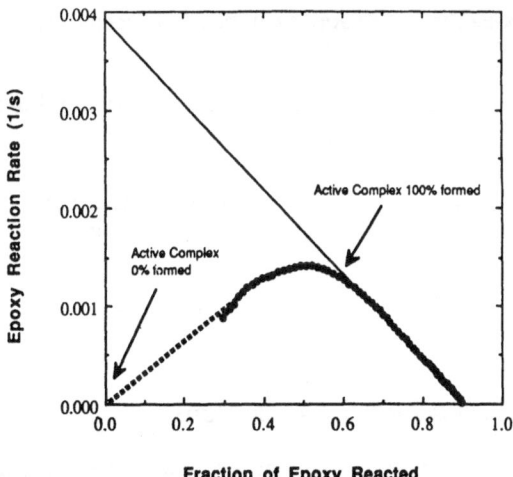

Figure 26.   Epoxy reaction rate as a function of the epoxy fractional conversion for the 1.00 stoichiometry at 150°C. The difference between the first-order extrapolated rate and the actual rate scales with the fraction of active complex formed.

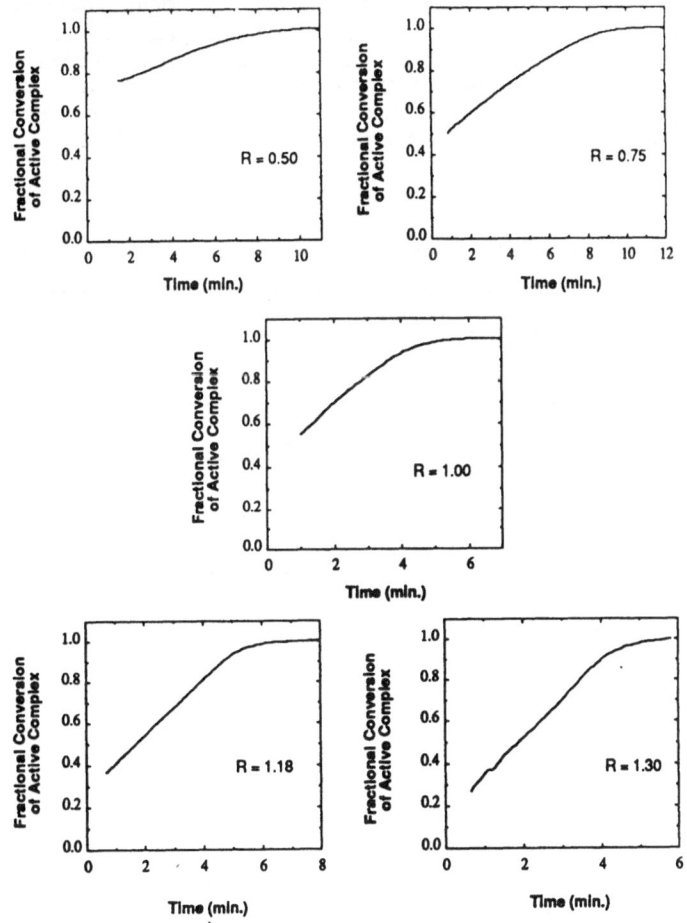

Figure 27.   Fraction of active complex formed for several stoichiometric ratios, R.

time zero. This indicates that some active complex is formed during sample preparation and may be an equilibrium product.

The general form of the zeroth-order rate law describing this behavior is:

$$x_{ac} = x_{ac,0} + k_{ac}\, t \qquad \text{for} \quad t \leq \frac{1 - x_{ac,0}}{k_{ac}} \tag{53}$$

$$x_{ac} = 1 \qquad \text{for} \quad t > \frac{1 - x_{ac,0}}{k_{ac}}$$

Determination of Zeroth-Order Rate Constant

An analysis for x_{ac} with the same data set as was used for determination of the first-order rate constant, k_1, allows the functional relationship of the reaction rate on the catalyst concentration to be clarified further. Zeroth-order rate constants are fitted to the linear regions of the x_{ac} curves. These zeroth-order rate constants, $k_{ac,0}$, were found to be a function of the product of the stoichiometric ratio, R, and the initial TPP concentration, C_{t0}, for all temperatures investigated; this is shown in Figure 28. Once again, the results show that a fitted line passes through the origin. These results are summarized in Table 4, where q_{ac} is the slope of the least squares line and r^2 is its correlation coefficient. These data do not yield as high a correlation coefficient as was achieved for the "steady-state" propagation rate constant, k_1. This may be due to the accumulated degree of error which resulted from the numerical processing entailed in deriving the activated complex data from the epoxy reaction rate data.

Table 4. Temperature Dependency of Zeroth-Order Rate Constant.

Temperature (°C)	q_{ac} (min^{-1} mole-^{1})	r^2
130	2688.5	0.999
150	8954.4	0.945
170	23,530	0.979
200	48,330	0.988

Figure 29, the Arrhenius plot of these data, yields an adequate exponential fit with a correlation coefficient of 0.973. The fit is neither as good nor as reliable as that of the first-order reaction rate because fewer points were obtainable for either the 200°C reactions - where initiation occurs very rapidly, making measurement of the active complex formation difficult - or for the 130°C reactions - where few data points had been collected because of the sluggishness of the reaction at this temperature.

The zeroth-order kinetic constant for the formation of the active complex may thus described as:

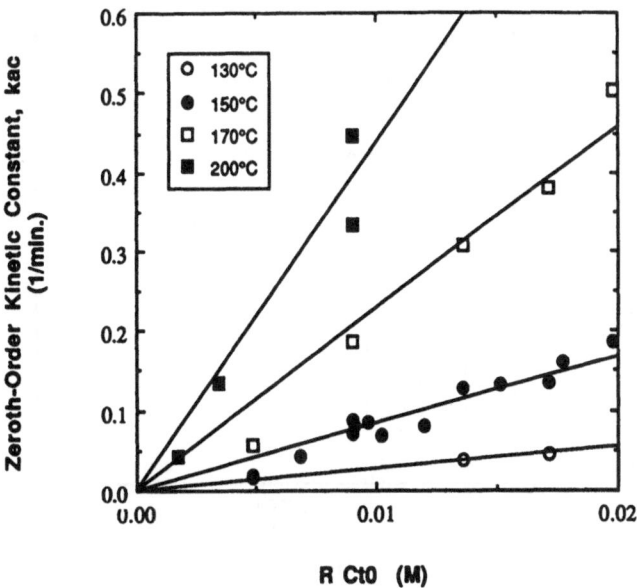

Figure 28.   Correlation of zeroth order kinetic constant, k_{ac}, with R C_{t0} for the range of temperatures 130-200°C.

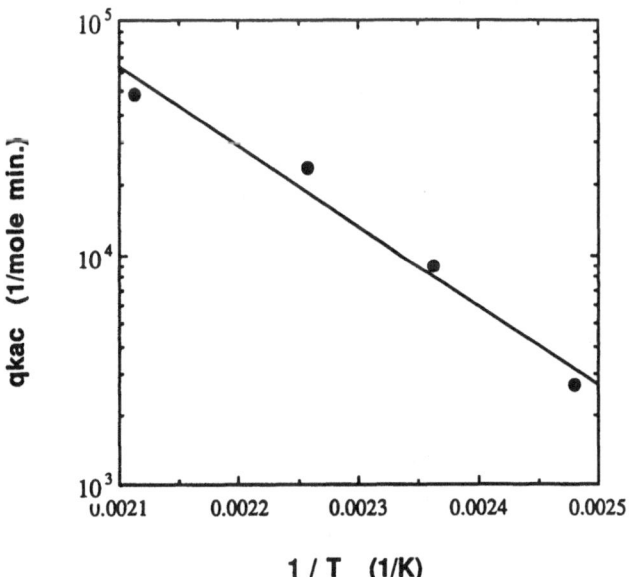

Figure 29.   Arrhenius plot of q_{kac}, used in calculation of the temperature dependency of the first-order rate constant, k_{ac}.

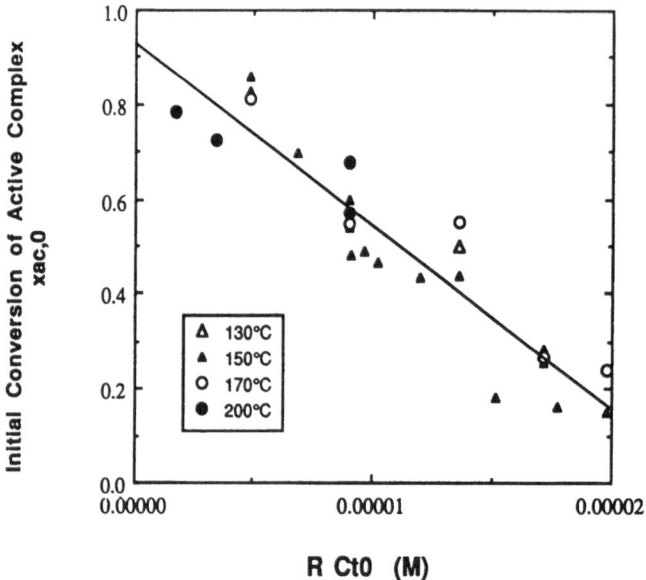

Figure 30.   Scaling of the initial fractional conversion of the activated complex
with $R\, C_{t0}$ for the range of temperatures 130-200°C.

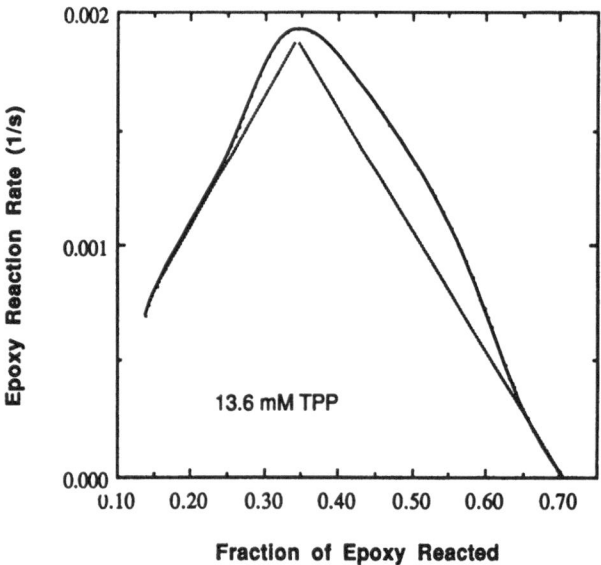

Figure 31.   Enhancement in rate of epoxy reaction for 2.00 stoichiometry, which
may be due to a secondary branching reaction. (150°C, 13.6 mM TPP)

$$k_{ac} = m_{kac} \, R \, C_{t0} \, e^{-E_{kac}/R_g T} \tag{54}$$

where $m_{kac} = 1.14 \times 10^9$ min^{-1} mole^{-1}, $E_{kac} = 65.6$ kJ/mole, R is the stoichiometric ratio (C_{e0}/C_{p0}), R_g is the ideal gas constant (8.314 J/mole °K), and T is absolute temperature. Thus, the activation energy for activated complex formation is very similar to that which was seen for the first-order epoxy reaction propagation. This similarity implies that the initiation and propagation steps of the reaction may share similar rate-limiting steps.

Determination of Initial Active Complex Fraction

The initial fraction of the active complex, $x_{ac,0}$, is plotted in Figure 30 as a function of the product between the stoichiometric ratio and the initial TPP concentration, ($R \, C_{t0}$). This functionality was found to give the best overall fit to the data. Lesser correlations were observed individually for C_{e0}, C_{p0}, and C_{t0}. There is considerable scatter in the data, in part because of the error induced by the repeated extrapolations. No functional dependency on cure temperature is discernable. A least squares fit to the data yields

$$x_{ac,0} = 0.932 - 38.8 \text{ mole}^{-1} \, (R \, C_{t0}) \tag{55}$$

with a correlation coefficient of 0.881. Based on a visual inspection, there is an apparent scatter of ±0.12 around the best fit line. This dependency on initial composition implies that some equilibrium amount of the active complex is formed during the mixing step.

SUMMARY OF EQUATIONS

In summary, the mathematical expressions describing the kinetics of the epoxy-phenol reaction are:

Epoxy Reaction Rate

$$\frac{d \, x_e}{d \, t} = k_1 \, (x_{e,max} - x_e) \tag{56}$$

where x_e is the fraction of epoxy reacted, $x_{e,max}$ is the maximum fraction of epoxy reacted at the given stoichiometry and temperature, and k_1 is the first-order kinetic constant; and

$$k_1 = m_{k1} \left(\frac{C_{ac}}{C_{p0}} \right) e^{-E_{k1}/R_g T} \tag{57}$$

where $m_{k1} = 2.14 \times 10^8$ s^{-1}, $E_{k1} = 68.0$ kJ/mole, R_g is the ideal gas constant (8.314 J/mole °K), and T is absolute temperature. $x_{e,max}$ is the maximum conversion which can be achieved by reaction before the reaction is quenched by diffusion limitations, and it has values of 0.87-1.00 in the temperature range 150-200°C.

Active Complex Formation

The concentration of active complex is $C_{ac} = x_{ac} C_{t0}$, and the fraction of active complex, x_{ac}, formed is:

$$x_{ac} = x_{ac,0} + k_{ac} t \qquad \text{for} \quad t \leq \frac{1 - x_{ac,0}}{k_{ac}} \tag{58}$$

$$x_{ac} = 1 \qquad \text{for} \quad t > \frac{1 - x_{ac,0}}{k_{ac}}$$

where

$$x_{ac,0} = 0.932 - 38.8 \text{ mole}^{-1} (R \, C_{t0}) \tag{59}$$

$$k_{ac} = m_{kac} R \, C_{t0} \, e^{-E_{kac}/ R_g T} \tag{60}$$

where $m_{kac} = 1.14 \times 10^9 \text{ min}^{-1} \text{ mole}^{-1}$, $E_{kac} = 65.6 \text{ kJ/mole}$, R is the stoichiometric ratio (C_{e0}/C_{p0}).

EVALUATION AND SELECTION OF MODELS

Many observations about the bulk-state ECN-PN reaction have been made and can be used to determine the best mechanisms amongst those reviewed in Part I. The first observation concerns diffusional effects on the reaction rate. The second pertains to the existence of first-order kinetics. The third deals with the initiation period, which plateaus into some sort of steady state; the beginning of the steady-state most likely signals the complete conversion of the TPP into an active complex. The fourth is that the activation energies for initiation and propagation are essentially the same. The fifth is the failure to observe oxidation of the TPP. The sixth is the branching reaction occurring in the 2.00 stoichiometry. Many others could be listed, but these will suffice for the task of analyzing the mechanisms currently under consideration.

The sluggishness of the non-catalyzed reaction in comparison to the catalyzed reactions has been well-demonstrated in the ionic conductivity experiments. These non-catalyzed reactions become important only at very high temperatures (>200°C) which are held for extended periods of time (several hours). Under these conditions, oxidation of the polymer might also become significant. The studies considered here were done under conditions generally unfavorable to the noncatalyzed reactions, since they are performed over relatively short time intervals (2 to 40 minutes) and at lower temperatures. It must be kept in mind, however, that non-catalyzed reactive pathways *do* exist. These pathways can have significant influence on the reaction distribution under the right conditions, such as very low catalyst loading, conditions in which the catalyst is rapidly oxidized (or deactivated in some other way), or through extended aging of the polymer at high temperatures.

Initation Period

The initiation period is *not* explained by any of the mechanisms in which TPP directly catalyzes the epoxy-phenol reaction. A directly catalyzed reaction would be at its most rapid rate at the beginning of the reaction, when all reactant concentrations are highest. The result would be some type of nth order decay in reaction rate from the initial and most rapid reaction rate. On the contrary, initiation periods of 3 to 10 minutes in duration are seen for the ECN-PN formulation. The mechanisms used by Banthia (Rxns. 16-17), one of those used by Sorokin and Shode (Rxns 12-15), that of Romanchick (Rxns. 18-20), and the main reaction proposed by Gagnebien (Rxns. 24-26) do not account for this initiation period.

These reactions would also be expected to show a proportional scaling between the reaction rate and the *product* of the TPP concentration and phenolic hydroxyl concentration ($C_{t0} C_{p0}$), but the reverse is seen for the ECN-PN system: the reaction rate scales with the *ratio* (C_{t0}/C_{p0}).

Diffusion of Catalytic Complex or Catalyst

The epoxy-phenol reaction is controlled by the activation energy in its later curing stages; this is shown by the linearity of the Arrhenius plots throughout the range of temperature, stoichiometric ratio, and conversion. This fact is surprising given the tremendous change in viscosity of the thermoset as it cures. The inverse of the ionic conductivity ($1/\sigma_{ion}$) can be interpreted as a measure of local viscosity seen by molecular size segments. As this viscosity increases by several orders of magnitude, however, the reaction exhibits first-order dependency on the epoxy concentration for the duration of the reaction. Very few of the mechanisms allow for this.

A very bulky quaternary catalytic complex, such as that envisioned by Sorokin and Shode, should exhibit severe diffusional and steric constraints. The steric constraints would be enormous because the complex between epoxy, phenol, and TPP would literally be chained to the polymer network through the epoxy and phenolic novalacs. Upon gelation and crosslinking, severe reaction inhibition would be expected due to the decreasing mobility of the complex as its ties to the network become more restricted. Diffusional limitations would also be substantial for a molecule so large. This particular mechanism is implausible because the complex would have to orient itself into a favorable position with both another epoxy *and* a phenol in order to catalyze the reaction.

Those mechanisms in which the TPP directly catalyzes the reaction can be called into question by the diffusion issue as well. Direct catalysis would require the bulky TPP to diffuse through the mixture as it polymerizes. The ionic conductivity decreases by many orders of magnitude as cure progresses, and the mobility of TPP decreases correspondingly. These mechanisms require that TPP orient with both epoxy and with phenol. Such favorable orientations between three reactants would become increasingly unlikely as the crosslink density increases and the resulting mobility decreases.

Charge Transfer Mechanisms

The mechanisms which seem most plausible for describing the bulk-state epoxy-phenol reaction are those which involve only a transfer of charge or proton between molecules. These would exhibit fewer diffusional limitations. Favorable orientation is necessary between only two reactants at a time: bimolecular reactions are much more feasible than trimolecular reactions in a highly crosslinked medium.

The zwitterion generated through the mechanisms of Shechter and Gagnebien is not capable of opening the epoxy ring directly;[4,11] however, it can enable reaction between epoxy and phenolic. This zwitterion can generate a phenoxide ion which may then react with the epoxy ring by extracting the proton from the phenolic hydroxyl. The alcoxide ion formed in this manner is capable of reacting with an epoxy; however, it is much more likely to extract a proton from the phenolic if sufficient phenolic is available. The transfer of protons and charge between alcoxide and phenoxide then repeats itself, and the reaction propagates. Thus, the core of the mechanisms 3-6 (Shechter) and 27,32-35 (Gagnebien) is essentially an ionic polymerization.

SYNTHESIS OF NEW MECHANISM

These two sets of mechanisms are similar, differing primarily in terminology and symbology. The Shechter model allows for equilibrium between the zwitterion and epoxy-amine. It also allows for equilibrium of both alcoxide ion and phenol with both secondary alcohol and phenoxide. The Gagnebien model allows for the alcoxide ion to react with epoxy, thereby enabling a secondary branching reaction. A synthesis of the two models is as follows; it includes an a second initiation mechanism stemming from the ionic nature of the TPP-phenol interaction:

Initiation Mechanism #1

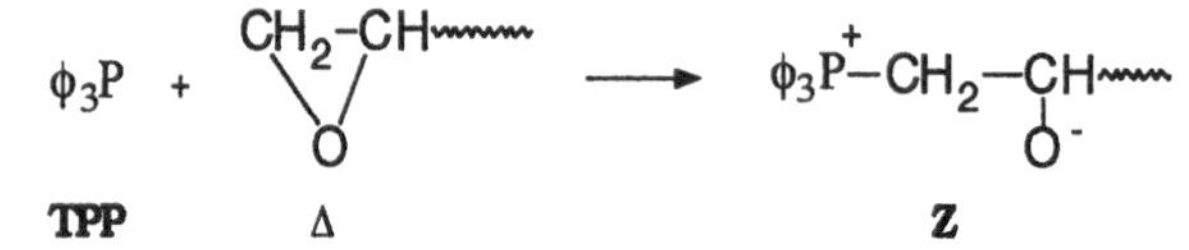

$$\phi_3P \;+\; \underset{\Delta}{\overset{CH_2-CH\!\!\sim\!\!\sim}{\triangle}} \;\longrightarrow\; \underset{Z}{\overset{+}{\phi_3P}-CH_2-\underset{O^-}{CH}\!\!\sim\!\!\sim} \tag{61}$$

$$\underset{Z}{\overset{+}{\phi_3P}-CH_2-\underset{O^-}{CH}\!\!\sim\!\!\sim} \;+\; \underset{\phi OH}{\sim\!\!\sim\!\!\langle O\rangle\!-\!OH} \;\rightleftharpoons\; \left[\underset{IP_2}{\sim\!\!\sim\!\!\langle O\rangle\!-\!O^-,\;Q^+}\right] \tag{62}$$

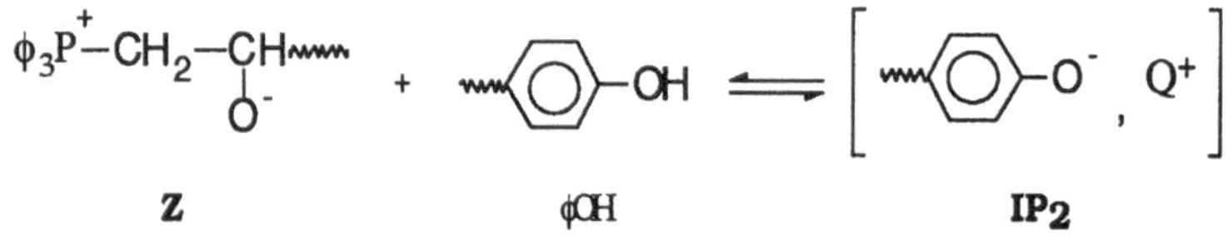

Initiation Mechanism #2

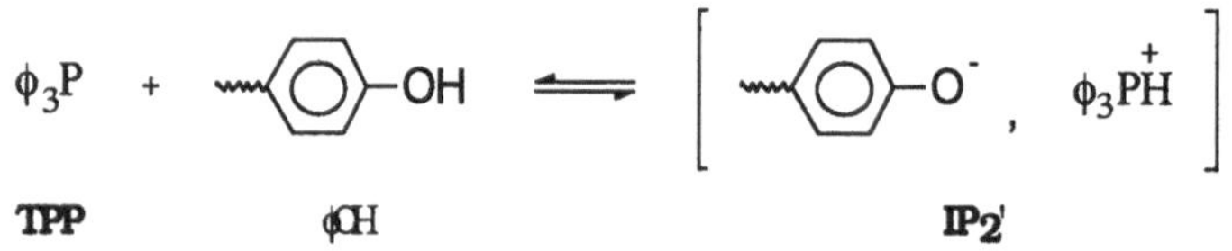

$$\underset{TPP}{\phi_3P} \;+\; \underset{\phi OH}{\sim\!\!\sim\!\!\langle O\rangle\!-\!OH} \;\rightleftharpoons\; \left[\underset{IP_2'}{\sim\!\!\sim\!\!\langle O\rangle\!-\!O^-,\;\overset{+}{\phi_3PH}}\right] \tag{63}$$

Propagation Mechanism (Primary Epoxy Consumer)

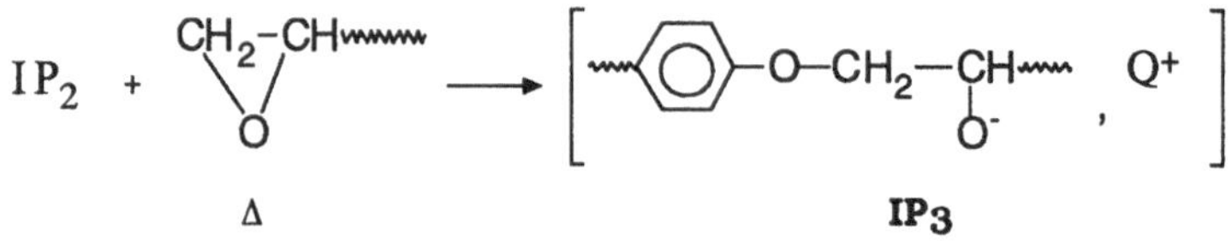

$$IP_2 \;+\; \underset{\Delta}{\overset{CH_2-CH\!\!\sim\!\!\sim}{\triangle}} \;\longrightarrow\; \left[\underset{IP_3}{\sim\!\!\sim\!\!\langle O\rangle\!-\!O\!-\!CH_2-\underset{O^-}{CH}\!\!\sim\!\!\sim,\;Q^+}\right] \tag{64}$$

$$\underset{IP_3}{\sim\!\!\sim\!\!\langle O\rangle\!-\!O\!-\!CH_2-\underset{O^-}{CH}\!\!\sim\!\!\sim} \;+\; \underset{\phi OH}{\sim\!\!\sim\!\!\langle O\rangle\!-\!OH} \tag{65}$$

$$\rightleftharpoons\; \underset{P}{\sim\!\!\sim\!\!\langle O\rangle\!-\!O\!-\!CH_2-\underset{OH}{CH}\!\!\sim\!\!\sim} \;+\; \underset{IP_2}{\sim\!\!\sim\!\!\langle O\rangle\!-\!O^-}$$

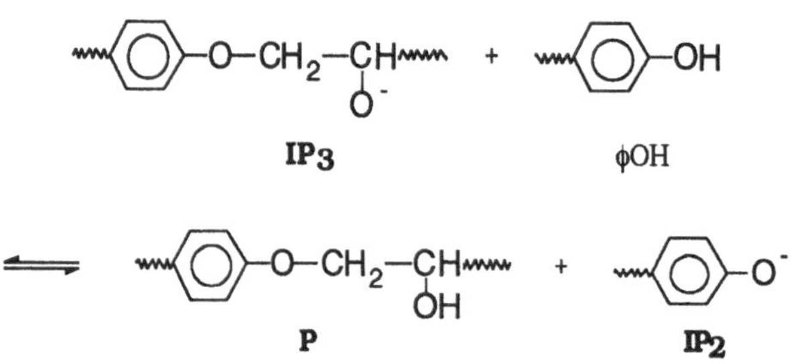

Branching Reaction (Secondary Epoxy Consumer)

$$IP_3 \; + \; \text{(epoxy)} \; \xrightarrow{\Delta} \; \left[\text{IP}_3' \right] Q^+ , \tag{66}$$

IP$_2$ and IP$_2$' are both ion pairs capable of reacting with epoxy. These would be very difficult to isolate; and they are considered equivalently reactive (IP$_2$ = IP$_2$') since they contain the same reactive species, namely a phenoxide ion. Their actual reactivity may differ based on the degree of ionization in the respective acid-base pairs. Similarly, IP$_3$ and IP$_3$' possess comparable reactivity to epoxy.

This ionic polymerization mechanism offers the most convincing explantation for the data on the ECN-PN cure. Such a polymerization mechanism would show diffusion limitations only very late in the cure because the charges are highly mobile. It is also preferred because the reaction requires coordination between only two molecules at a time, rather than the three or more molecules required in the other mechanisms.

An ionic reaction mechanism such as this may be expected to cause charging problems on an electrode surface. This is the case, in fact, as was seen in the ionic conductivity experiments of Part II. An increase in ionic conductivity was also noted by both Luston and Antoon while studying epoxide-anhydride-tertiary amine reactions.[19,20]

This reaction mechanism also allows for an interpretation of the the complex reaction rate data of the 2.00 stoichiometry reaction. As noted earlier, two parts epoxy are reacting with one part phenolic for this stoichiometry. The reaction is expected to quench at 50% reaction of the epoxy if epoxy reacts only with phenolic. Instead, up to 70% reaction of the epoxy is actually observed by DSC experiments. Ionic conductivity experiments have shown that the noncatalyzed reactions occur very slowly in comparison to the catalyzed reactions. Therefore, the only reasonable explanation is a catalyzed epoxy branching reaction, such as in Rxn. 61.

The proximity of the charges in the zwitterion, Z, neutralizes its strength in opening epoxy. However, the electronegativity and charge on the oxygen are enough to extract the loosely held proton from the phenolic, thereby generating the phenoxide or IP$_2$. Phenoxides (IP$_2$ & IP$_2$') possess a strong enough charge (due to charge separation) to open the epoxy, which creates an alcoxide (IP$_3$). This alcoxide, IP$_3$, would then respond in a similar manner as Z might: either it opens an epoxide *or* it extracts a proton from available phenolic. Epoxide is much more electrically neutral than the acidic phenolic hydroxyl, and the alcoxide will therefore exhibit a greater propensity to extract the proton rather than open the epoxy ring.

The equilibrium between the alcoxide ions and the phenoxide ions (Rxn. 60) is strongly affected both by the secondary alcohol concentration and by the phenolic hydroxyl concentration:

$$K_{equilib} \frac{[\text{Alcoxide Ion}]}{[\text{Phenoxide Ion}]} = \frac{[\text{Secondary OH}]}{[\text{Phenolic OH}]} \tag{67}$$

The equilibrium is driven towards the alcoxide ion as phenolic hydroxyls are consumed by the reaction which generates secondary hydroxyls. Thus, the alcoxide ions possess longer lifetimes and their probability of reacting with epoxy is increased. A reactive formulation such as the 2.00 stoichiometry enhances the likelihood for the branching reaction because epoxy greatly outnumbers the amount of phenol present. An alcoxide ion would likely collide with a large number of epoxides before the more reactive collision with a phenolic hydroxyl might take place. The branching reaction (61) may occur in any of these collisions with an epoxy. Reaction 61 may occur repeatedly, consuming more epoxy as the branch grows, and this branch growth could continue until terminated by proton abstraction from a phenol.

The branching reaction can occur only subsequent to the primary propagation step. If it occurs on a similar timescale to the primary reaction, the branching reaction might be observed as an increase in the reaction rate of the epoxy; this is because it would cause epoxy to be consumed more rapidly than by only the primary reaction. This apparent enhancement in reaction rate would increase as phenol is depleted, thereby favoring the branching reaction.

The presence of unreacted phenol is necessary for the branching reaction to occur. [4] Romanchick noted that the TPP oxidizes soon after the phenol is depleted, and this terminates any branching reaction which may occur.[6] Based on these two considerations, the secondary reaction is expected to cease as, or even before, the phenol depletes.

This understanding can now be used for interpreting the isothermal cure of the 2.00 stoichiometry. The 150°C cure of the 13.65 mM TPP formulation is shown in Figure 31. It deviates from the linear, smooth rise and fall of the reaction rate observed for the other stoichiometric ratios. It also shows epoxy consumption beyond the 50% expected for epoxy-phenol reaction. Extrapolated dotted lines are inlaid to show an approximation of the reaction rate which would be found in only primary epoxy-phenol reaction. The increase in reaction rate is believed to be due to the secondary reaction, and it is noted that this increase in reaction rate first begins at approximately 25% epoxy consumption. The deviational increase in reaction rate becomes larger as epoxy (and phenol) are consumed. Finally, the epoxy reaction rate appears to return to the primary first-order mechanism at around 5-10% of the remaining reaction before the reaction fully terminates. This termination appears to result from phenol depletion, which causes cessation of both the primary epoxy-phenol reaction and secondary branching reaction. Similar behavior is observed over the range of TPP concentrations studied (8.60 mM to 17.7 mM).

Rate-Limiting Step

It is noteworthy that the activation energies for both reaction propagation and reaction initiation are very similar, 68.0 and 65.6 kJ/mole, respectively. A probable explanation for this is that both the primary epoxy-phenol reaction and the complex activation reaction have similar rate-limiting steps, which in all likelihood is the opening of the epoxide ring.
This would account for the first-order dependency on epoxy concentration if Rxn. 59, the opening of the epoxide ring by the phenoxide ion, is the rate limiting step of the main propagation mechanism. Rxn. 61, the opening of the epoxide ring by the alcoxide ion, probably occurs more slowly than Rxn. 59 because it is driven by the weaker alcoxide ion. If the rate limiting step in the initiation reaction is Rxn. 56, the opening of the epoxy ring by the TPP, then the energetics would be expected to be similar to those for Rxn. 59. Both the propagation reaction and the initiation reaction have the same activation energy, and this supports the argument that the opening of the epoxy ring in all cases is the rate limiting step.

Comment on Techniques Employed

Differential scanning calorimetry is not sensitive enough to allow the primary reaction to be separated from the secondary reaction in a quantitative manner, such as was the case for the 2.00 stoichiometry. A more species sensitive technique, such as NMR or FTIR, is necessary to discern the individual extents and rates of reaction of the various species. Such data could then be used to verify or refine the currently proposed model.

CONCLUSIONS

A semiempirical rate law has been developed which describes the epoxy-phenol reaction as catalyzed by triphenylphosphine. It involves a pseudo-zeroth-order activated complex formation step and a first-order propagation step. The activated complex appears to exist in an equilibrium state. The activation energies of the initiation and propagation reactions are similar, which suggests that these reactions share similar rate-limiting steps.

Accordingly, a reaction mechanism for the catalyzed epoxy-phenol reaction has been proposed. It is synthesized from the features of the models discussed in the literature which are consistent with the experimental observations of this dissertation. This mechanism is shown to have internal consistency by explaining the key features of the cure: the initiation period, the first-order propagation, the lack of diffusional limitations until quite late in the cure, the similar rate-limiting steps, and the branching reaction observed for the 2.00 stoichiometry.

Acknowledgment

This work has been supported by the Office of Naval Research via N0001487-K0211.

REFERENCES

1. Bauer, R. S. In *Epoxy Resins and Epoxy Resin Technology - Short Course*; Shell Development Company: Houston, 1989.

2. Dangayach, K. C.; Jablonski, B. B.; Biernath, R. W. In *Proc. of 3rd SAMPE Electronics Conf.*; SAMPE: Los Angeles, June, 1989; p 983.

3. Van den Bogert, W. F.; Belton, D. J.; Molter, M. J.; Soane, D. S.; Biernath, R. W. *IEEE Trans. CHMT* **1988**, 11(3), 245.

4. Shechter, L.; Wynstra, J. *Ind. Eng. Chem.* **1956**, 48(1), 86.

5. Lin, Y. G.; Sautereau, H.; Pascault, J. P. *J. Appl. Polym. Sci.* **1986**, 32, 4595.

6. Romanchick, W. A.; Sohn, J. E.; Geibel, J. F. In *ACS Symposium Series 221 - Epoxy Resin Chemistry II*; Bauer, R. S., Ed.; American Chemical Society: Washington, D. C., 1982, p 85.

7. Romanchick, W. A.; Geibel, J. F. *Polym. Mat. Sci. Eng.* **1981**, 46, 410.

8. Ricciardi, F.; Romanchick, W. A.; Joullie, M. M. *J. Polym. Sci.: Polym. Chem. Ed.* **1983**, 21, 1475.

9. Sorokin, M. F.; Shode, L. G. *Zh. Org. Khim.* **1966**, 2(8), 1447.

10. Sorokin, M. F.; Shode, L. G. *Zh. Org. Khim.* **1966**, 2(8), 1452.

11. Gagnebien, D.; Madec, P. J.; Marechal, E. *Eur. Polym. J.* **1985**, 21(3), 273.

12. Lunsford, D. J.; Banthia, A. K.; McGrath, J. E. *Polym. Prepr.* **1981**, 22(1), 194.

13. Banthia, A. K.; McGrath, J. E. *ACS Polym. Prepr. Div. Polym. Chem* **1979**, 20(2), 629.

14. 5Alvey, F. B. *J. Appl. Polym. Sci.* **1969**, 13, 1473.

15. Bauer, R. S. In *ACS Symposium Series #285 - Applied Polymer Science*; Tess, R. W. Poehlein, G. W., Ed.; American Chemical Society: Washington, D. C., 1985, p 931.

16. Son, P. N.; Weber, C. D. *J. Appl. Polym. Sci.* **1973**, 17, 2415.

17. Lin, S. C.; Andrejak, R. A. *Polym. Mat. Sci. Eng.* **1985**, 52, 350.

18. Uejima, A. *Nippon Kagaku Kaishi* **1981**, 3, 399.

19. Antoon, M. K.; Koenig, J. L. *J. Polym. Sci.: Polym. Chem. Ed.* **1981**, 19, 549.

20. Luston, J.; Manasek, Z.; Kulickova, M. *J. Macromol. Sci.-Chem.* **1978**, A12(7), 995.

21. Narracot, E. S. *Brit. Plastics* **1953**, 26(4), 120.

22. Baba, H.; Matsuyama, A.; Kokubue, H. *J. Chem. Phys.* **1964**, 41, 895.

23. Hale, A. Ph.D. Dissertation, University of Minnesota, 1988.

24. Wittig, G.; Haag, W. *Chem. Ber.* **1955**, 88, 1654.

25. Sheppard Jr., N. F.; Day, D. R.; Lee, H. L.; Senturia, S. D. *Sensors & Actuators* **1982**, 2, 263.

26. Senturia, S. D.; Sheppard Jr., N. F. *Adv. Polym. Sci.* **1986**, 80, 1.

27. Day, D. R. Dielectric Properties of Polymeric Materials. *Course Notes.* 1987.

28. Day, D. R. Personal Communication. 1988.

29. Narracot, E. S. *Brit. Plastics* **1953**, 26(4), 120.

30. Heise, M. S.; Martin, G. C. *J. Appl. Polym. Sci.* **1990**, 39(3), 721.

31. Sickfeld, J.; Mielke, W. *Prog. Org. Coat.* **1984**, 12, 27.

32. Prime, R. B. In *Thermal Characterization of Polymeric Materials*; Turi, E. A., Ed.; Academic Press: New York, 1981, p 435.

33. Oleinik, E. F. *Pure & Appl. Chem.* **1981**, 53, 1567.

34. Oleinik, E. F. *Adv. Polym. Sci.* **1986**, 80, 49.

35. Chern, C.-S.; Poehlein, G. W. *Polym. Eng. Sci.* **1987**, 27(11), 788.

36. Huguenin, F. G. A. E.; Klein, M. T. *Ind. Eng. Chem. Prod. Res. Dev.* **1984**, 24, 166.

37. Sorokin, M. F.; Shode, L. G. *Zh. Org. Khim.* **1966**, 2(8), 1447.

38. Gagnebien, D.; Madec, P. J.; Marechal, E. *Eur. Polym. J.* **1985**, 21(3), 273.

39. Shechter, L.; Wynstra, J. *Ind. Eng. Chem.* **1956**, 48(1), 86.

40. Luston, J.; Manasek, Z. *J. Macromol. Sci.-Chem.* **1978**, A12(7), 983.

41   Antoon, M. K.; Koenig, J. L. *J. Polym. Sci.: Polym. Chem. Ed.* **1981**, 19, 549.

42.   Romanchick, W. A.; Sohn, J. E.; Geibel, J. F. In *ACS Symposium Series 221 - Epoxy Resin Chemistry II*; Bauer, R. S., Ed.; American Chemical Society: Washington, D. C., 1982, p 85.

A THEORETICAL TREATMENT OF THE STRUCTURAL ASPECTS

OF THE TOPOCHEMICAL POLYMERIZATION OF DIACETYLENES

S. E. Zutaut, M. Jalali-Heravi[#], and S. P. McManus[*]

Department of Chemistry
University of Alabama in Huntsville
Huntsville, AL

ABSTRACT

A theoretical model for predicting which diacetylene monomers will undergo topochemical polymerization has been developed. The semiempirical SCF-MO methods MNDO, AM1, and PM3 have been used to carry out calculations based on this model. These methods were used to minimize the monomer geometries and the intermolecular distances and angles ($\underline{R}$ and $\underline{\alpha}$) between the monomer pairs. They have been applied to reactive and unreactive derivatives as well as some unknown derivatives. The results from these calculations suggest that this method may be applicable to a large variety of substituted diacetylene monomers. Specific examples of cases which show both the utility and the limitations of the model are discussed.

INTRODUCTION

Some substituted polydiacetylenes have shown promise as materials for use in optical switching devices due to their third-order nonlinear optical (NLO) properties[1]. A great deal of research has gone into finding substituents which can be added to diacetylene which will produce polymers with the best properties. Some of these substituted diacetylene monomers have been synthesized and crystallized, only to discover that they will not undergo the topochemical polymerization (by 1,4' addition, as shown in Figure 1) that will produce the desired polymer single crystal. The mechanism for the solid-state polymerization was discussed many years ago by Hirshfeld and Schmidt[2] and Wegner[3]. This mechanism, which is diffusionless and totally lattice-controlled, involves the rotation of the substituted diacetylene monomers to bring the 1 and 4' carbons of adjacent molecules close enough to react. It has been shown experimentally that there are two important parameters, $\underline{d}$ and $\underline{\gamma}$ (Figure 2) which determine whether or not the molecules are close enough to react. The desired values for polymerization in the solid-state are $\underline{d} \sim 5$ Å and $\underline{\gamma} \sim 45°$ [4,5,6]. Unfortunately, these values can only be obtained for a given substituted diacetylene after it has been synthesized and crystallized. Once this stage is reached the quickest way to find out if polymerization will occur is to just try it. The goal of our

---

[#] Visiting professor from Department of Chemistry, University of Kerman, Kerman 76175, Iran.
[*] To whom inquiries should be addressed.

Contemporary Topics in Polymer Science, Vol. 7., Edited by
J.C. Salamone and J. Riffle, Plenum Press, New York, 1992

research is to find a relatively inexpensive way to predict the likelihood of polymerization for a given substituted diacetylene monomer before the time and money is invested in synthesis. This model has been briefly presented previously[7].

The purpose of this model is calculate the parameters that are important to predicting the polymerizability of the substituted diacetylenes in a monomer crystal. Polymerization can occur in the solid state when the backbones of the diacetylene molecules along the stacking axis are parallel and close enough to allow simultaneous rotation and translation along that axis. This must occur to bring the $C_{(1)}$ and $_{(4')}$ carbons of the adjacent molecules close enough to react. This is called the least motion principle[4,5]. "Close enough" is defined by two parameters, $\underline{d}$ and γ, where $\underline{d}$ is the distance between each two monomers and γ is the angle of each backbone with respect to the stacking axis. These parameters are illustrated in Figure 2.

It had been shown in experimental plots of $\underline{d}$ vs. γ obtained from crystallographic data[6] that the substituted diacetylenes that underwent solid-state polymerization tended to

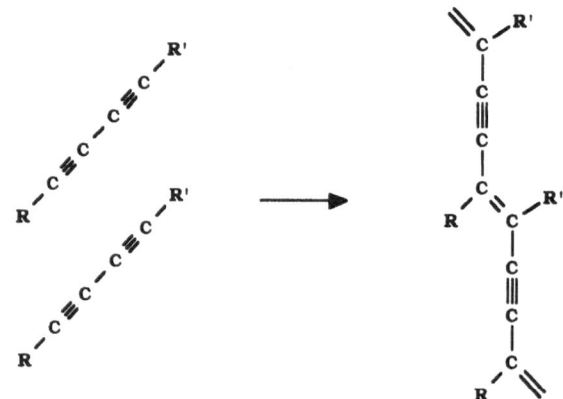

Figure 1. 1,4' addition of diacetylene to form polymer.

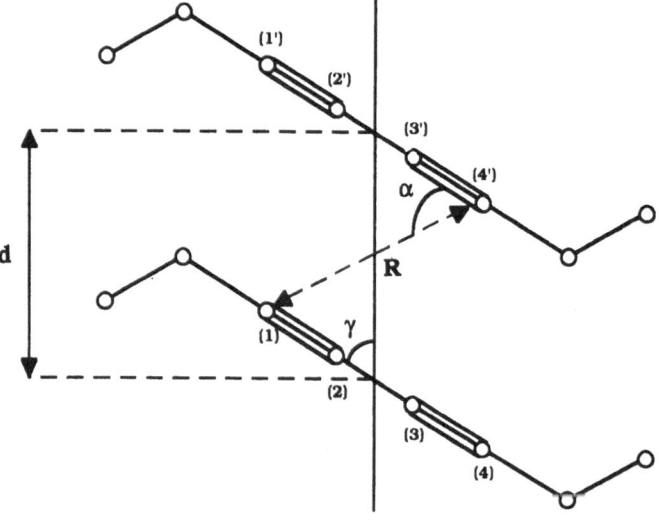

Figure 2. Schematic representation of the model for calculations of gas phase parameters of diacetylene monomers.

lie in a cluster around $\underline{d} \sim 5$ Å and $\gamma \sim 45°$, as stated above. Although we did not expect the values of $\underline{d}$ and γ to be the same in our gas-phase model and the solid-state experimental values, we hoped that we might see similar trends.

In order for this model to be useful, programs must exist which can provide reasonable geometries for substituted diacetylenes and must properly model intermolecular interactions between monomer pairs. Ab initio programs are unsuitable for this project due to the size of the molecules involved. Since we are also interested in electronic properties as well as geometric properties, it was decided that the most suitable candidates available were the semiempirical SCF-MO methods: MNDO[8], AM1[9], and PM3[10]. These methods are all found in the program MOPAC[11]. It has been determined that these methods do indeed provide reasonable geometries for substituted diacetylenes[12] but only the latter two properly model intermolecular interactions.

RESULTS AND DISCUSSION

Model for Calculations

It is clear that finding a theoretical model for calculation of the packing parameters which is practical and matches all essential conditions in the solid -state is difficult. To simplify the problem one needs to apply some approximations. In order to calculate the gas phase packing parameters for some diacetylenes, the following approximations are applied:

(I)   The symmetry of the monomers is retained, with the carbon backbones $C_{(1)}$ - $C_{(4)}$, and $C_{(1')}$ - $C_{(4')}$ forced to be linear (Figure 2).
(II) Only one monomer pair is considered. Therefore, only their interaction is considered.
(III) The distance $\underline{R}$ between $C_{(1)}$ and $C_{(4')}$, and the angle α that $\underline{R}$ makes with the backbone (see Figure 2) are optimized along with the monomer geometries, but the backbones are forced to remain parallel to each other.
(IV) Translational symmetry is assumed initially.
(V) The monomers are initially aligned so that the steric effects of the substituents are minimized.

Assumption (I) is necessary because we have not currently found a way of determining beforehand how crystal packing will affect the shape of the monomers. Existing X-ray data suggests that the backbone will remain essentially linear in the solid state, so that part of the assumption should be fair. We have used symmetry options where necessary to relate the two monomers to avoid having flexible substituents move away from each other during optimization because the substituents would not really have anywhere to go in the solid state, since the molecules in the solid state are completely surrounded by other molecules. Otherwise, we might get unrealistically low values for $\underline{d}$.

Assumption (II) implies that the effects of the other molecules that would normally surround the monomer pair do not affect the equilibrium distance and angle and would not participate in the solid-state reaction. Clearly the other molecules surrounding the monomer pair do affect them, which is why we do not expect to get the same values for $\underline{d}$ and γ as the solid-state values. However, if the effect of the surrounding molecules is essentially substituent independent, then we should get similar trends. For cases where substituents can participate in hydrogen bonding or are very large, this assumption may not work as well. Hydrogen bonding may cause the substituents to be so tightly bound to each other that the backbones cannot rotate enough for the reaction to occur, even though both $\underline{d}$ and γ are reasonable, or even if the backbones can rotate, the strain may eventually cause the hydrogen bonds to break, giving a broken crystal. Conversely, if the substituents are flexible, the backbones may be able to rotate without destroying the hydrogen bonds. The assumption that the other molecules do not participate in the reaction is generally good, especially for large substituents, but may not be for small substituents because a given monomer may be able to react with more than just two neighbors. This would lead to an amorphous polymer.

Assumption (III) is necessary because in the solid state, the monomer pair would be on a stacking axis, so the backbones must be parallel. We optimize $\underline{R}$ and $\underline{\alpha}$ because we want to control the starting geometry with respect to $C_{(1)}$ -$C_{(4')}$ geometry to insure that the local optimum geometry we find is the best with respect to the reactive carbons.

Assumption (IV) is made because we do not know how to predict the space group in which a given diacetylene derivative will pack. This is especially complicated by the fact that the same substituted diacetylene can have different crystal structures depending on the way in which it was made. One (or more) structure(s) may be reactive and the other(s) may not. As it turns out, for many substituted diacetylenes, this is in fact the case, although the monomer pairs along a reactive axis may be related by inversion symmetry or mirror symmetry. It should be mentioned here that although all of the calculations presented here started from a translational symmetry relation for the monomer pair, in some cases we have allowed the substituents to optimize into a form where the monomer pair does not have translation symmetry (such as inversion symmetry where R=R') and that although our plots only show data for that asumption, we have carried out calculations for other symmetry relations where the specfic substituents warranted it.

Finally, it was found that assumption (V) was needed to adequately model systems where the substituents are bulkier in one direction than another, i.e. rings. In many of the crystal structures of substituted diacetylenes, this assumption is valid. Certainly it should be for the reactive diacetylenes because if the rings were parallel to the stacking axis, the monomers would be pushed farther apart, leading to large $\underline{d}$ values.

Calculations

All substituted diacetylene monomer geometries were initially optimized using MNDO, AM1, and PM3 from Version 5.04 of the MOPAC package running on a CRAY X-MP/24. These optimized geometries were then used to create a monomer pair, with $\underline{R}$ = 4.0 Å and $\underline{\alpha}$ = 90°. The choice for $\underline{R}$ was based on the fact that the experimental solid-state data for $\underline{d}$ and γ in the reactive part of the plot suggests that $\underline{R} \leq 4.0$ Å for reactive substituted diacetylenes [6]. Symmetry and restrictions are defined as described previously. The monomer pair is then optimized using MNDO, AM1, and PM3. The optimized values of $\underline{R}$ and $\underline{\alpha}$, together with the bond lengths $C_{(1)}$-$C_{(2)}$, $C_{(2)}$-$C_{(3)}$, and $C_{(3)}$-$C_{(4)}$, are then used to calculate the values for $\underline{d}$ and γ. Table 1 shows a selection of the substituents we have calculated. The calculated values of $\underline{d}$ vs. γ are plotted for MNDO, AM1 and PM3 in Figures 3-5, respectively.

The heat of formation of the optimized monomer and monomer pair (ΔH_M and ΔH_{MP}, respectively) were checked in each case to insure that the pair of monomers was more stable than the two monomers individually (i.e. $\Delta H_{MP} - 2\Delta H_M = \Delta H_{MA} < 0$). This was always true for the AM1 and PM3 methods, but not for MNDO. This confirms conclusion drawn by Dewar and Zoebisch[13] that MNDO overestimates repulsive interactions between atoms at large internuclear separation and predicts purely repulsive interactions between neutral molecules, which moreover are already large at the Van der

Table I. Diacetylene substituents for Figures 3 - 5.

No.	R=	R'=	No.	R=	R'=
1	-H	R	11	-phenyl	R
2	-H	-Cl	12	-CH$_3$	-C≡C-CH$_3$
3	-Cl	R	13	-pyrrolyl	R
4	-CH$_3$	R	14	-CH$_2$-pyrrolyl	R
5	-CH$_3$	-H	15	-carbazolyl	-CH$_2$OH
6	-H	-OCH$_3$	16	-CH$_2$NHCOCH$_3$	R
7	-CH$_2$OH	R	17	-pyrrolyl	-CH$_2$OH
8	-CH$_2$CH$_2$OH	R	18	-CH$_2$-pyrrolyl	-CH$_2$OH
9	-COCH$_3$	R	19	-CH$_2$-carbazolyl	-CH$_2$OH
10	-CHOHCH$_3$	R			

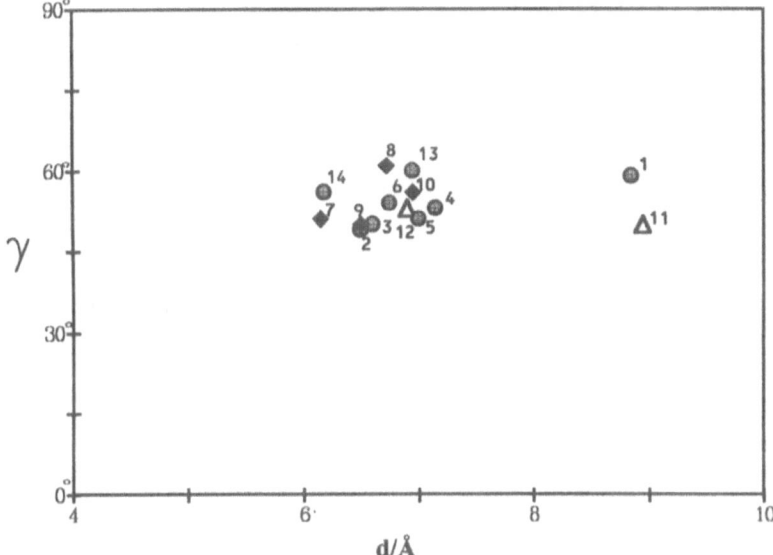

Figure 3. Plot of MNDO calculated d̲ vs. γ. △: Inactive structures; ◆: reactive structures; ◉: reactivity is unknown.

Waals' distance. This problem has been overcome in AM1 and one may expect the heats of monomer association values calculated by this method to be reasonable. Because of deficiency in the MNDO method, it was not used in later calculations. When hydrogen bonding was involved between the monomer pairs, the heat of monomer association was much more negative than otherwise. Unfortunately, this value would not be affected by hydrogen bonding that occurs between monomer pairs that are not along the stacking axis, since they are not being modeled.

Gas-Phase Packing Parameters

The relevance of the model considerations can be tested using calculated packing data for a number of diacetylene monomers experimentally known to be reactive toward polymerization or not. From molecules included in Table I, the reactivity of molecules 1-6,13-14,and 17-18 are unknown; molecules 7-10 and 15-16 are reactive and species 11 and 12 are inactive and the experimental form of 19 [6] is inactive.

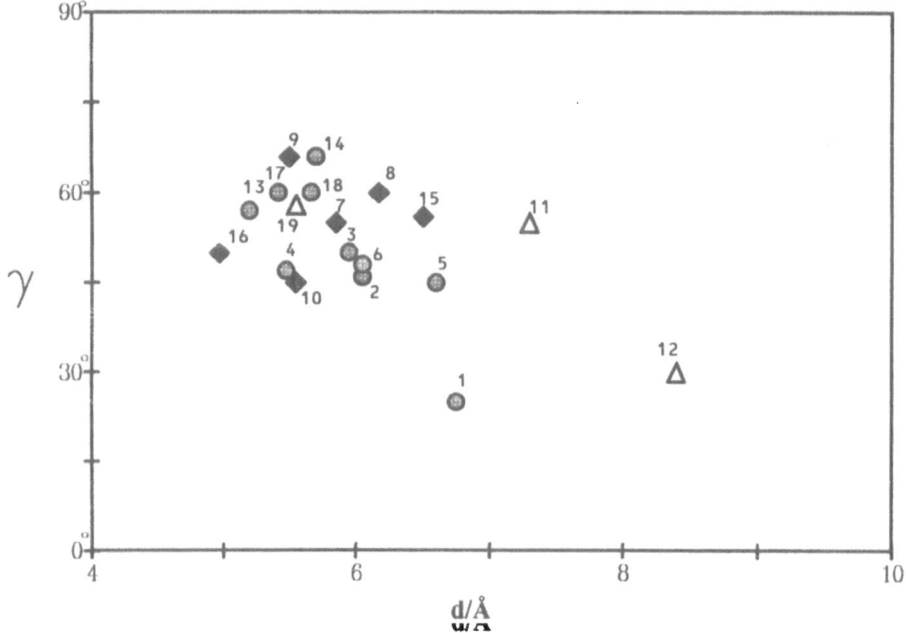

Figure 4. Plot of AM1 calculated $\underline{d}$ vs. γ. △: inactive structures; ◆: reactive structures; ◉:reactivity is unknown.

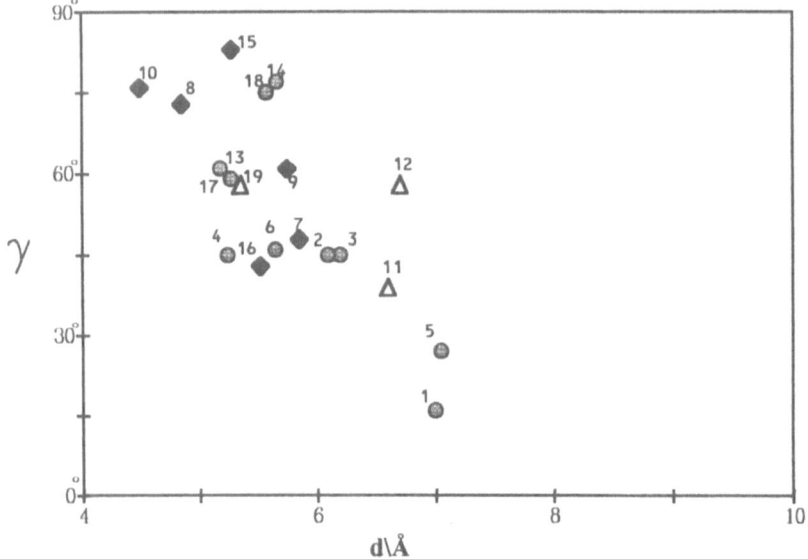

Figure 5. Plot of PM3 calculated $\underline{d}$ vs. γ. △:Inactive structures; ◆: reactive structures; ◉: reactivity is unknown.

As discussed earlier, MNDO was found to be unsuitable for calculating diacetylene monomer interactions. An example of this failure can be seen for molecule #11, which has similar for $\underline{d}$ and γ values to the reactive compounds.

The calculated AM1 parameters $\underline{d}$ and γ vary between 4.97 Å - 6.50 Å and 45 - 66 degrees respectively for reactive monomers. In the case of the PM3 method, the parameter $\underline{d}$ shows a 1.35 Å variation and angle γ has values between 43 and 83 degrees for reactive molecules. For the inactive compounds, both AM1 and PM3 predict 11 and 12 properly. Molecule 19 was predicted to be reactive by both methods. One may conclude that the AM1 and PM3 methods predict the reactivity of diacetylene monomers in a reasonably satisfactory manner, with AM1 being better because PM3 gives a much larger range of γ for the reactive molecules. The unknown molecules were calculated to provide a means to test the model. From data obtained by the AM1 method, one would predict that mono- and dichlorosubstituted diacetylene monomers (molecules 2 and 3) are reactive. Molecule 6, which forms by substitution of hydrogen in diacetylene monomers by a methoxy group may also show some reactivity. Work is currently underway to synthesize some of the substituted diacetylenes which are predicted to be reactive and may have interesting NLO properties.

Both solid-state and gas-phase models show that the reactivity is largely controlled by the monomer packing and not by the chemical nature of the substituents. It should be emphasized at this point that packing parameters for solid-state or gas phase both estimate the relative reactivity and give no absolute scale for the reaction rate.

Examples

3,5-octadiyne-1,8-diol - Molecule #8. Molecule #8 is a good example of the possible benefits or problems of hydrogen bonding. The heat of monomer association of the monomer pair in the conformation shown above is very small (ΔH_{MA} = -0.179 kcal/mol for AM1, ΔH_{MA} = -1.196 kcal/mol for PM3), but given the substituents, one might expect that hydrogen bonding would be important in the crystal structure. As it turns out, it is [14]. In this case, the hydrogen bonding favors polymerization because it allows polymerization along the axis perpendicular to the hydrogen bonding. Without the hydrogen bonding, polymerization could have occurred along more than one axis, which would have led to an amorphous material. Also, the flexibility caused by the methylene groups allow rearrangement without forcing the hydrogen bonds to break. In systems where there is not much flexibility (i.e. 2,4-hexadiyn-1-ol), the hydrogen bonding works against getting a good polymer crystal.

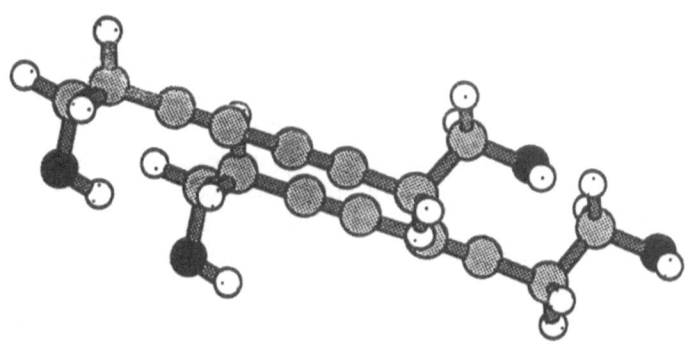

It should be mentioned that the initial selection for the geometry of the monomer for molecule #8 came from the crystal coordinates [14]. These monomer coordinates were optimized and then a monomer pair was optimized. The picture is rotated so that we are almost looking down the stacking axis.

1-(N-Carbazolyl)hexa-2,4-diyn-6-ol - Molecule #19. This molecule has been shown to be unreactive [6], but our plots for AM1 and PM3 show it to be in the reactive regions. A full crystallographic analysis was not carried out on the molecule, so no values for $\underline{d}$ and γ are available. Before we decided to add assumptions (IV) and (V) to the model we used several different symmetries when optimizing the monomer pairs and assumed that the one with the lowest ΔH_{MA} was the "best" structure[7]. We later decided that this assumption was inconsistent with a solid state system. As an example, one would expect that the monomer pairs that are coupled by hydrogen bonding would have lower heats of monomer association, but hydrogen bonding does not always occur along the stacking axis (molecule #8), which is the pair we are trying to model. To avoid this problem, we have chosen to use the translational symmetry criterion instead. The reason that the experimental data shows molecule #19 as being inactive may be that those crystals did not pack so that the monomers along the stacking axis (assuming there is one) have the same relationship as that used in our model. It is possible that some other crystallization method would lead to a packing arrangement that would favor polymerization. Molecule #18, which has a pyrrolyl group instead of a carbozolyl group, was affected in the same way by the change in assumptions. Our group is currently in the process of synthesizing molecule #18 in an attempt to verify the prediction that it has some reactivity.

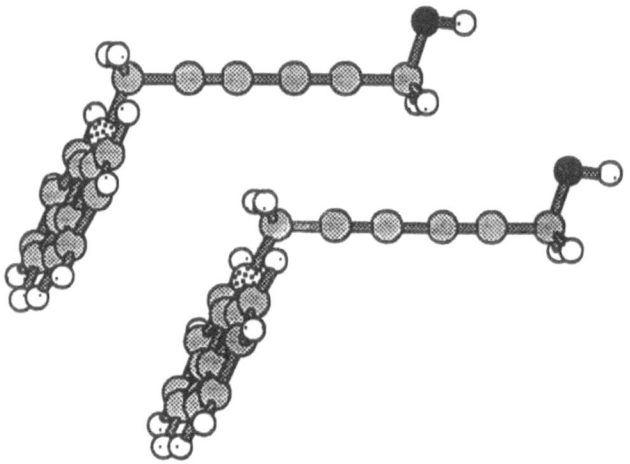

CONCLUSIONS

The main goal of this work is to find a theoretical model to assess the relative reactivity of diacetylene monomers. In the solid-state, the packing parameters $\underline{d}$ and γ play the major role in solving this problem. It is obvious that finding a theoretical model for gas phase calculations which matches all requirements in the solid-state is difficult. By applying approximations, we have developed a model for calculation of gas phase packing parameters. This model, in conjunction with MNDO, AM1 and PM3 semiempirical methods, is used to calculate the packing parameters of a number of reactive, inactive and unknown diacetylene monomers. It is found that the theoretical gas phase model qualitatively mimics the solid-state model. Among three semiempirical methods used in this work, AM1 and PM3 predict the reactivity of diacetylene monomers in a reasonably satisfactory manner. On the other hand, MNDO, which overestimates the repulsive interaction between atoms with large separation, is not good for investigation of the interactions between diacetylene monomers.

The method appears to be useful for a large number of substituents. It has the same qualitative features that the solid-state plot[6] gives. If a given substituted diacetylene monomer is calculated using the AM1 or PM3 method and values for $\underline{d}$ and γ fall outside the cluster of reactivity, then it quite probably will not polymerize. If $\underline{d}$ and γ the values fall within the acceptable range for reactive molecules, then one should consider the possible effects that might occur from hydrogen bonding, substituent flexibility and steric effects which may prevent the monomers from moving enough to react. One should also look at the possibility that hydrogen bonding for inflexible substituents may cause the crystal to crack during polymerization and at the possibility that polymerization may occur along more than one axis. Values for $\underline{d}$ and γ do not give any indication of the potential for these problems, either in the solid-state plot or the gas-phase calculations. Also, the fact that the model predicts that polymerization can occur if the monomers in the crystal are arranged like the one being modeled (i.e. translation, inversion, or mirror plane), it does not say anything about how one should perform the crystallization in order to get that desired stacking arrangement.

ACKNOWLEDGEMENTS

A generous allocation of computer time by the Alabama Supercomputer Network and financial support by the National Aeronautics and Space Administration (Contract NAGW-812) and the NASA Space Grant Fellowship Program is gratefully acknowledged.

REFERENCES

1. Reviews: G. R. Meridith, MRS Bulletin, 13(8):24-29, (1988) ; A. J. Heeger and D. R. Ulrich, "Nonlinear Optical Properties of Polymers", North-Holland Press, New York, (1988); P. N. Prasad and D. A. Ulrich, "Nonlinear Optical and Electroactive Polymers", Plenum, New York, (1988); D. J. Sandman, "Solid-State Polymerization", American Chemical Society, Washington, DC, (1987); D. J. Williams, "Nonlinear Optical Properties of Organic and Polymeric Materials", American Chemical Society, Washington, DC, (1983).
2. F.L. Hirshfeld and G. M. J. Schmidt, J. Poly. Sci.:Part A, 2:2181, (1964).
3. G. Wegner, Z. Naturforsch, 24b:824, (1969).
4. R. H. Baughman, J. Polym. Sci., Poly. Phys. Ed. 12:1511, (1974).
5. R. H. Baughman and K. C. Yee, J. Poly. Sci, Macromal. Rev. 13:219, (1978).
6. V. Enkelmann, Structural Aspects of the Topochemical Polymerization of Diacetylenes, in: "Advances in Polymer Science 63: Polydiacetylenes", H. J. Cantow, ed., Springer-Verlag, New York, (1984).
7. M. Jalali-Heravi, S. E. Zutaut, S. P. McManus, Polymer Preprints, 32(1):78, (1991).
8. M. J. S. Dewar, W. Thiel, J. Am. Chem. Soc. 99:4899, (1977).
9. M. J. S. Dewar, E. G. Zoebisch, E. F. Healy, J. J. P. Stewart, J. Am. Chem. Soc., 107:3902, (1985).
10. J. J. P. Stewart, J. Comp. Chem., 10:209, (1989).
11. Stewart, J. J. P.; MOPAC, A Semiempirical Molecular Orbital Program, QCPE 455 (1983), Version 5.04, (1989).

12. M. Jalali-Heravi, S. P. McManus, S. E. Zutaut, and J. K. McDonald, <u>Macromolecules</u>, 24:1055, (1991).

13. M. J. S. Dewar, E. G. Zoebisch, <u>J. Molecular Structure (Theochem)</u>, 180:1, (1988).

14. D. A. Fisher, D. J. Ando, <u>Acta Crystallogr., Sect. B</u>, B34(12):3799, (1978).

CREATING CONDUCTING MATERIALS THROUGH SOLUTION BLENDING OF

CONDUCTING POLYMERS WITH COMMERCIAL POLYMERS

David MacInnes, Jr.

Chemistry Department, Guilford College
Greensboro, NC 27410

ABSTRACT

The soluble conducting polymer poly-o-ansidine, or PANIS, was blended
with various other polymers to give conducting materials. The
conductivity of the blends can be controlled by the relative amounts of
polyanisidine and host polymer and is one of the few times that the
conductivity of a conducting polymer has been modified without chemical
doping. In this study, polyansidine was been blended with a number of
commercial polymers to form free-standing, stable, flexible films having
electrical resistance values dependent upon the concentration of PANIS.

INTRODUCTION

Polyanisidine (PANIS) is a new and novel electrically conductive
polymer (1) that is generating much interest due to it's combination of
air stability, electrical conductivity, and solubility in organic
solvents. This unique blend of properties has been heretofore
unavailable in synthetic conductors. PANIS is synthesized by either
chemical or electrochemical means. The polymer may exist in four unique
states, each a function of the level of oxidation and protonation. Each
state is characterized by a unique set of electrical and optical
properties, solubility and air stability. The states are easily
convertible by simple chemical and electrical means. The most conductive
and stable form of PANIS is the oxidized acid form shown in Figure 1.

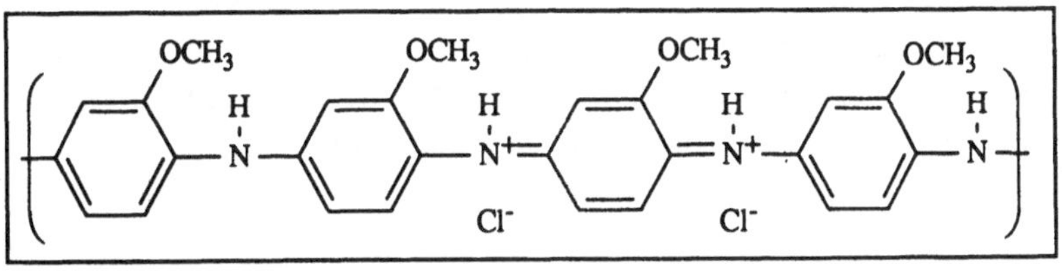

Figure 1. Polyanisidine as prepared from 1 M HCl

Contemporary Topics in Polymer Science, Vol. 7., Edited by
J.C. Salamone and J. Riffle, Plenum Press, New York, 1992

The oxidized acid form of PANIS is present as a dark green crystalline polymer having a molecular weight of approximately 2200 and a conductivity of 13.0 S/cm (1). PANIS does not have a definite melting point and begins to degrade at approximately 130 deg C. The presence of the pendent ortho-methoxy groups allows dissolution in various common solvents, such as N,N-dimethylformamide, trifluoroacetic acid, and N-methyl-pyrrolidinone (1,2). The potential applications of PANIS are widely varied and span the range from the utilization of the polymer's electrical properties, (e.g. polymeric electrodes and thin film resistors), to the optical properties of the polymer, (e.g. optical switching devices). The fabrication of such articles, however, requires that the polymer be processable utilizing the current state-of-the-art industrial technology. This study was one of the first steps in developing such technology.

EXPERIMENTAL

Chemical Synthesis of Polyanisidine

The compound o-anisidine was polymerized chemically in air by oxidative polymerization. A 21.60 g sample (0.17 mol) of o-anisidine (Aldrich) was dissolved in 1 M hydrochloric acid and cooled in an ice bath at 5 deg C. Ten grams of ammonium peroxydisulfate (0.4 mol, Fisher) were dissolved in 200 mL of 1 M hydrochloric acid at 5 deg C and placed in a separatory funnel above the o-anisidine solution to facilitate drop-wise addition. The ammonium peroxydisulfate solution was slowly introduced into the o-anisidine solution over an approximately 10 minute period with constant stirring. After 1 hour of stirring at 5 deg C, the solid PANIS was filtered out, rinsed with 1 M hydrochloric acid, rinsed with distilled water, and dried under dynamic vacuum at 60 deg C. The yield of dark, emerald green crystalline product was 3.6 g (16.7%). A pressed pellet of PANIS powder exhibited a two probe resistance of less than 5 ohms when checked with an ohmmeter.

Testing Methods

UV-Visible Spectra

Each form of PANIS has been shown to posses a unique color (1). In particular, the oxidized acid form is dark green (emeraldine) and shows a strong absorption peak in the range from 825 nm to 875 nm (2). Therefore, UV-visible spectra can be used to indicate the presence of the proper form of PANIS. UV-visible spectra were recorded on a Hitachi Model U-2000 spectrophotometer scanning between 320 nm and 1100 nm which clearly show the absorption peaks of the material.

Most of the tested samples had resistivities of magnitudes not easily measured using the standard four probe method. Due to the negligible amount of contact resistance compared to the total resistance of the sample, two-point methods were utilized for high resistivity films while the standard four point probe method was used for the more conductive films. Two-point resistance values were obtained on a FLUKE digital ohmmeter. A simple method were developed to measure resistance through the thickness of a sample. In this method the film sample was placed on a piece of gold foil. The circuit was formed by placing one gold plated probe (diameter 1/16 inch) in contact with the top surface of the sample (a 100 g weight was used to standardize contact pressure) and resting another probe on the gold foil. Very small samples were easily checked with this method. The following equation was used to calculate the conductivity of the sample:

$$s = 4t/pd^2R$$

where s = conductivity in (ohm-cm)-1 or S/cm

R = measured resistance in ohms

t = film thickness in cm

d = diameter of sample probe in cm

p = pi

A second method was developed to measure the resistance along the plane of the film. Ohmmeter leads were inserted in two small pools of mercury (contained in 3 mm I.D. O-rings) and the resistance was recorded.

The 4-point probe resistance meter used was a Loresta-FP made by Mitsubishi Petrochemical Co., Ltd. The four pin probe is an AS probe with brass pins each separated by 0.5 cm. The pins are spring loaded to a constant 100 g of pressure. The meter is capable of reading from 0.0001 ohm to 1.999 M ohm. All measured resistances were converted to surface resistivity by multiplying the result by 4.532 (p.3, instruction manual). Volume resistivity for thin films whose thickness can be measured is the surface resistivity multiplied by the sample thickness in cm.

Most of the film samples had conductivities (or resistances) of magnitudes not easily measured using a four probe method. For this reason, the two probe method was used on these films and was found to give the most consistent and, it is felt, reliable readings. A comparison between two and four probe methods can be seen by looking at the conductivity of pure PANIS. When PANIS was dissolved in TFA and cast into a film, the two probe technique gave a conductivity of 0.06 S/cm. Measurement of the same sample using the four probe technique gave a conductivity of 13 S/cm, identical to that found for pressed powder samples of polyansidine. Thus the conductivities reported for the more conductive films are really low by a factor of 216 (or log S = +2.3).

Film Appearance

A Nikon stereomicroscope fitted with a polaroid camera was used to produce photomicrographs at magnifications of 100X, 200X, and 400X. Films were also viewed with a typical laboratory stereomicroscope. At the relatively low magnification of 100X, the dark brown or green films appeared to have a fairly uniform mottled appearance. At magnifications of 200X and 400X, regions of clear host polymer were seen, evenly dispersed with fibrils of the darker polyanisidine.

Solubilities of Polyanisidine

Unblended Polyanisidine Films

   a. Sulfuric acid: The polyansidine polymer dissolves easily and completely in concentrated sulfuric acid to give a thick black solution. The best films were prepared by casting a film in air,

TABLE 1. PANIS - ACID, OXIDIZED FORM

	Anion Present		
Solvent	BENZENESULFONATE	SULFATE	TRIFLUOROACETATE
DMF	soluble	soluble	sol
DMSO	soluble	sol	sol
TOLUENE	insol	insol	sol
CHCl$_3$	s.sol	s.sol	s.sol
CH$_3$CN	s.sol	s.sol	sol
ACETONE	insol	s.sol	s.sol
METHANOL	insol	insol	insol
ETHANOL	insol	insol	insol
NMP	sol	sol	sol
ETHYL ACETATE	insol	insol	s.sol
THF	insol	insol	s.sol

TABLE 2. PANIS - BASE, OXIDIZED FORM

NMP	soluble, bluish black
THF	insoluble
Toluene	insoluble
Acetonitrile	insoluble
DMF	soluble
DMA	soluble
CHCl$_3$	s.soluble

treating with water and drying overnight in an oven at 110 deg C.   The final result was a shiny dark film which was somewhat brittle, no piece larger than 0.5 cm resulted.

      b.   Dimethylformamide: As previously reported, polyanisidine can be dissolved in N,N-dimethylformamide (1).   The resulting solution is dark brown in color. The solution can be cast onto a glass plate and dried under an infrared heat lamp.   Upon solvent evaporation in air, under the infrared lamp, a dark brown powder was produced. The powder exhibited poor film properties as it was brittle and would not form a free-standing film.

      c.   Trifluoracetic acid: A 10% (w/v) solution of PANIS in trifluoroacetic acid was prepared.   The dark, emerald green solution was swabbed onto a clear glass slide and allowed to air dry.   The resulting film was dark green, adhered well to the glass surface (impossible to remove intact), smooth, and had no visible pinholes or voids.

      d.   N-methylpyrrolidinone: Polyanisidine in its base, oxidized form was dissolved in N-methylpyrrolidinone (NMP), filtered, and the thin solution poured into a petri dish.   After drying overnight at 90 deg C a thick dark blue film was formed which adhered tightly to the glass.   The measured resistance was greater than 2 M ohm. The film was allowed to sit in 1M HCl for 2 days after which it had a green color on reflection yet blue in transmission.   The film's four probe surface resistivity was now 0.9 ohm and the film came off the glass easily.   The volume resistivity was calculated to be 0.009 ohm-cm.   Surface conductivity is thus 1.1 S/cm and volume conductivity is 110 S/cm^2.   The material is relatively strong and is probably cross linked by action of the solvent, since it is now insoluble in NMP.

SOLUBILITIES OF COMMERCIAL POLYMERS AND CONDUCTING POLYMERS

      The solubilities of various commercially available polymers were tested in a number of organic and inorganic solvents. Conducting polyanisidine containing different anions were tested in the same solvents. Results show that the most useful solvents for all types of polymers used are, in descending order of utility, trifluoroacetic acid (TFA), 1-methyl-2-pyrrolidinone (NMP), dimethylformamide (DMF), chloroform, tetrahydrofuran (THF), and ethyl acetate.

Easily available commercial polymers which show promising solubilities are polyvinyl chloride, polystyrene, Nylon 6T (polyterphthalamide), polyethyleneoxide and polyisoprene. Some low molecular weight casting resins were examined. The polymer resins Acyloid B72 and B66, polyester PE 20 were soluble in tetrahydrofuran and dimethylformamide but suitable films could be cast from THF. A coating powder was also tested (nylon 11) but was found to be insoluble in all solvents. Solvent mixtures (TFA/DMF, TFA/CHCl$_3$ and others were tried. These gave good films for the pure polymer and could be used to cast films of the polymer blends. Of the various forms of polyanisidine, those containing the benzenesulfonate ion and the trifluoroacetate ion were soluble in DMF, dimethylsulfoxide (DMSO), and NMP. They were slightly soluble in chloroform, acetonitrile, acetone, methanol, ethanol, ethyl acetate and THF.

Polymer Blends

The term polymer blend as used here refers to the physical mixing of two polymers in an attempt to utilize the unique properties of each polymer in the resulting mixture. Although there are various approaches to such combinations, the work discussed here deals exclusively with processes where polymers are blended through solution mixing. Due to the low thermal stability of polyanisidine, melt blending was not attempted.

Preparation of Solution Blends

PANIS/Polystyrene Film

A ten percent (w/w) solution of polystyrene in N,N-dimethyl-formamide was prepared. Two grams of the solution were mixed with 2 grams of one-half percent PANIS in N,N-dimethylformamide (w/w). Such mixture yields a dried film consisting of approximately 5% PANIS. The resulting mixture was applied by spin casting onto a glass microscope slide and then drying under an infrared heat lamp. This procedure yielded a strong film that could be readily peeled from the glass substrate. The polyanisidine, however, did not appear to be well dissolved, as the film had a granular, particulate appearance and was quite chalky. A UV-visible absorption at 841 nm indicates that the conductive form of PANIS was present in the polystyrene; however, these results could not be consistently duplicated. The two-probe surface

resistance tests showed infinite resistance. An alternate solvent
(trichloroethylene) for the polystyrene solution was used in an attempt
to get better film formation. Films were formed by dissolving expanded
polystyrene in trichloroethylene and then adding various amounts of a
solution of PANIS dissolved in N,N-dimethylformamide. Films containing
up to 90% PANIS were formed, none of which showed any conductivity when
tested for electrical resistance.

In response to the unsuccessful PANIS/polystyrene blends, other
polymers were examined. In addition, in order to reduce the particle
size and improve the consistency of the powder, PANIS and the host
polymer were ground to a fine powder. The polymer was placed into a
small plastic rotary tumbler with nickel shot. Ten to twelve grams of
powder were masticated for approximately thirty minutes, resulting in a
very uniform powder. The powder was then separated from the shot using a
fine mesh screen (approximately 42 mesh or 0.350mm opening).

PANIS/Poly(Trimethyl Hexamethylene Terephthalamide)

Films of various concentrations of PANIS in poly(trimethyl hexamethylene
terephthalamide), also known as Nylon 6T, were prepared by dissolving
both polymers in trifluoroacetic acid and spreading the solutions onto
glass slides by passing them under a stationary blade to achieve a wet
film thickness of approximately 0.2 mm. Films were also prepared using
N-methylpyrrolidinone as the solvent. Surface resistance and resistance
through the thickness of the film were measured. The results are shown
in Table 3 and in Figure 2.

Table 3. Percent PANIS versus log (Conductivity)
for Several PANIS/Nylon Blends.

SAMPLE	%PANIS	CONDUCTIVITY (1)
NMP(2)/water	25	0.00015
NMP/water	50	0.0003
TFA(3)/water	25	0.0001
TFA/water	50	0.004
TFA/water`	75	0.013
TFA/methanol	25	0.00013
TFA/methanol	50	0.0017
TFA/methanol	75	0.013
TFA/methanol	100	0.010
TFA/methanol	0	.00001
TFA/methanol	100	0.06 (4)

1. Siemens/cm (ohm-cm)-1 by two point probe
2. N-methylpyrridinone. 3. Trifluoroacetic acid.
4. Conductivity = 13 S/cm for 4 probe.

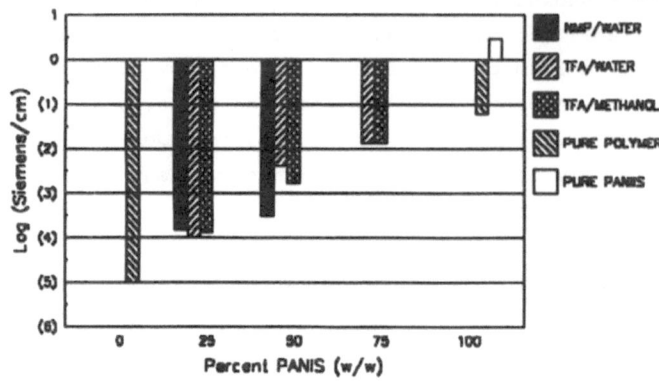

Figure 2.   Percent PANIS versus log (Conductivity)
for Several PANIS/Nylon Blends.   Data from Table 3.

Solutions of Other Polymers

    Blends of PANIS and solutions of Celanese Nylon 6-6, Valox
(Natural) and Celanex polyester PBT plastics were prepared.   Ten percent
(w/w) solutions were prepared.   The two solutions were mixed in ratios to
give mixtures which contained 5%, 25%, 50%, 75%, and 95% PANIS content.
The total solids content of the solutions ranged from 2% to 9%.   These
mixtures were then used to cast thin films.   Films prepared in this
manner were uniformly poor due to the hydroscopic nature of trifluoroacetic acid.   Only those blends with greater than 75% PANIS showed any
significant conductivity and these films were brittle.

CONCLUSIONS

    This study demonstrates, for the first time, the ability to
manipulate the resistance of a conductive polymer without doping.
Techniques were developed to allow blending conductive PANIS with
commercially available polymers, and methods were devised to measure the
resistance of the resulting blends. Polymer preparation, solvent
selection and solvent removal have been shown to be critical in producing
the desired physical, electrical and optical properties.
    The films had very low conductivity, for the most part and blends
with concentrations of polyanisidine higher than 43% were very brittle.
The commercial polymers investigated were relatively insoluble which
meant that unsatisfactory solvents like trifluoroacetic acid had to be
used to dissolve and cast the films.   While TFA did give nice looking
films, their strength and conductivity were not what we were looking for.
When lower molecular weight polymers were tried, the results were not
encouraging.   Attempts to use solvent blends resulted in poor films due
to phase separation.

178

The results of this work was encouraging. It was established that PANIS can be blended with other polymers to yield electrically conductive films. Although the blended polymers showed considerably higher electrical resistance than the pristine polyanisidine powder, the improved processability and physical properties should more than compensate for the partial loss of conductivity.

From the work done so far with solution blending, it appears that a good solvent or solvent mixture is yet to be found. The best looking films are those prepared from Poly(trimethylhexamethylene terphthalamide, Nylon 6,6 or Celanese Nylon. The stronger films have less than 43% PANIS content which gives them very high electrical resistances. Films with higher PANIS content show lower resistance but are very brittle. Most films cast from TFA show no phase separation and came out as shiny evenly black films. Blends made using solvent mixtures such as polycarbonated / PANIS blends all showed strong phase separation. As a result, these were brittle, uneven films with very high electrical resistivity. Thus more work needs to be done on finding solvents suitable for solution casting of PANIS blends.

REFERENCES

1. David MacInnes, Jr. "Poly-o-methoxyaniline: A New Soluble Conducting Polymer," Synthetic Metals, 25(1988):237.
2. Research results, Guilford College Laboratory.
3. Ibid., p 236.

PREPARATION OF SUBSTITUTED POLYPHENYLENES AND

CATALYZED AROMATIZATION OF POLYCYCLOHEXADIENEDIOL DERIVATIVES

Donald R. Wilson[*], Hamsa Jathavedam and Norman W. Thomas

Celgene Corporation
7 Powder Horn Drive
Warren, New Jersey 07059

INTRODUCTION

Polyphenylene (I) has been the subject of considerable research effort due to its unique chemical structure. The all aromatic nature and the absence of functional groups in either the main chain or on pendant groups result in important advantages over most other classes of organic polymers. These include, high thermal stability, low moisture sensitivity and stability against the environment and many chemicals.

Potential applications envisioned for these polymers include:

1) Electrical Conductors - In the doped state they are among the highest conductivity reported.
2) Dielectric Materials - They have an excellent combination of properties for packaging and interconnection in electronic devices owing to the very high thermal stability, low moisture sensitivity and low dielectric constant.
3) 3rd Order Nonlinear Optical Materials - Extended conjugation along the polymer chain should result in excellent properties.
4) Structural Materials - With functional end-groups the rigid-rod nature of oligomers should make them excellent building blocks for high performance polymers.

Up until the mid 1980's interest in polyphenylene was dampened by the polymers obtained having low molecular weight (typically n < 15), difficulty in fabricating the polymer into shapes and coatings, and the many impurities resulting from the synthetic method employed. These older synthetic methods[1,2,3] included:

Oxidative cationic polymerization of benzene;
Coupling of dibromobenzene (Grignard reaction, or electropolymerization); and,
Bromination followed by dehydrobromination of polycyclohexadiene.

A new synthetic method, which significantly reduced these deficiencies, was reported by ICI workers[4,5] and McKean and Stille[6] and is based on the heat treatment of poly(cyclohexadienediol) derivatives (III) as

---

[*] Consultant to Celgene Corporation from Advanced Polymer Technology, Inc., P.O. Box 221, Landing, NJ 07850.

Contemporary Topics in Polymer Science, Vol. 7., Edited by
J.C. Salamone and J. Riffle, Plenum Press, New York, 1992

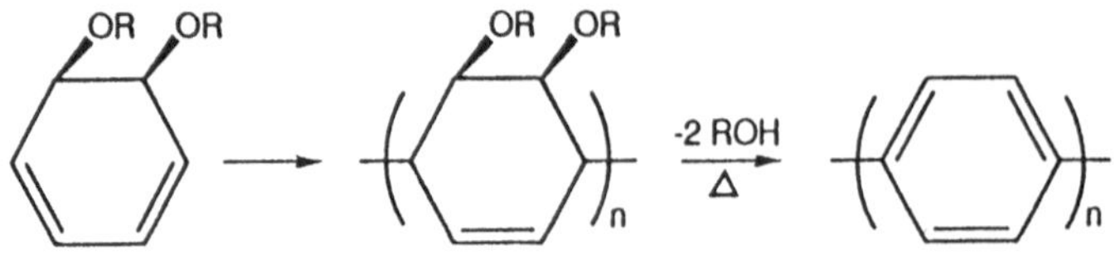

$$\text{II} \longrightarrow \text{III} \xrightarrow[\Delta]{-2 \text{ ROH}} \text{I}$$

shown above. Monomeric cyclohexadienediol derivative (II) is free radical polymerized to high molecular weight III which is soluble in many organic solvents and can be cast to films and coatings. The volatile organic leaving groups result in few impurities from the aromatization process as opposed to the metals and halides employed in the older methods. However, the polymer is not all 1,4-linked as shown but contains a significant fraction (about 15 %) of 1,2-linkages in polymerizations reported to date.

The objectives of work described in this paper were:

- Incorporation of functional groups onto polyphenylenes for improvements in adhesion, solubility, etc.; and
- Aromatization process improvements such as rate, yield, degree of completion and product stability.

A Celgene mission is to employ biotechnical approaches to synthesize monomers difficult or uneconomical to obtain by strictly chemical means.

PRECURSOR POLYMERS

Monomer Synthesis

cis-5,6-Dihydroxy-1,3-cyclohexadiene (CHD) was microbially produced from benzene using a bacterial strain of Pseudomonas putida[4],[5]. cis-1,6-Dihydroxy-2,4-cyclohexadiene-1-carboxylic acid (Carboxy-CHD) was prepared by the biological transformation of benzoic acid using a strain of Alcaligenes eutrophus[7]. CHD-Diacetate (IV), mp 35-36°C, lit.[5] 40°C, and CHD-Bismethylcarbonate (V), mp 34-36°C, lit.[5] 36°C were prepared by the appropriate reactions with acetic anhydride and chloromethyl formate. The methyl ester of cis-1,6-dihydroxy-2,4-cyclohexadiene-1-carboxylic acid was prepared by reaction with diazomethane and then the acetate derivative (VI), mp 91-93°C, lit.[7], and cycliccarbonate derivative (VII), mp 93-95°C, were prepared by reactions with acetic anhydride and triphosgene respectively.

Homopolymerizations

Free radical polymerization of CHD derivatives IV and V are reported in the literature[4],[5],[6] and very high MW polymers (10^5 to 10^6) are claimed by the ICI workers[4],[5]. Polymer derived from CHD derivative V has been analyzed by a combination of NMR and IR spectroscopy to be about 85 % 1,4-linked with the remainder 1,2-linked[5],[6]. Our polymerization of both of these derivatives resulted in polymers with inherent viscosities of about 0.4 (in chloroform c=2.5g/100 ml) similar to literature values of intrinsic viscosities but GPC MW's were less than 100,000.

Free radical polymerization of the diacetate carbomethoxy-CHD derivative (VI) under the best conditions using benzoyl peroxide at 90°C resulted in low MW polymer (about 5000). The cycliccarbonate carbomethoxy-CHD derivative also did not yield a high MW product. In the latter case only low MW oligomers in low yield in addition to 30 % of a dimeric product were obtained. Higher temperature polymerizations (necessitated by higher monomer

IV	V	VI	VII

melting point) or the presence of solvents at lower temperatures probably contribute to the difficulty in achieving high MW. However, a larger factor, as discussed in the copolymer studies below, is probably due to lower propagation and/or initiation rates.

Copolymerizations

Copolymers from carbomethoxy-CHD monomers with both the diacetate and bismethylcarbonate CHD monomers were prepared by free radical polymerization in the melt with the results shown in Tables 1 and 2. Reactivity of the carbomethoxy-CHD monomers are less than that of the unsubstituted CHD monomers as evidenced by the mole % incorporated versus the mole % charged. One also sees that MW is reduced by the presence of carbomethoxy-CHD monomers as evidenced by decreasing solution viscosities. These results are consistent with slower propagation rates for these monomers allowing time for more chain transfer to occur. The degree of incorporation of comonomers is slightly less with the CHD-bismethylcarbonate monomer which probably reflects the known higher rates of propagation for this monomer versus the CHD-diacetate[5]. The copolymers prepared with a 25 mole % charge of comonomer were of sufficient MW to give good and uniform coatings when cast onto glass plates.

AROMATIZATION TO POLYPHENYLENES

Uncatalyzed

Polycyclohexadienediol derivatives (esters and carbonates) are known to thermally aromatize to polyphenylenes[4,5,6]. The reactions are ini-

Table 1. Copolymers of CHD-Diacetate (IV) With Carbomethoxy-CHD Derivatives

Carbomethoxy-CHD Comonomer	Mole % Comonomer		Ninh
	Charged	Incorporated	
None	0	0	0.4
Diacetate	25	14	0.29
Diacetate	50	37	0.18
Cycliccarbonate	25	14	0.23
Cycliccarbonate	50	?	0.18

Table 2. Copolymers of CHD-Bismethylcarbonate (V) With Carbomethoxy-CHD Derivatives

Carbomethoxy-CHD Comonomer	Mole % Comonomer		Ninh
	Charged	Incorporated	
None	0	0	0.34
Diacetate	25	11	0.25
Diacetate	50	?	0.05
Cycliccarbonate	25	10	0.20
Cycliccarbonate	50	?	0.17

tially first order in monomer units with half-lifes on the order of 5-20 minutes at 300°C[5]. It has also been shown that significant MW degradation occurs during this uncatalyzed process[8].

CHD-Diacetate copolymers also exhibit aromatization rates which are initially first order in monomer units and the rates increase with increasing comonomer content as shown in Figure 1. However, the weight yield is significantly lower than theory as determined by TGA and shown in Table 3. The higher than theoretical weight loss is believed due to loss of oligomers and/or depolymerization. Infrared spectra of the pyrolyzed copolymers show the presence of an aromatic ester and is discussed later. Elemental analysis indicates aromatization is only 94 % complete for the homopolymer (even though the infrared spectra is essentially free of carbonyl absorption) and 96-97 % complete for the copolymers. Difficulty in achieving good elemental analyses on polyphenylenes has been noted by other workers[1,6].

CHD-bismethylcarbonate copolymers exhibit a slower rate of aromatization than for homopolymer in the presence of carbomethoxy-CHD diacetate comonomer and the rate is unaffected by cycliccarbonate comonomer. As seen in Table 3 the weight yields approximate theoretical values.

Base Catalyzed

Aromatization of CHD-bismethylcarbonate derivatives are reported to be catalyzed by bases (metal salts, amines, amide solvents) resulting in sig-

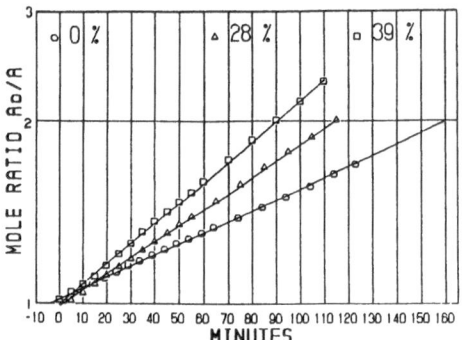

Fig. 1. First Order Rate Plots for the Aromatization of CHD-Diacetate Copolymers. Effect of Mole % Carbomethoxy-CHD Comonomer.

Table 3. Weight Yield Upon Thermal Aromatization of CHD Copolymers Under Nitrogen By Thermal Gravimetric Analysis

Monomer (Mole %)		Weight %	
CHD	Carbomethoxy-CHD	Yield	(Theory)
Diacetate (100)	---	36.4	(38.2)
Diacetate (86)	Diacetate (14)	34.8	(41.4)
Diacetate (86)	Cycliccarbonate (14)	35.8	(42.9)
Bismethylcarbonate (100)	---	33.5	(33.4)
Bismethylcarbonate (89)	Diacetate (11	34.6	(35.7)
Bismethylcarbonate (90)	Cycliccarbonate (10)	35.9	(36.4)

nificantly faster rates (e.g., half-life of 2 min at 200°C)[4,5,8,9]. A significant side benefit is that the MW degradation discussed above for the uncatalyzed process is not observed[8]. Base catalysis of aromatization for the organic acid derivatives such as the diacetate is not observed[5].

Base catalyzed aromatization of the CHD-bismethylcarbonate copolymers was shown to occur in the presence of trioctyl amine.

Acid Catalyzed

A new method of catalyzed aromatization was identified for the CHD-diacetate polymers and copolymers in the presence of strong acids. As seen in Figure 2 the addition of 3.7 mole % p-chlorobenzenesulfonic acid reduces the temperature of thermal aromatization by greater than 100°C. The order of acid catalyst effectiveness observed is,

3,4-Dichlorobenzenesulfonic = 1,5-Naphthalenedisulfonic >

p-Chlorobenzenesulfonic > 5-Sulfosalicylic > 2-Naphthalene-

sulfonic > p-Toluenesulfonic > Sulfuric > Phenylphosphonic,

and is probably a function of acidity and volatility. Catalyst activity decreases as aromatization proceeds, probably due to a combination of volatilization and condensation of the acid with the polymer. For p-chlorobenzenesulfonic acid with CHD-diacetate polymer, Cl and S analyses indicate

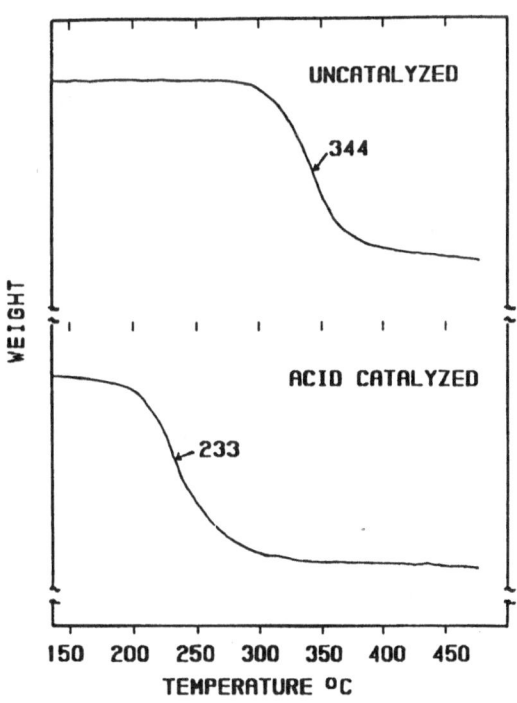

Fig. 2. Thermogravimetric Analyses of CHD-Diacetate Homopolymer in Nitrogen at 10°C/min. (a) Uncatalyzed; (b) With 3.7 wt.% p-Chlorobenzenesulfonic Acid

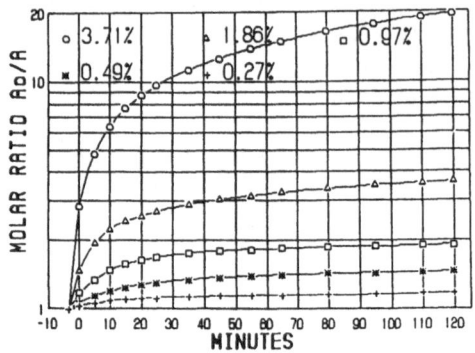

Fig. 3. Aromatization Rate Plots of CHD-Diacetate Homopolymer
at 239°C. Effect Wt.% p-Chlorobenzenesulfonic Acid

Table 4.   Weight Yield From Acid Catalyzed Aromatization of CHD-
Diacetate Polymers and Copolymers Under Nitrogen By
Thermal Gravimetric Analysis

Mole % Monomer		Weight %	
CHD	Carbomethoxy-CHD	Yield	(Theory)*
Diacetate (100)	---	40.5	(37.4–40.7)
Diacetate (86)	Diacetate (14)	40.7	(39.6–43.4)
Diacetate (86)	Cycliccarbonate (14)	43.4	(41.3–44.6)

*Range represents acid volatilization versus condensation.

about 79 % volatilization. Figure 3 shows the decrease of catalyst activity
with time as a function of acid concentration.

Weight yield after aromatization of homopolymer and copolymers is much
closer to theory as shown in Table 4. This probably indicates less or no MW
degradation as is observed in the base catalyzed processes. Infrared spectra and degree of aromatization is essentially the same (with the more
volatile acids) as for the uncatalyzed aromatizations.

For the CHD-bismethylcarbonate polymers and copolymers acid catalysis
is only effective for the first 11–18 mole % of the reaction and then the
rate becomes equivalent to the uncatalyzed reaction. This anomaly suggests
a different structural entity for at least 11 mole % of the CHD-bismethylcarbonate homopolymer.

AROMATIZED POLYMER PROPERTIES

Films (Coatings)

Aromatized films were prepared by casting precursor polymers onto glass
plates then heating under nitrogen from 175–340°C at 25°C/min followed by 4
hours at 340°C. A useful casting solvent was acetic acid containing 10 %

acetic anhydride. Aromatized films are pale yellow to reddish brown to almost black depending on thickness and catalyst. Acid catalyzed samples from diacetate monomers have the most color. All aromatized films are continuous at thicknesses of 0.7 to 1.5 microns. Only polymers from acid catalyzed aromatization are continuous at greater thicknesses and can be removed from glass plates and handled as free films. Continuous free films up to 0.8 mil thick of such films were obtained. At larger thicknesses wide cracks (high shrinkage) were observed.

Infrared Spectra

All aromatized polymers have the reported absorptions[10] at about 3030, 1600, 1480, 1000, 805, 765 and 695 cm^{-1}. Aromatized carbomethoxy copolymers have a carbonyl absorption at 1725 cm^{-1} attributed to an aromatic ester.

Saponification of copolymer resulted in carboxylate absorptions at 1600 and 1390 cm^{-1}. Neutralization resulted in carboxylic acid carbonyl absorption at 1720 cm^{-1} and hydroxy at 3450 cm^{-1}. Heating aromatized films at 340°C was necessary to remove residual carbonyl absorptions of the diol-ester groups.

Thermal Stability In Nitrogen

Thermal stability of aromatized polymers in nitrogen are shown in Table 5. A particular advantage of acid catalyzed aromatizations is a higher thermal stability as compared to polymers from both uncatalyzed aromatizations and from base catalyzed aromatizations. The weight loss of CHD-bismethylcarbonate polymers appear to be unaffected by catalysis. The observed weight loss is presumably from aromatized oligomers.

Ultraviolet Spectra

UV absorption maxima are indicative of degree of conjugation in poly-phenylenes and reach a maximum of 380-390 nm[1,6]. Absorption maxima of films on glass plates are shown in Table 6. The absorption is apparently unaffected by catalyst and the observed maxima for CHD-diacetate derived polymer is close to the highest reported value. A slightly lower maxima for CHD-bismethylcarbonate polymer may be the result of a slightly higher 1,2- versus 1,4-linkage in the polymer. The copolymers have a significantly lower absorption maxima indicating less extensive conjugation. The presence of a carbomethoxy group likely reduces coplanarity of adjacent rings due to steric hindrance. The degree of 1,2- versus 1,4-linkage could also be effected.

Table 5. Thermal Stability of Aromatized CHD Polymers and Copolymers In Nitrogen By Thermal Gravimetric Analysis From 383 to 479°C. Effect of Monomers and Catalysts

Precursor Polymer – Monomer (Mole %)		% Wt. Loss	
CHD	Carbomethoxy-CHD	Uncat.	Cat.
Diacetate (100)	---	20.3	4.4[a]
Diacetate (86)	Diacetate (14)	27.3	5.9[a]
Diacetate (86)	Cycliccarbonate (14)	24.8	8.3[a]
Bismethylcarbonate (100)	---	7.5	8.8[b]
Bismethylcarbonate (89)	Diacetate (11)	16.2	13.9[b]
Bismethylcarbonate (90)	Cycliccarbonate (10)	14.7	14.6[b]

(a) Acid catalyzed.   (b) Base catalyzed.

Table 6. Ultraviolet Spectra of Aromatized CHD Polymers and
Copolymers. Effect of Monomers and Catalysts.

| Precursor Polymer - Monomer (Mole %) | | Absorb. Max.(nm) | |
CHD	Carbomethoxy-CHD	Uncat.	Cat.
Diacetate (100)	---	388	388[a]
Diacetate (86)	Diacetate (14)	357	360[a]
Diacetate (86)	Cycliccarbonate (14)	366	364[a]
Bismethylcarbonate (100)	---	384	384[b]

(a) Acid catalyzed.   (b) Base catalyzed.

Table 7. Adhesive Strength of Polyphenylenes from Acid-Catalyzed
Aromatization of Diacetate Precursor Polymers

Base	Carbomethoxy Group Mole %	Adhesive Strength (psi)	Surface Failure Base/Epoxy
Glass	None	1200	70/30
Glass	14	>1830*	None
Copper	None	620	50/50
Copper	14	890	5/95

*Limit of test was 1830 psi.

Adhesion Data

Coating pull-off strengths were measured for polyphenylenes derived
from the acid-catalyzed aromatizations of CHD-diacetate polymers on both a
glass and a copper surface using a PATTI Adhesive Tester. In this test an
aluminum pull tab is cemented to the exposed film surface with epoxy resin
and the tab is pulled from the surface in a perpendicular direction[11].
The results in Table 7 show the superiority of adhesion with the carbomethoxy containing polyphenylene. By comparison, previously reported adhesion data for polyphenylene on silica using a similar test was only 0.34
N/m^2 (equivalent to 49 psi assuming m=mm)[12].

SUMMARY

cis-5,6-Dihydroxy-1,3-cyclohexadiene (CHD) derivatives with and without
a carbomethoxy substituent have been copolymerized. Copolymers have been
converted to useful polyphenylenes containing 10-15 mole % carbomethoxy
groups. A new method of acid catalysis for aromatization of CHD derived
polymers was identified which results in polyphenylenes with improved yield,
thermal stability and film integrity. The adhesive strength to glass and
copper of polyphenylenes containing a carbomethoxy substituent were greatly
improved. Presence of a carbomethoxy substituent reduces conjugation between adjacent phenylene rings in the substituted polyphenylene.

REFERENCES

1. Kovacic, P., Jones, M.B., Dehydro Coupling of Aromatic Nuclei by Cata-
lyst-Oxidant Systems: Poly(p-phenylene), Chem. Rev., 87, 357-379,
(1987).
2. Elsenbaumer, R.L., Shacklette, L.W., Phenylene-Based Conducting Polymers, in: Handbook of Conducting Polymers, Vol.1, Skotheim, T.A.,
ed., Marcel Dekker, Inc., New York, (1986).

3. Wegner, G., Polymers With Metal-Like Conductivity - A Review of Their Synthesis, Structure and Properties, Ang.Chem.Int.Eng., 20, 361-380, (1981).

4. Ballard, D.G.H., Courtis, A., Shirley, I.M., Preparation of Aromatic Polymers from Cyclohexa-3,5-diene Polymers, U.S. Pat. 4,476,296, October 9, 1984.

5. Ballard, D.G.H., Courtis, A., Shirley, I.M., Taylor, S.C., Synthesis of a Polyphenylene from a cis-Dihydrocatechol, a Biologically Produced Monomer, Macromol., 21, 294-304, (1988).

6. McKean, D.R., Stille, J.K., Electrical Properties of Poly(5,6-dihydroxy-2-cyclohexen-1,4-ylene) Derivatives, Macromolecules, 20, 1787-92, (1987).

7. Reinecke, W., Otting, W., Knackmuss, H.-J., cis-Dihydrodiols Microbially Produced from Halo and Methyl Benzoic Acids, Tetrahedron, 34, 1707-14 (1978).

8. Ballard, D.G.H., Holmes, P.A., Nevin, A., Twose, D.L., Shirley, I.M., Polyaromatics, Eur.Pat.Appl. 0 243 065, April 13, 1987.

9. Cheshire, P., Elimination Process, U.S. Pat. 4,454,307, June 12, 1984.

10. Marvel, C.S., Hartzell, G.E., Preparation and Aromatization of Poly-1,3-cyclohexadiene, J.Am.Chem.Soc., 81, 448-452, (1959).

11. Pull-Of Strength of Coatings Using Portable Adhesion Tester, ASTM D 4541-85, American Society for Testing and Materials, Philadelphia, PA (1985).

12. Abbott, S.J., Shirley, I.M., A Coated or Encapsulated Product, Eur.Pat.Appl. 0,125,767, November 21, 1984.

RADICAL CATION SALTS OF SIMPLE AROMATICS -
A FAMILY OF ORGANIC METALS

V. Enkelmann

Max-Planck-Institut f. Polymerforschung
P.O.Box 3148, D-6500 Mainz/ West-Germany

C. Kröhnke

CIBA-GEIGY AG, Additives Division
R-1038.5.09, CH-4002 Basle/ Switzerland

Radical cation salts of pure aromatic hydrocarbons can be prepared by anodic electrocrystallization from solutions of the corresponding arene in solvents like methylenechloride, THF, 1,1,2-trichlorethane or chlorobenzene in the presence of appropiate electrolytes like $(NR_4)^+ X$ (R : e.g. ethyl, n-butyl ; X^- : BF_4- , ClO_4^- , PF_6^- , AsF_6^-, SbF_6^-,...). Hereby electrically conducting crystals of the composition $(Ar_{2-y})^{.+} X^-$ (0 < y < 1) are obtained. They can be considered as a family of organic conductors. The aromatic rings in most of these radical cation salts are arranged in staggered structures with the counterions in the channels between the stacks. Due to the close packing found between the adjacent aromatic rings within one stack compounds of this class of materials can be regarded as models for the interchain inter- actions in conducting polymers e.g. doped polyacetylene and doped poly-p-phenylene. Of considerable influence on the electrical and magnetic properties is the molecular overlap which changes in a metal-insulator transition near 200 K. Detailed investigations has been made with the phase transitions on radical cation salts of fluoranthene (FA) - In contrast to most of the simple arenes 2,6-di- methylnaphthalene can be characterized as a two-or three-dimensional organic metal.

INTRODUCTION

Conditions for the electrical conductivity in organic crystals

In general the electrical conductivity of a material is determined by the amount of charge carriers and their mobility. One possibility to create charge carriers of sufficient mobility is by charge transfer between donator-and acceptor-molecules. Another way is the creation of unpaired electrons which are delocalized within a defined , regular structure ; moreover a weak electron-electron- repulsion is advantageous.

PREPARATION OF RADICAL CATION SALTS BY ELECTROCRYSTALLIZATION

As it was first shown on naphthalene[1] radical cation salts of various aromatic hydrocarbons can be prepared by anodic oxydation and subsequent crystallization in solvents like methylenechloride, chlorobenzene , 1,1,2-trichloroethane or THF if an electrolyte like NR_4X with suitable anions X^- is present (R: n-butyl ; X^-: BF_4^-, ClO_4^-, PF_6^-, AsF_6^-, SbF_6^-,....)[2,3].
Three steps describe the electrochemical process which lead to the crystal formation:

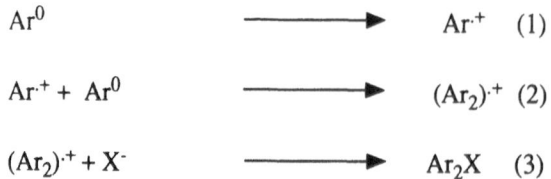

$$Ar^0 \longrightarrow Ar^{\cdot +} \quad (1)$$

$$Ar^{\cdot +} + Ar^0 \longrightarrow (Ar_2)^{\cdot +} \quad (2)$$

$$(Ar_2)^{\cdot +} + X^- \longrightarrow Ar_2X \quad (3)$$

It should be emphasized that in all cases the ratio arene : counterion is 2:1 or less. In deta the composition and morphology of the electrochemically grown crystals are influenced by parameters like the nature of the counterion, solvent, concentration(-ratio), temperature, voltage, current-density, size, shape and material of the electrodes and the arrangement of the electrochemical cell. Referring to pyrene products with different composition can be obtained.

Corresponding to the electrochemical oxydation of the aromatic hydrocarbons the chemic or electrochemical oxydation or reduction of conjugated polymers - called "doping"-occurs.

STRUCTURAL PRINCIPLES OF RADICAL CATION SALTS

Fig.1 shows the crystal structure of the fluoranthene (FA)-salt FA_2AsF_6 at 300K with the projection in stack direction.The aromatic molecules form stacks which are pseudohexagonal arranged leaving channels in which the counterions are located; here anions of different shape at size can be incorporated causing onlly small changes of the lateral packing of the FA-columns . Neighboured FA- rings with very short interplanar distances of 3.2- 3.3 Å are alternately rotated 180° .Some atoms of adjacent molecules are in such a close contact that a strong interaction of tl corresponding π-orbitals in stack direction takes place.

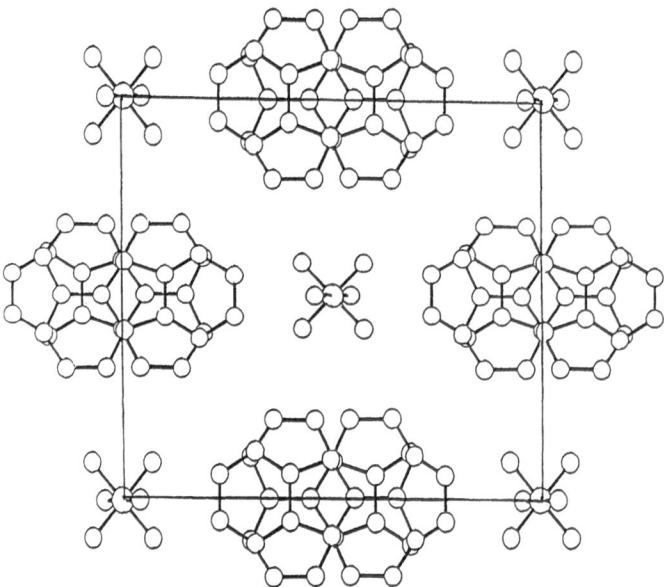

Fig. 1. Crystal structure of FA_2AsF_6 at room temperature; projection in stack direction.

The angle φ between two neighboring aromatic rings seems to be typical for each individı compound ; naphthalene forms columns with φ= 90° whereas in case of pyrene φ= 0° , 60° and 9 are found. In relation to the crystal symmetry all FA rings are identical with the consequence that radical cation state is delocalized along the stack.

In many cases radical cation salts undergo phase transitions near 200 K correlated with a change from metallic to semiconducting behaviour. Fig.2 demonstrates the difference in the cryst structure of FA_2AsF_6 at room temperature and at 120 K which has been studied in some detail. A

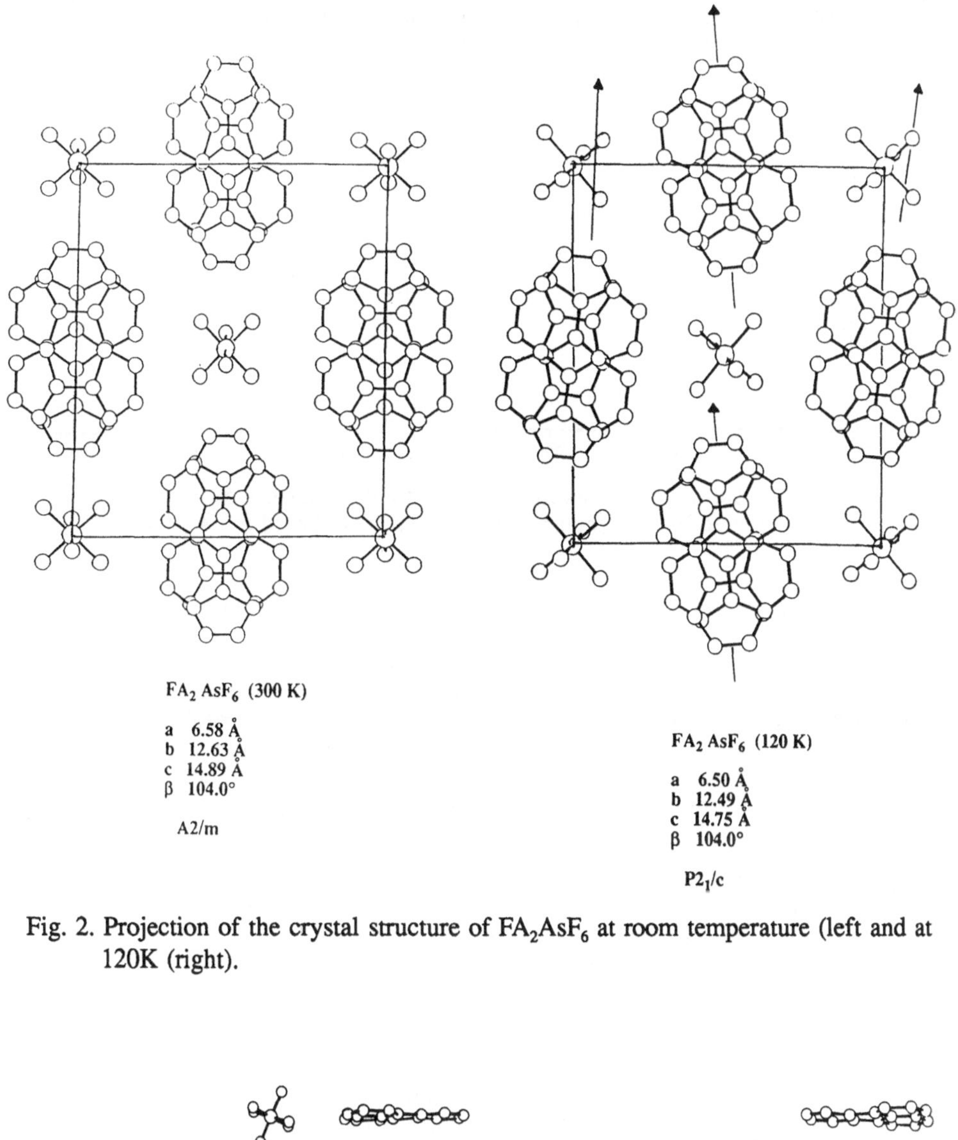

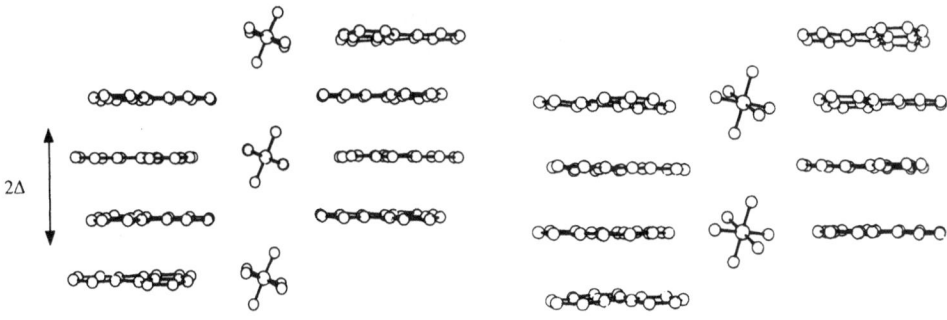

FA₂ AsF₆ (300 K)

a 6.58 Å
b 12.63 Å
c 14.89 Å
β 104.0°

A2/m

FA₂ AsF₆ (120 K)

a 6.50 Å
b 12.49 Å
c 14.75 Å
β 104.0°

P2₁/c

Fig. 2. Projection of the crystal structure of FA₂AsF₆ at room temperature (left and at 120K (right).

2Δ

Fig. 3. Projection of crystal structures perpendicular to the stack axis:
left: FA₂PF₆ at 295K; right: FA₂AsF₆ at 110K.

result of this investigation one has to recognize that the interplanar spacings are slightly different (see fig.3) although the overlap of adjacent molecules is strictly symmetrical. A spacing of $\Delta_1 = ?$ Å follows next a spacing of $\Delta_2 = 3,33$ Å at roomtemperature; the corresponding values for the l temperature modification are 3,22 Å and 3,28 Å (110 K) .The temperature-depending curve of th difference δ shows a loss in continuity at the temperature of the phase transition (T= 217K) (fig. A corresponding plot for FA_2PF_6 with an incontinuity at the phase transition (206 K) is shown ir fig.4 B.

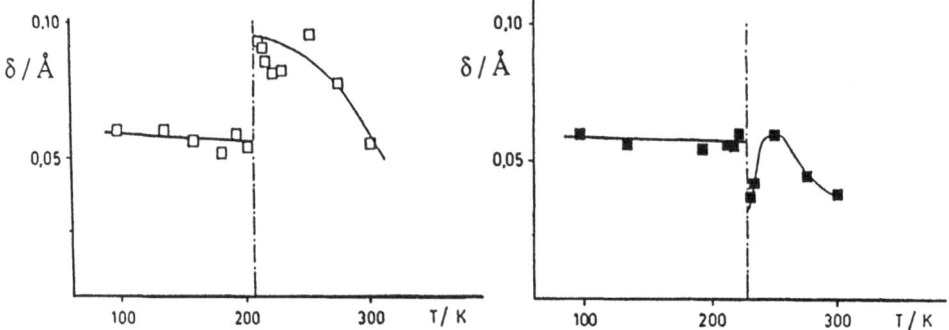

Fig. 4. Temperature-dependent difference δ as for the distances Δ_1 and Δ_2 between three neighboured FA-molecules within a stack; $\square$: FA_2PF_6 ; $\blacksquare$: FA_2AsF_6

In the high temperature structures of the mentioned FA-salts with a space group A2/m the orientation of both the counterions and the FA-rings are fixed by the symmetry. The FA-molecul are located on mirror planes ,the anions on sites with the symmetry (2/m).

In the course of the phase transition the space group A2/m changes into its subgroup $P2_1/$ loosing certain symmetry elements (lattice centering , mirror plane, twofold axis). Below the pha transition the FA-rings as well as the counterions gain a limited freedom for rotations.

The phase transition described above can be characterized as a second order transition. However, the nature of the metal-insulator-transition in FA_2X-salts is still an unsolved question a a matter of continuing research. In DSC experiments[4] two transitions at approximately 200 K an 180 K are observed of which the first one is the transition which can be monitored crystallographically. However, the electronic properties change abruptly at 180 K. The assumptic of the existence of a lower transition is also supported by some results of ESR-[5] and NMR-experiments[6] with a pronuonced line broadening at and below the phase transition around K as well as by measurements of the static susceptibility [7,8] and the electrical conductivity. At thi temperature no anomalies in the temperature dependence of the crystal structures can be detected

Preliminary experiments [9,10] indicate the possibility of a charge density wave in FA_2PF_6 which locks in at this temperature.

Earlier ESR-measurements[11] show for the high temperature phase of FA_2PF_6 an signal w an extrem small linewidth ($\delta H_{pp}= 10$mG)(fig.5) which corresponds to 1/100 of the linewidth of diphe-　nylpicrylhydrazin (DPPH) and 1/1000 of that of TTF/TCNQ.

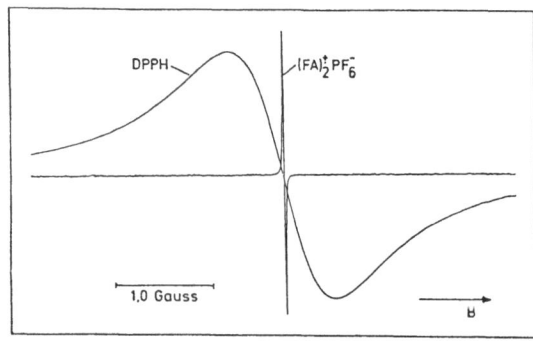

Fig. 5. ESR-linewidth of FA_2PF_6 in comparison with diphenylpicrylhydrazine (DPPH)

Due to interaction found between the aromatic rings
the radical cation salts can be regarded as model compounds for the interchain interactions in conducting polymers like polyacetylene. The analogy is illustrated in fig.6; at the right side the stack-forming elements represent segments of the oxidized ("doped") polymer main chain. A section of such a polymer is represented e.g. by quaterphenyl (QP) a low molecular weight analogon of poly-p- phenylene. Fig. 7 shows the crystal structure of $QP_3QP(SbF_6)_3$ in two different projections. According to the projection in stack direction the QP molecules are more or less aligned in one direction whreas the counterions are located in separated layers. In addition the projection perpendicular to the stacks elucidates an additional (neutral) QP-molecule placed in the anion layer. In contrast to most other radical cation salts each QP molecule within the stack carries one positive charge on the average .

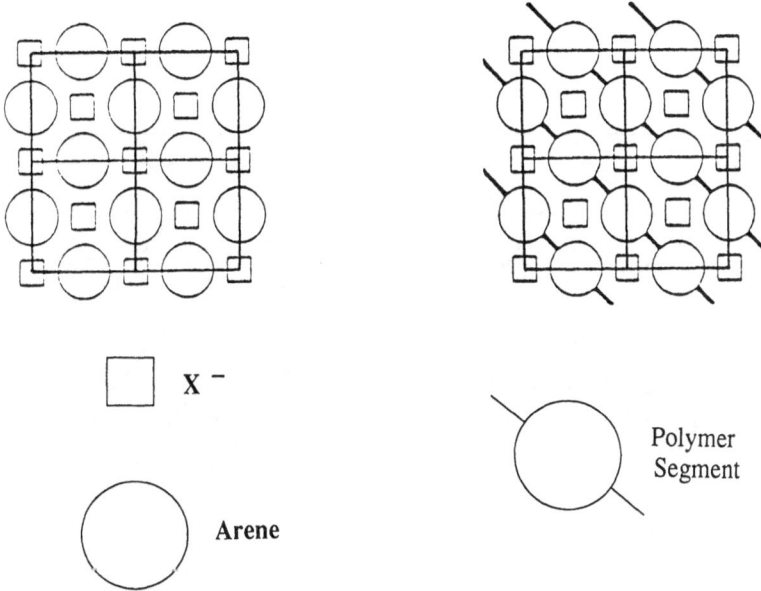

□ X⁻

Arene

Polymer
Segment

Fig. 6. Packing-analogy in radical cation salts (left) and a proposed structural model for conducting polymer salts e.g. polyacetylene (right).

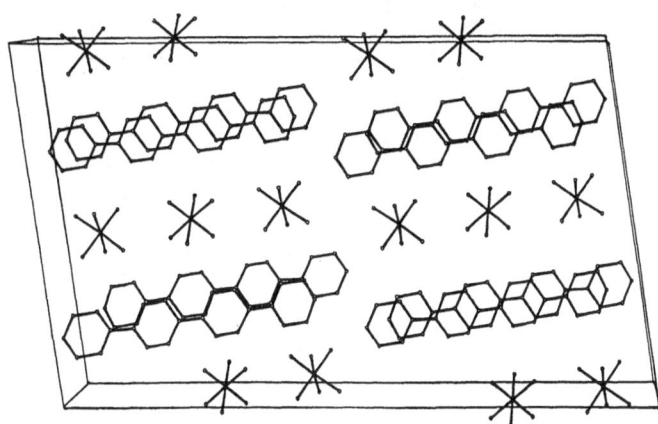

Fig. 7. Crystal structure of $QP_3QP(SbF_6)_3$; projection in stack direction (left) and perpendicular to the stack (right)

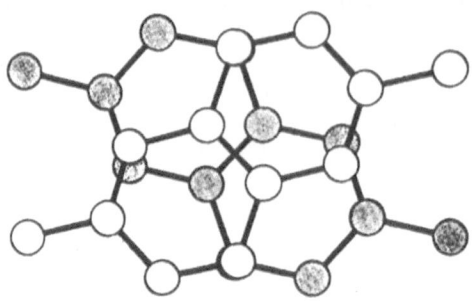

Fig. 8. Molecular overlap of DMN-rings in $DMN_3(SbCl_2)_2(CH_2Cl_2)$ at room temperature by projection in stack direction.

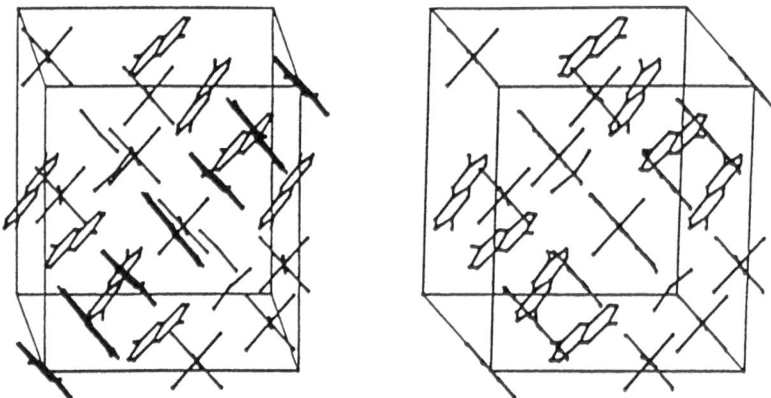

Fig. 9. Stereoscopic projection of the crystal structure of $DMN_3(SbCl_2)_2(CH_2Cl_2)$

As knowm for typical one-dimensional organic metals most radical cation salts of arenes are arranged in columnar structures with the crystallization of planar arenes in segregated stacks and crystallographically uniform spacing within the stack elements which are present in a partial oxydized form. Therefore anisotropies of at least factor 100 have been reported together with specific electrical conductivities within the stack direction of nearly 1000 S/cm[12].

In contrast to those 1-d-systems radical cation salts of 2,6-dimethylnaphthalene (DMN) like $DMN_3(SbF_6)_2(CH_2Cl_2)$ are the first examples with higher dimensionality[13]. The molecular overlap of DMN-rings is shown in fig.8 whereas fig.9 presents a stereoscopic projection of the crystal structure with two different stacks oriented perpendicular to each other. Parallel columns are packed in layers so that the stack direction alternates along the c-axis. Between the stacks anions and solvent molecules are placed; anions may compensate charges in stacks of different orientation. The interplanar arene spacing change periodically between extremly short (3,06 Å) and "normal" (3,32 Å).

The specific conductivity ($\sigma = 10^{-1}$ S cm^{-1}) is equal along all three crystallographic axis which can be interpreted as a charge transfer along the stacks as well as perpendicular to them in direction to the c-axis.

REFERENCES

1. H.P.Fritz, H.Gebauer, P.Friedrich, P.Ecker, R.Artes and U.Schubert, Z.Naturforsch.(b) **33**,498 (1978) ; H.P.Fritz,H.Gebauer, P.Friedrich and U.Schubert, Angew.Chem. **90**, 305 (1978).
2. C.Kröhnke, V.Enkelmann and G.Wegner, Angew.Chem. **92**, 941 (1980)
3. V.Enkelmann, B.S.Morra, C.Kröhnke, G.Wegner and J.Heinze, Chem.Phys. **66**, 303 (1982).
4. Th.Schimmel, W.Rieß, G.Denninger and M.Schwoerer, Ber. Bunsenges. Phys. Chem., **91**, 901 (1987).
5. E.Dormann and G.Sachs, Ber.Bunsenges.Phys.Chem. **91**, 879 (1987).
   G.Sachs, E.Pöhlmann and E.Dormann, J.Magn.Magn.Mater., **69**, 131 (1987).
   G.Denninger, Mol.Cryst.Liq.Cryst. **171**, 315 (1989)
6. G.Sachs and E.Dormann, Synth.Metals **25**, 157 (1988)
7. U.Köbler, J.Gmeiner and E.Dormann, J.Magn.Magn.Mater., **69**, 189 (1987).
8. E.Dormann, Synth. Metals, **27**, B529 (1988).
9. H.Eichele, M.Schwoerer, C.Kröhnke and G.Wegner, Chem.Phys.Lett. **77**, 311 (1981).
10. V.Enkelmann, Synth.Metals , in press
11. W.Rüß and E.Dormann, ibid., in press
12. H.J.Keller, D.Nöthe, H.Pritzkow, D.Wehe, M.Werner, P.Koch and D.Schweitzer, Mol.Cryst.Liq. Cryst. **62**, 181 (1981)
13. V.Enkelmann and K.Göckelmann, Ber.Bunsenges.Phys.Chem. **91**, 95 (1987).

APPROACHES TO ORTHOGONALLY FUSED CONDUCTING

POLYMERS FOR MOLECULAR ELECTRONICS[1]

James M. Tour,*[,2] Ruilian Wu, and Jeffry S. Schumm

Department of Chemistry
University of South Carolina
Columbia, South Carolina 29208

Molecular electronics-based computing instruments possess tremendous technological potential. There is the hope of developing single molecules that could each function as a self-contained electronic device. Thus, one can envision computing systems with molecular-sized electronic elements and operational efficiencies far exceeding that of present systems.[3] Recently, Aviram of the IBM Corporation has suggested that molecules which contain a pro-conducting (non-doped or non-oxidized system, hence insulating) polymer which is fixed at a 90° angle via a non-conjugated sigma bonded network to a conducting (doped or oxidized system) should exhibit properties which would make it suitable for interconnection into future molecular electronic devices.[4] These devices may be useful for the memory, logic, and amplification computing systems. The molecule **1** (in doped form) is an example of this pro-conducting/sigma/conducting type of molecule.

1

We have undertaken the synthesis of several molecules which fit the structural requirements of this electronic model. From the synthetic standpoint, several aspects are challenging. First, there must be a one spiro-fused junction separating two potentially conducting chains with a tetrahedral bonding atom at the center to maintain the 90° angle via a sigma bonded network. Secondly, all four conducting chains originating from the central segment must be *identical* in length. These requirements prohibit the use of any random polymerization methods. Initial reports suggested conducting chains ~50 Å long (from end to end rather than from end to core) would fulfill the model.[4]

Our initial approach to these systems involved the synthesis of the key spiro core **2** from which we envisioned selective oligomerization to the target molecule **1**. A retrosynthetic analysis is shown in eq 1.

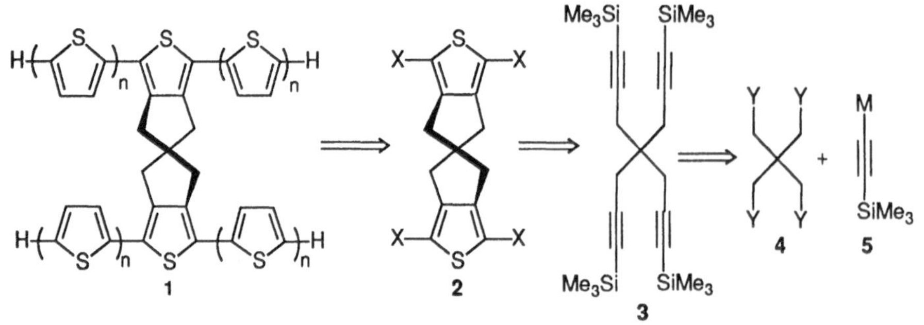

(1)

Though substitutions on pentaerythrityl tetrahalides involves reactions on a neopentyl system, exhaustive substitution has been accomplished using oxygen, nitrogen, and sulfur nucleophiles.[5] Attempted formation of **3** using 1-metallo-2-(trimethylsilyl)acetylenes **5** and pentaerythrityl tetrahalides and tosylates **4** proved to be very difficult even though we tried numerous coupling procedures (M= MgBr, Li, ZnCl, Cu, AlR$_2$ with and without Pd and Ni catalysis). In several cases, we obtained the cyclopropyl system **6**.[6] In an effort to overcome these difficulties while maintaining the required sigma-

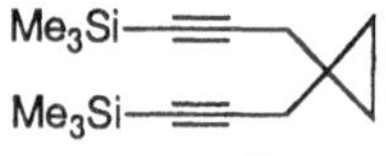

bonded tetrahedral spiro junction, we turned our attention to the use of silicon as the central atom. Accordingly, treatment of SiCl$_4$ with the silyl protected propargyl Grignard reagent cleanly afforded the tetra(alkyne) **7**.[6] Treatment of **7** with a zirconocene equivalent, generated *in situ* from zirconocene dichloride and butyllithium, and quenching with sulfur monochloride afforded the trimethylsilyl-spiro core **8** (eq 2).[6,7] To our knowledge, use of

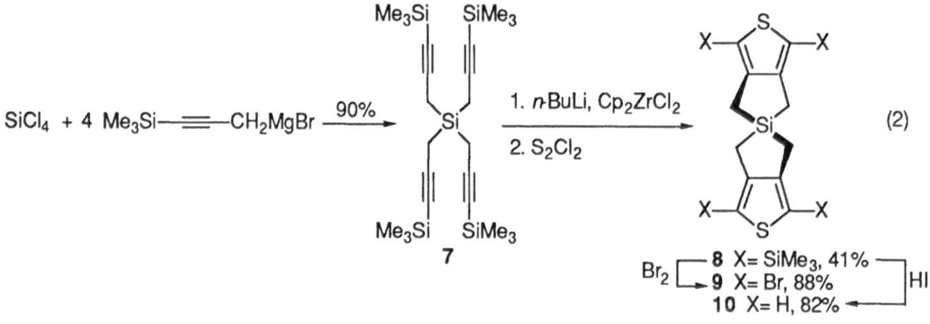

(2)

8 X= SiMe$_3$, 41%
9 X= Br, 88%
10 X= H, 82%

this group IVA coupling procedure for a *bis*(bicyclization) has never before been demonstrated. The trimethylsilyl core (**8**) was converted to the tetrabromide (**9**) and parent core (**10**) under electrophilic substitution conditions.[6,8] Remarkably, no attack on the pseudo allylic central silicon atom was observed.

Likewise, we have synthesized another key core segment based on a *p*-polyphenylene[9] conducting unit which fits the general electronic architectural requirements. Conversion of 2-aminobiphenyl to the corresponding iodide under Sandmeyer[10] conditions followed by lithium halogen exchange and quenching with fluorenone afforded the alcohol **11** Acid treatment to close the spiro system[11] followed by reaction with bromine and $FeCl_3$ gave the tetrabromide **12** in excellent yields (eq 3).[6] Bromination occurred only at the positions para to the second ring in the chain as one

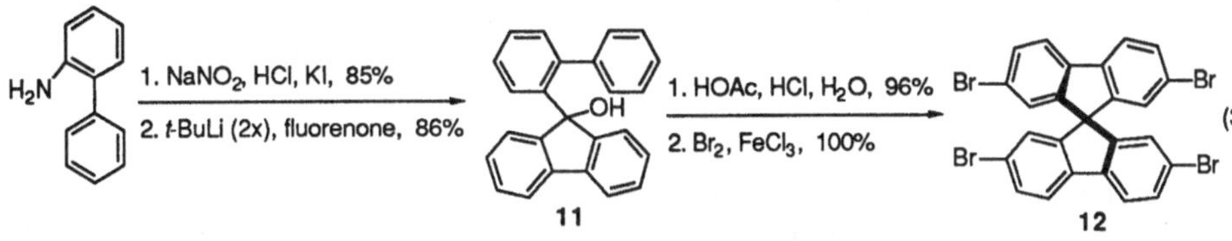

(3)

would expect by resonance stabilization arguments of the ionic intermediate. It is imperative that the bromination take place at the para position since a 4-substituted moiety is essential to afford a highly conducting system.[9]

With two key core units in hand, we then addressed methods to selectively and equally extend the chains in all four directions. Coupling **9** and **12** with **13** and **14**, respectively, using transition metal catalysis[12] would allow for the selective introduction of a known number of units.[13] Additionally, the terminal trimethylsilyl group in **13** and **14** would allow for

$M= R_3Sn, XZn, XMg$

selective bromination at those sites and, hence, a position for further coupling if necessary.[8a] Accordingly, treatment of **9** with **13** (M= Bu_3Sn, n= 1) in the presence of catalytic $Pd(PPh_3)_4$ afforded **15** in 41% yield. Likewise, treatment of **12** with **14** (M= ClZn, n= 1) under similar catalytic conditions afforded **16** in 40% yield.[6] Thus the methodology for attachment to the cores worked

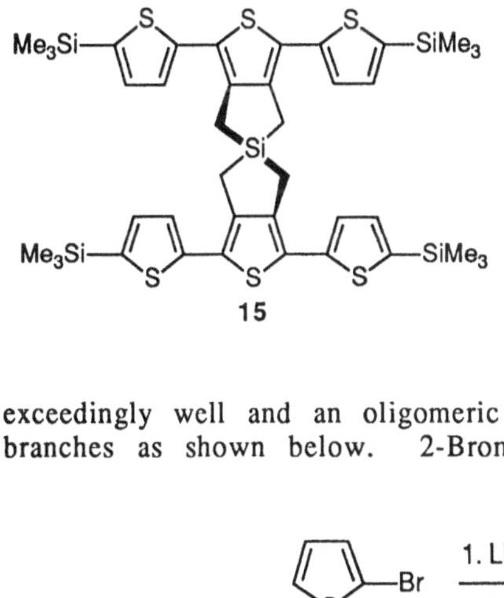

exceedingly well and an oligomeric system was prepared in order to extend the branches as shown below. 2-Bromothiophene was lithiated with lithium

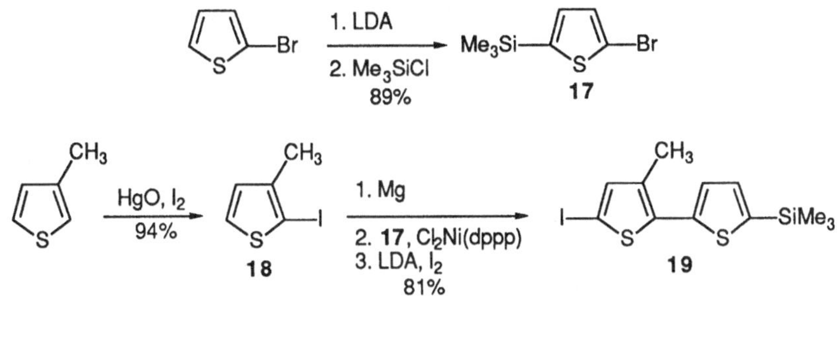

$$18 \xrightarrow[\substack{2.\ \mathbf{19},\ Cl_2Ni(dppp) \\ 3.\ LDA,\ n\text{-}Bu_3SnCl \\ 72\%}]{1.\ Mg}$$

20

diisopropylamide (LDA) and quenched with chlorotrimethylsilane to give **17**. 3-Methylthiophene was iodinated exclusively at the 2-position with mercury(II) oxide and iodine to give **18**[15] which was then converted to its Grignard reagent. Treatment of the Grignard reagent of **18** with the bromide **17** and (diphenylphosphino)-propanenickel(II) dichloride catalysis[16] afforded the dimer which was deprotonation with LDA and quenched with iodine to give the dimer iodide **19**. Treatment of the Grignard reagent of **18** with **19** under nickel-catalyzed conditions afforded the trimer which was metalated to give **20** in 72% yield for the final three steps (based on recovered **19**). When the silylated thiophene unit in **20** had a methyl substituent in the position α to silicon, desilylation was extremely facile, even upon silica gel chromatography. This desilylation also occurred with a triethylsilyl group in place of the trimethyl substituent. Thus, we chose to keep the terminal thiophene unit de-alkylated. Note that alkylation of several of the thiophene units was essential in order to maintain a soluble orthogonally fused system as describe below.

With the trimer **20** in hand, we could then address the final coupling reaction. Treatment of the core **9** with excess **20** in the presence of tetrakis(triphenylphosphine)palladium(0) catalyst[12] afforded the target orthogonal system **21** that is approximately 25 Å from end to end, one half of the length necessary to fulfill the initial Aviram model.

21

In summary, we describe a convergent approach to the orthogonally fused macromolecules that may function as molecular-sized components in future computing devices.

Acknowledgements

This research was funded by the Department of the Navy, Office of the Chief of Naval Research, Young Investigator Program (N00014-89-J-3062), the National Science Foundation (RII-8922165), and the University of South Carolina Venture Fund.

References

1. a. Tour, J. M.; Wu, R.; Schumm, J. S. *J. Am. Chem. Soc.* **1990**, *112*, 5662. b. Tour, J. M.; Wu, R.; Schumm, J. S. *Polym. Preprints* **1990**, *31*, 408.

2. Recipient of the Office of Naval Research, Young Investigator Award (1989-92).

3. a. Bowden, M. J. in *Electronic and Photonic Applications of Polymers*; Bowden, M. J.; Turner, S. R., Eds; (Advances in Chemistry, 218) American Chemical Society: Washington DC, 1988. b. *Molecular Electronic Devices*; Carter, F. L., Ed.; Marcel Dekker: New York, 1982. c. *Molecular Electronic Devices II*; Carter, F. L., Ed.; Marcel Dekker: New York, 1984. d. Third International Symposium on Molecular Electronic Devices, Washington DC, October, 1986; Roland Etvos Physical Society, Satellite Symposium on Molecular Electronics, Budapest Hungary, August 1987. e. Krummel, G.; Huber, W.; Mullen, K. *Angew. Chem. Int. Ed. Engl.* **1987**, *26*, 1290.

4. Aviram, A. *J. Am. Chem. Soc.* **1988**, *110*, 5687 and references therein.

5. a. Padias, A. B.; Hall, H. K., Jr.; Tomalia, D. A.; McConnell, J. R. *J. Org. Chem.* **1987**, *52*, 5305. b. Fujihara, H.; Imaoka, K.; Furukawa, N. *J. Chem. Soc. Perkin Trans. I* **1986**, 465.

6. All new compounds were fully characterized spectroscopically and the elemental composition was established by high resolution mass spectrometry and/or combustion analysis. All reported yields pertain to isolated homogeneous materials which were purified by recrystallization or chromotography.

7. a. Negishi, E. Holmes, S. J.; Tour, J. M.; Miller, J. A.; Cederbaum, F. E.; Swanson, D. R.; Takahashi, T. *J. Am. Chem. Soc.* **1989**, *111*, 3336. b. Fagan, P. J.; Nugent, W. A. *J. Am. Chem. Soc.* **1988**, *110*, 2310.

8. a. Chan, T. H.; Fleming, I. *Synthesis*, **1979**, 761. b. Utimoto, K.; Kitai, M.; Nozaki, H. *Tetrahedron Lett.* **1975**, *33*, 2825.

9. For a discussion of polyphenylene, see: a. Elsenbaumer, R. L.; Shacklette, L. W. in *Handbook of Conducting Polymers*, Skotheim, T. A., Ed.; Dekker: New York, 1986. For related syntheses, see b. Yamamoto, T.; Hayashi, Y.; Tamamoto, A. *Bull. Chem. Soc. Jpn.* **1978**, *51*, 2091. c. Kovacic, P.; Oziomek, J. *J. Org. Chem.* **1964**, *29*, 100.

10. Heaney, H.; Millar, I. T. *Org. Synth.* **1960**, *40*, 105.

11. Clarkson, R. G.; Gomberg, M. *J. Am. Chem. Soc.* **1930**, *52*, 2881.

12. a. Negishi, E.; Baba, S. *J. Chem. Soc. Chem. Commun.* **1976**, 596. b. Negishi, E.; Takahashi, T.; Baba, S.; Van Horn, D. E.; Okukado, N. *J. Am. Chem. Soc.* **1987**, *109*, 2393. c. Stille, J. K. *Angew. Chem., Int. Ed. Engl.* **1986**, *25*, 508. d. Stille, J. K. *Pure Appl. Chem.* **1985**, *57*, 1771.

13. For a discussion of polythiophene and its derivatives, see: Tourillon, G. in ref. 9a.

14. For a discussion of mixed thiophene/phenylene semiconductors, see, for example: Pelter, A.; Maud, J. M.; Jenkins, I.; Sadeka, C.; Coles, G. *Tetrahedron Lett.* **1989**, *30*, 3461.

15. Uhlenbroek, J. H.; Bijloo, J. D. *Rev. Trav. Chim.* **1960**, *79*, 1181.

16. Tamao, K.; Kodama, S.; Nakajima, I.; Kumada, M.; Minato, A.; Suzuki, K. *Tetrahedron* **1982**, *38*, 3347.

POLYPYRROLES FROM ISOPOROUS ALUMINA MEMBRANES

R.P. Burford, S.N. Atchison, R.P. Chaplin and T. Tongtam*

University of New South Wales

University of NSW, Kensington, NSW Australia
*Royal Thai Naval Science Dept., Bangkok, Thailand

INTRODUCTION

Since 1970, polymers which are intrinsically conductive have been synthesised[1,2]. These have a conjugated unsaturated polymeric backbone, and typically contain a low molecular weight species counter-ion or dopant. Several hundred chemically distinctive classes of polymers have been shown to have intermediate to high conductivities[3,4]. Generally they are synthesised using highly active catalysts or are made by electrodeposition, although a new class has emerged in which dehydrohalogenation by controlled thermal degradation is performed[5].

One major class of conducting polymers is based on cyclic monomers, in particular pyrrole and thiophene. During the past decade, over a thousand papers and patents have appeared, reflecting intense interest which can be attributed to several factors. Polypyrroles can possess high thermal stability, and are far more durable than polyacetylenes. They are relatively simple to synthesise, and can be made with a wide range of conductivities, depending on dopant type and level, and other factors. The major limitations with these polymers are their poor ductility, lack of processability and opacity.

The potential usefulness of conducting polymers as switching devices, catalytic supports and carriers for active ingredients has brought with it a need to form substrates with a high and predictable surface area and controlled morphology. As these polymers are not generally able to be dissolved, despite recent attempts to develop soluble precursors[6,7] attention has been directed towards *in situ* polymerization, either using chemically based oxidants such as $FeCl_3$ or by electrodeposition from the monomer.

One route for producing fibrillar polypyrroles is to employ isoporous polycarbonate (PC) membranes. This was first reported by Martin *et al*[8] and additional refinements have been provided by the present authors[9]. Recently tubular products using ferric based catalysts have been reported[10].

The previous work[8,9] using PC membranes suffer in that filaments are widely spaced, although length can be controlled[8,11].

The use of membranes as a template to produce fibrillar polypyrroles has recently been extended by Martin *et al*[12] who used an Austec γ-alumina membrane as the template. In this work a thin skin grows <u>over</u> the gold coated alumina substrate, and in their case non-conductive composites based on divinyl benzene and ethylvinyl benzene were emphasised. Narrow polypyrrole fibres have also been illustrated recently using both Nuclepore and Anopore membranes, with high conductivities ($\approx$ 7500 Scm^{-1}) claimed[12].

Contemporary Topics in Polymer Science, Vol. 7., Edited by
J.C. Salamone and J. Riffle, Plenum Press, New York, 1992

In this work we show how lightly controlled filamentary polypyrroles can be formed by growing through γ-alumina from both 0.2 and 0.02 μm Anotec materials.

2.   EXPERIMENTAL

2.1 Materials

All chemicals, if not otherwise mentioned, were reagent grade and used without further purification. Polymerization solutions consisted cf pyrrole monomer (Sigma Chemical Company) dopant electrolyte (tetraethylammonium-*p*-toluene) sulphonate, Alfa/Aldrich Chemical Company) and distilled water. The ratio of pyrrole:dopant concentration was maintained at 3:1, as this has previously been found to be optimum[13], and the initial pyrrole concentration was 0.75 moll^{-1}.

2.2 Polymerization using Anotec - alumina membrane hosts

Anotec γ-alumina membranes are prepared by an adonizing process[14], and have the morphology shown in Figures 1 and 2[15]. Hexagonal pores with a nominal diameter of 0.2 μm persist through the membrane, with some constriction at the top and bottom surfaces. The membranes used in these experiments were 45 μm thick.

As the γ-alumina is resistant to organic solvents, the first experiments employed a conductive film prepared by painting colloidal silver suspended in amyl acetate. Subsequent experiments used a · conductive gold film, with an anode assembly as described in Figure 3. Here single-sided tape was applied to the gold layer, so that an air gap remained between the Anotec membrane and the glass slide facilitating subsequent manipulation of the product. As the 0.02 μm Anotec membrane is quite anisotropic, separate polymerizations were prepared with gold coated either on the constricted or open side.

Electrodeposition time for this type of electrode was typically 1 h. Unlike the previous experiments with polycarbonate membranes, leaching of the membrane host requires more aggressive conditions.

To separate the γ-alumina without destroying the polypyrrole, screening experiments were performed to discover the optimum level of exposure to sodium hydroxide leachant. From the Anotec literature[15] it is evident that γ-alumina dissolves readily in NaOH, but it is also known that conductivities and structure of polypyrrole are affected by exposure to bases.

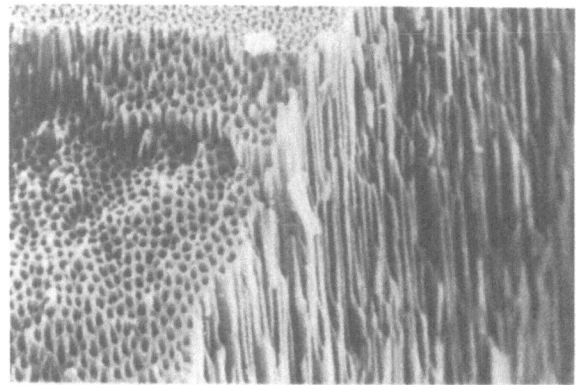

0.2 micron ANOPORE

Figure 1.   Morphology of 0.2 μm Anotec γ-alumina
membrane. Pores are homogeneous.

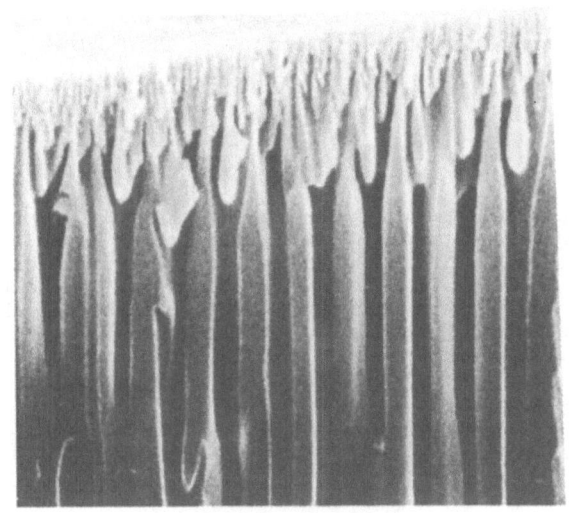

0.02 micron ANOPORE

Figure 2(a). Morphology of 0.02 μm Anotec γ-alumina
            membrane.

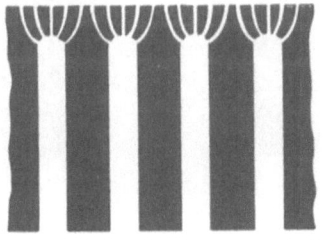

Asymmetric Pores

Figure 2(b). Schematic of 0.02 μm Anotec γ-alumina
            membrane.

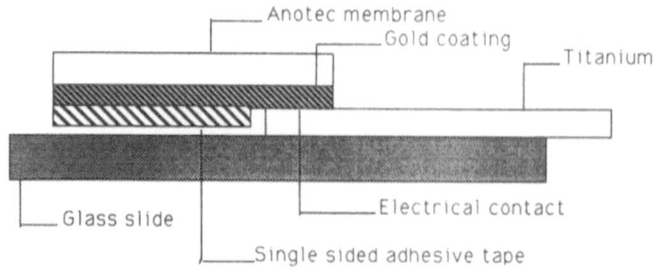

Anotec membrane
Gold coating
Titanium
Glass slide
Electrical contact
Single sided adhesive tape

Thus in separate experiments 300 μm thick, polypyrrole films prepared on titanium electrodes were exposed to 1 M aqueous NaOH for times from 0.5 to 48 h. The back surface of polypyrrole stripped from the electrode is shown prior to exposure (Figure 4a) and after 0.5 h (Figure 4b). It can be seen that some modest microscopic changes have occurred, attributable to changes in counter-ion levels, but that the matrix has essentially survived with little alteration.

From those experiments and the observation that total leaching of the γ-alumina is accomplished in less than 1 h by 1 M NaOH, removal of the polypyrrole product from the alumina host was performed by immersion for 40 min.

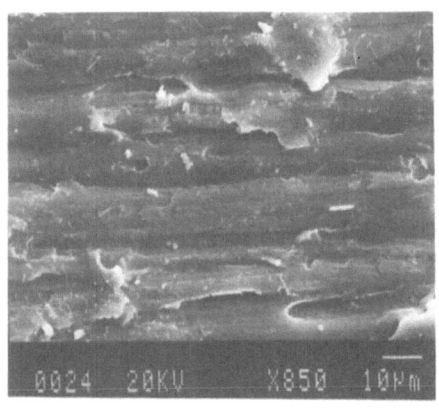
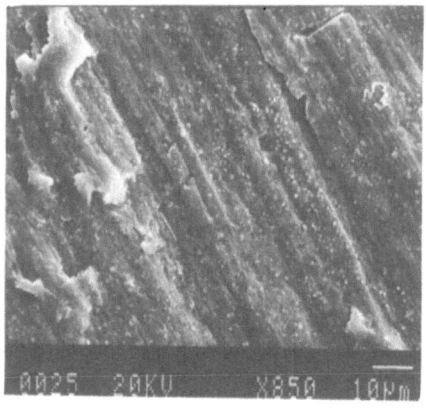

(a)   (b)

Figure 4.   Polypyrrole surface (a) before exposure (b)
            after 0.5h exposure to 1M NaOH.

2.4 Microscopy

Electrodes and filamentary polypyrroles were routinely examined at various stages using a Nikon SMZ stereozoom optical microscope. Morphologies of both uncoated and gold coated polypyrroles were recorded using a Cambridge Stereoscan 360 SEM.

RESULTS AND DISCUSSION

The type of morphology commonly encountered after isoporous polycarbonate (PC) based membrane hosts are removed is indicated by Figure 5. With the techniques described previously length of filaments can be kept constant (i.e. the thickness of the polycarbonate), thus allowing some degree of structural control. However, as can be seen, the density of fibres is low. With PC membranes having large diameter pores (5 μm), the spacing between each stump is large and the walls of resulting fibres are jagged. As pore diameter decreases to 1 μm or less, the density of pores (and associated polypyrrole filaments) increases and the sides of each fibre are considerably smoother.

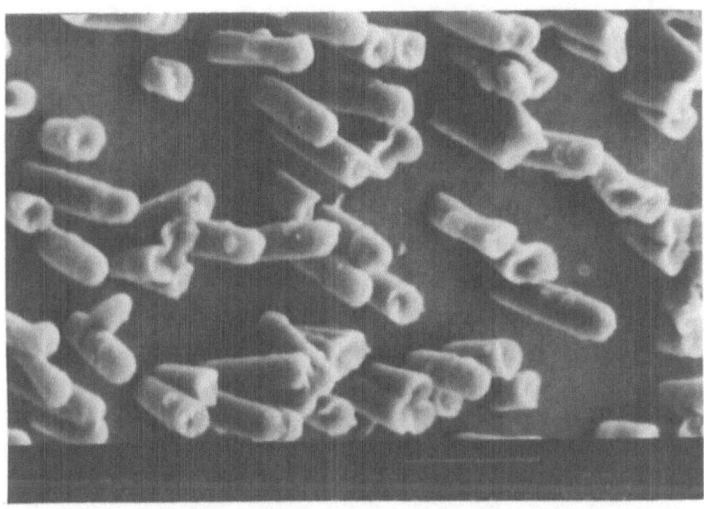

Figure 5. Polypyrrole prepared from an isoporous polycarbonate membrane.

With the 0.2 μm γ-alumina membranes, a much higher density of fibres, reflecting the close hexagonal packing of the membrane, are formed, to give a uniform carpet, as shown in Figure 6. At higher magnification it is seen that each fibre is more of less hexagonal and the mean diameter is of the same order as the original pores (i.e., ≈ 0.2 μm). It should be noted that the Anotec membrane is not comprised of uniformly sized nor perfectly shaped polygons, and so these irregularities are replicated in the polypyrrole product. Notwithstanding this, a quite regular range of hexagonal/pentagonal fibres with a reasonably narrow distribution of diameters is seen, and this level of order may make the product useful for modelling of uptake or release of adsorbants, for example.

The products from the 0.02 μm Anotec membrane can be less rigorously correlated with the structure of the host. When electropolymerization of the pyrrole appears to be complete, as indicated by a black layer across the face of the γ-alumina, and extraction is performed, two types of product (depending upon whether the Anotec is coated on the "up" or "down" side) are found, although in neither case do they appear to be as thick as the host. In one case, a uniform hexagonal array of fibres is found, with few finer filaments extending (Figure 7). In this instance the diameter is again of the order of 0.2 μm, and indeed this agrees with the main cross-sectional size of the pores, below the skin of the membrane.

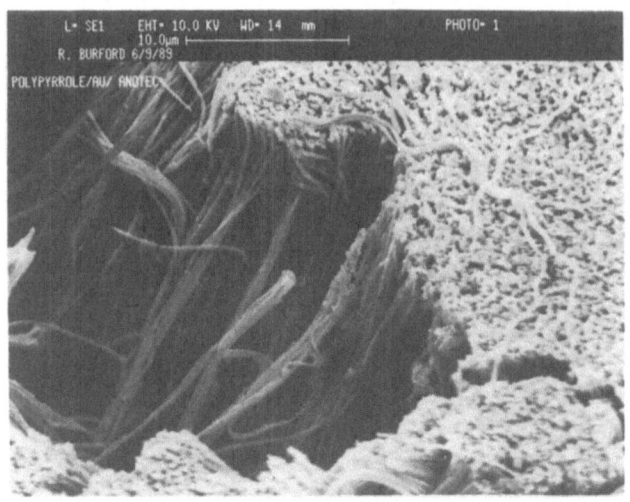

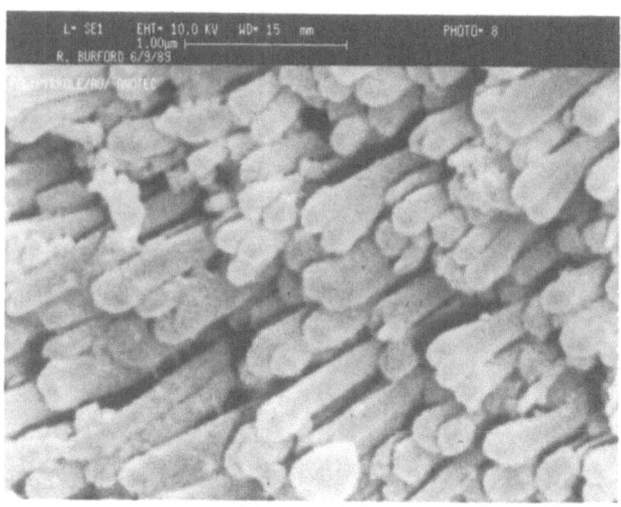

Figure 6.   Polypyrrole prepared from the 0.2 μm
            Anotec membrane.

Figure 7. Polypyrrole prepared from 0.02 μm Anotec membrane.
Grown from the "upper" side.

When the polypyrrole is formed from the other side, again a
hexagonal sub-layer is found, but in this instance many simpler fibres
extend, typically for a further 5 μm (Figure 8). However, it is noted
that the total thickness of the "carpet" is about 10 μm, much less than
the 45 μm specified for the Anotec host. Secondly, in no case have we
yet observed the very fine (i.e. 0.02 μm) tufts of polypyrrole "hairs"
which would be predicted to form at the extremities. There are several
possible reasons for this. First, it is possible that much longer
electrodeposition times may be required due to the low rates of monomer
diffusion through the narrow constrictions. Secondly, fouling of the
membrane skin by prematurely formed polymer or contaminments may arise.
Finally, the fibres themselves are predicted to be very fragile and so
may be mechanically or chemically disrupted during subsequent work-up.

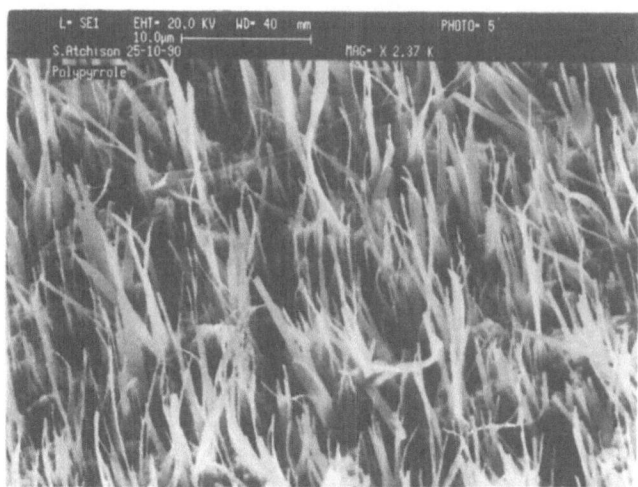

Figure 8. Polypyrrole prepared from 0.02 μm Anotec membrane.
Grown from the "lower" side.

Another factor which limits our success in correlating polypyrrole structure with 0.02 μm Anotec morphology is that we have yet to obtain a reliable complete cross-sectional profile of the membrane. However, we are currently pursuing further high resolution studies of both γ-alumina membranes and extended electropolymerizations to better understand this aspect of the work.

REFERENCES

1. A.G. MacDiarmid and A.J. Heeger, Synthetic Metals, **1**, 101 (1980).
2. G. Wegner, Agnew Chem., **93**, 352 (1981).
3. G.B. Street and T.C. Clark, IBM J. Res. Dev., **25**, 51 (1981).
4. R.A. Pethrick and M.G.B. Mahbourbian-Jones, in "Polymer Yearbook 2" (Harwood Academic Publishers, London, 1985), p.201.
5. T.M. Keller, Chemtech., **18**, 635 (1988).
6. E.W. Meijer, S.Hijhuis and E.E. Havinga, Polymers for Electronics and Photonics, in: Integration of Fundamental Polymer Science and Technology, Vol.3, P.J. Lemstra and L.A. Kleintjens, eds., Elsevier Applied Science, Barking (1989).
7. W.J. Feast, Applications of Metatthesis in the Synthesis of New Materials, in Polymer 91, Melbourne, 10-14 Feb. 1991, 70-71.
8. R.M. Penner and C.R. Martin, J. Electrochem. Soc., **133**, 2206 (1986).
9. R.P. Burford and T. Tongtam, Conducting Polymers with Controlled Fibrillar Morphology, J. Materials Sci., In Press.
10. C.R. Martin, L.S. Van Dyke, Z. Cai and W. Liang, J. Am. Chem. Soc., **112**, 8976 (1990).
11. C.Liu, M.W. Espenscheid, W-J. Chen and C.R. Martin, J. Am. Chem. Soc., **112**, 2458 (1990).
12. Cited in M.G. Kanatzidis, Chem. Eng. News, **68**, (48) 49 (1990).
13. B.K. Moss, R.P. Burford and M. Skyllas-Kazacos, Materials Forum, **13**, 35 (1989).
14. European Patent 0178831, Porous Films of Anodic Aluminium Oxide, to R.C. Furneaux, Alcan Int. Ltd., 23 April, 1986.
15. Anotec - Inorganic Membrane Technology, Altech Associates Inc.,Illinois, USA, Bulletin #136,1988

SYNTHESIS OF CONDUCTING COMPOSITE POLYMER BEADS

Gia Y. Kim, Ronald Salovey, and John J. Aklonis

Loker Hydrocarbon Research Institute
Department of Chemistry and Department of Chemical
Engineering, University of Southern California
Los Angeles, CA 90089-1661

ABSTRACT

Composite conducting polymeric beads were prepared by taking advantage of ionic interaction between anionically charged beads and cationically charged conducting polymer precursors. Monodisperse crosslinked polystyrene (PS) beads in the fractional micron range were sulfonated in the gas phase above fuming sulfuric acid to yield surface activated monodisperse polystyrene sulfonic acid (PSSA) beads. Three conducting polymers, polyaniline (PA), polyphenylene vinylene (PPV) and poly (2,5-dimethoxy phenylene vinylene) (PDPV), were then coated on these beads. For PA, one conductive form of the polymer is a cation and was directly mixed with the beads; for PPV and PDPV, soluble cationic precursor polymers were mixed with the beads. Two different PPV precursors, poly (p-xylyidene tetrathiophenium chloride) (I) and poly (p-xylylene-α-dimethylsuphonium chloride) (II) and one PDPV precursor, poly (2,5-dimethoxy-1,4-xylene-α-diemthylsulfonium chloride) (III) were utilized. Chemical doping with AsF_5 for PPV and I_2 for PDPV, of pellets prepared by pressing the coated beads resulted in conductivities as high as 10^{-1} S/cm. The emeraldine base form of PA was self doped.

INTRODUCTION

Polymers have been employed in many fields where their applications range from common daily use to sophisticated aerospace applications. They are primarily used for insulators in the electrical and electronic industries. Recently, however, conjugated unsaturated systems have received special attraction due to their ability to conduct electricity. Doped polyacetylene, the first well characterized and studied conducting polymer, possesses conductivities as high as 10^5 S/cm.[1] Such values are comparable to metal conductors.

Unlike common metal conductors, organic conductors can display a wide range of conductivities. Other attractive features of polymers such as plasticity, superior strength-to-weight ratios, low cost, ease of fabrication and long term stability are well known.

One of the major problems associated with utilization of conducting polymers is processibility. Since these polymers are unsaturated, they tend to be insoluble and also infusible which makes shaping difficult. Moreover, air stability of conducting polymers after doping is a problem.[2] Thus, there is a need to develop air stable, processible conducting polymers.

This paper describes a method to prepare composite polymer beads which are moderately conducting and stable. These may be added to ordinary insulating polymers to yield conducting processible polymer systems. Our procedure utilizes monodisperse crosslinked PS beads which have been treated with SO_3 above fuming sulfuric acid to sulfonate the bead surfaces thus rendering the beads anionic. We have found gas phase sulfonation to be particularly suitable since the swelling of the beads that occurs in solution is avoided.[3] These anionically charged beads are subsequently added to a dilute aqueous solution of cationically charged precursor polymer which results in individually coated beads (Figure 1); they become conductive upon subsequent heating and doping.

Precursor polymers, (I) and (II) of PPV and (III) of PDPV were synthesized by the method of Karasz[4-7] et al. This involves the polymerization of a bis-sulphonium salt monomer to a water-soluble sulphonium salt polyelectrolyte precursor polymer which can subsequently easily be converted to PPV or PDPV by thermal elimination. In their work, metal-like conductivities were observed when these polymers were doped with AsF_5. Compared to PPV, PDPV has improved stability and can be doped under milder conditions.

Polyaniline was first discussed almost 100 years ago[8] and has been shown to be an unusually soluble and stable conducting polymer in the last decade.[9] Much of this recent work has been conducted by MacDiarmid et al.[10,11] A wide range of attractive electrical, electrochemical and optical properties, coupled with good stability, make PA potentially useful in electronic applications.[12]

Both the salt and the base forms of PA are soluble at room temperature up to polymer concentrations of more than 20 % w/w in several concentrated strong protonic acids. Thus, we have used H_2SO_4 as the solvent for a cationic PA salt (VII) to form a solution in which sulfonated beads were coated (VIII). These PA coated beads are attractive since they do not require chemical doping to be electrically conductive and are stable in air indefinitely.

EXPERIMENTAL

Monodisperse crosslinked polystyrene beads (453 nm) were synthesized via emulsifier-free emulsion polymerization[13] as described previously. The PS beads were then sulfonated in the gas phase using an enclosed glass chamber. Within the chamber, a small glass container held the PS beads above fuming sulfuric acid (containing 16-24 % SO_3) in the bottom of the larger glass chamber. Slow continuous rotation for three days at room temperature exposed all the beads. The white PS beads became yellowish after this treatment. The sulfonated beads were washed with distilled water several times and dried in a vacuum oven at room temperature for four days. X-ray Photoelectron Spectroscopy (VG Scientific ESCALAB MKII instrument) and Auger (Perkin Elmer) analysis were used to analyze the surfaces of the sulfonated beads. Elemental analysis (Controlled Equipment Corporation 240XA at 970°C) was done by Oneida Research Services.

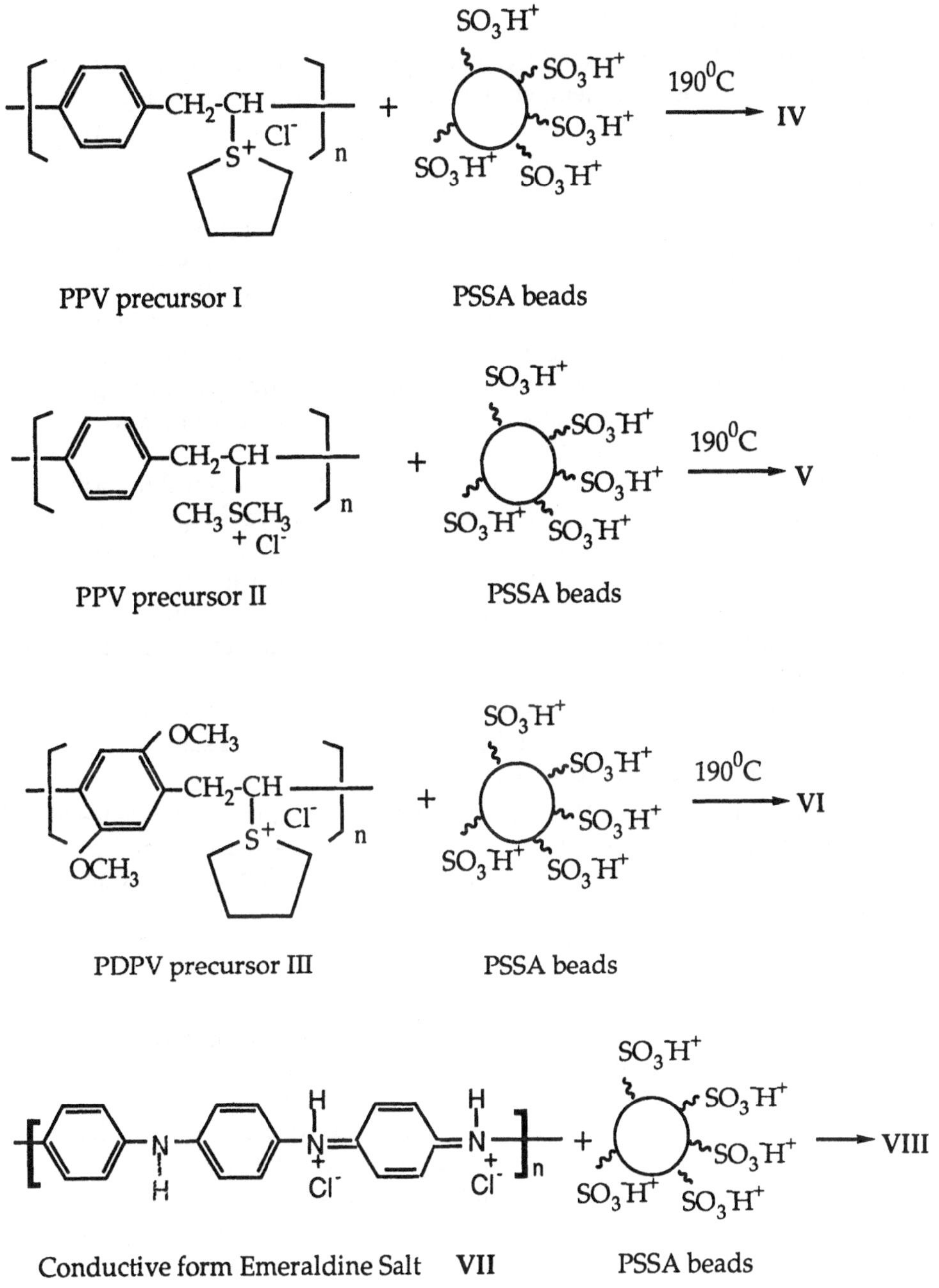

PPV precursor I   PSSA beads

PPV precursor II   PSSA beads

PDPV precursor III   PSSA beads

Conductive form Emeraldine Salt   VII   PSSA beads

Figure 1

Two different PPV precursor polymers, poly(xylylidene tetrathiophenium chloride), I, and poly(p-xylylene-α-dimethylsulphonium chloride), II, were used to coat the sulfonated PS beads. PPV precursors I and II were prepared as described in the literature.[14,15] The elemental analysis of the bis-sulphonium monomers and the infrared spectra of the PPV films obtained from I and II were found to agree with published data.

Sulfonated PS beads were mixed with aqueous 0.2 M PPV precursor I by stirring for 30 minutes at room temperature. Small aggregates were dispersed either by mechanical agitation or sonication. The mixture was then filtered and dried in a vacuum oven at room temperature for 4 days. The resulting slightly yellowish beads were heated to 190°C for 3 days under vacuum and turned yellow which we took as an indication of the formation of composite PPV polymer beads, IV (Figure 1). These composite beads were then pressed between PTFE sheets in a hydraulic press at 170°C and 2500 lb for 20 minutes to form a pellet. Two pressed pellets were doped with AsF_5 at approximately 100 torr for various periods of time and conductivities were measured. These measurements were done in air immediately after removal of the samples from the doping chamber and utilized a Keithley Model 197 Auto-ranging Microvolt DMM connected to an Alessi C4S and C4R 4-Point Probe.

The same synthetic procedure was employed with precursor II to form composite polymer beads, V except that the doping period was longer (20 days). After measuring the conductivity in the usual way, this sample was exposed to the laboratory atmosphere for 24 hours and its conductivity was remeasured. Then this doped pellet was broken into small pieces and repressed in a hydraulic press at 150°C and 2500 lb for 15 minutes. It was subsequently redoped with AsF_5 for 4 days at 100 torr.

The fractured surfaces of pellets made from beads IV and V were observed by SEM (Cambridge Instruments) and the distribution of As in a sample doped for 8 days was determined via EDXS (Cambridge Instruments).

The synthesis of the PDPV precursor, poly (2,5-dimethoxy-1,4-xylene-α-dimethylsulfonium chloride) also followed the technique reported by by Karasz et al.[16] The bead coating procedure was identical except that doping was done above solid I_2 at room temperature for 12 days.

Polyaniline was synthesized according to the procedure reported by the Heeger et al.[17] The reagent-grade aniline was vacuum distilled before use; 40 ml of this aniline in a three necked round bottom flask was added to 460 ml of 1 N HCl. The oxidant solution, 190 ml of 1.5 N HCl and 50 g of $(NH_4)_2S_2O_8$ was added slowly from a dropping funnel while stirring the mixture at –5°C. The reaction mixture was stirred for 4 hours. The precipitated polyaniline was collected from the reaction vessel, was filtered and then washed with distilled water until the water became colorless. Subsequently, the precipitated polyaniline was washed with several portions of methanol followed by diethyl ether. The washed polyaniline was dried in a vacuum oven at room temperature for 5 days. Dried polyaniline was extracted with tetrahydrofuran (THF) using a Soxhlet extrator followed by washing with methanol and ether. The polyaniline was again dried in vacuum oven at room temperature for 5 days.

The polyaniline synthesized in this way was dissolved in concentrated sulfuric acid and sulfonated polystyrene beads were added. The mixture was stirred for 1 hour at room temperature. The mixture was then filtered and the composite beads were washed with water in several portions. The green colored beads were dried in vacuum for several days. The conductivities of polyaniline beads were measured after pressing the powder beads at room temperature at 1000 lbs for 30 minutes with a hydraulic press.

RESULTS AND DISCUSSION

The degree of sulfonation of the PSSA beads was obtained from elemental analysis, XPS and Auger analysis. Assuming sulfonation of each phenyl ring in a PS bead, the elemental composition (by mass) would be 52.2% C, 4.3% H, 26.1% O and 17.4% S. Elemental analysis results were 84.72% C, 7.60% H, 4.59% O and 2.89% S which indicated that about one in every eight monomer units was sulfonated. Auger analysis confirmed the presence of sulfur and oxygen on the bead surfaces. XPS analysis, which ignores hydrogens, showed the elemental atomic composition to be 92.8% C, 6.0% O and 1.2% S. This technique also shows a trace of oxygen on the surface of "pure" PS beads which is probably due to a modest amount of surface oxidation. The unexpected 5 to 1 oxygen to sulfur ratio is probably due to this complication. Further studies on the sulfonation of PS beads are in progress.

The thermal elimination temperatures needed to form PPV from I and II are quite different. Precursor II was used in an attempt to make conducting composite beads but complete thermal elimination was impossible due to the high elimination temperature required to form PPV from II and the fact that PSSA beads decompose at the relatively modest temperature of 300°C. Elimination at temperatures lower than optimal prevents formation of the fully conjugated PPV polymer. Total elimination was possible for precursor I. From the data in Table 1, we observe that it was easier to dope IV than V; furthermore, higher conductivities were observed with longer doping times. The spatial heterogeneity of the conductivity of the pellets made from V was considerable (measured conductivities varied by about an order of magnitude).

When V was exposed to laboratory air for one day, its conductivity decreased by less than an order of magnitude. Upon repressing at 150°C, the conductivity dropped to below 10^{-8} S/cm. Subsequent redoping appeared to be more facile than the original doping.

A comparison of the properties of neat PPV films and conducting composite beads is interesting. The conductivity of a PPV film made from precursor I was 1.8 $\times 10^{-2}$ S/cm (Table 1) after doping for 8 days at 100 torr. The conductivity of the pellet prepared from the same precursor doped and measured under identical conditions was 2.4×10^{-4} S/cm. The decreased conductivity for the beads was probably due to gaps between the beads. Also, as expected, the conductivities increased as the doping period increased from 6 to 8 days for IV.

The fractured surface of a pellet made from IV was examined with SEM (Figure 2). The diameters of the beads increased from about 450 nm before coating to close to 480 nm. The SEM clearly shows that the beads retain their integrity and the PPV appears to be reasonably uniformly coated on their surfaces. EDXS showed the presence of arsenic atoms throughout the fractured surface.

Unlike PPV, PDPV can be doped with I_2 rather than AsF_5. However, the conductivity of PDPV coated PS beads was lower than PPV coated beads. After 12 days of doping, the conductivity was only 10^{-5} S/cm (Table 1).

Table 1. Conductivity measurements of PPV/PSSA and PDPV/PSSA composite beads.

Composite beads	Time of doping (days)	Thickness of pellet (cm)	Conductivity (S/cm)
	6	0.043	5.1×10^{-5}
IV	6	0.043	1.2×10^{-5}
	8	0.059	2.4×10^{-4}
	12	0.051	below 10^{-8}
	20	0.051	$1.4 \times 10^{-1} - 5.7 \times 10^{-2}$
V	20^a	0.051	1.1×10^{-2}
	20^b	0.051	2.0×10^{-4}

PPV film	8	0.0020	1.8×10^{-2}

PDPV	12	0.0035	1.4×10^{-5}

a exposured to air for one day
b repressed in hydraulic press then redoped for 4 days

Unlike PPV and PDPV coated beads, PA coated beads do not require chemical doping to conduct. In addition, PA coated beads are stable in air indefinitely. Unfortunately, the conductivities of composite beads based on PA are much lower than those prepared with PPV as shown in Table 2. These low conductivities may be due to uneven coverage of the surfaces of the beads which can be seen in Figure 3 which is an SEM picture showing aggregates presumably of PA on the beads surfaces. It is probable that since PA exists in both reduced and oxidized states, uncharged portions of the PA might not bind well to the sulfonated bead surface. The smooth coverage observed with PPV and PDPV is consistent with the fact in these cases every monomer unit of the chain contains a charge which can interact with the anionic beads.

A possible application of conducting composite beads may involve adding them to polymer matrices to enhance conductivities in systems which can be shaped using conventional means. The regular shape of the beads should make predictions and control of rheology straightforward.

Table 2. Conductivity measurements of PA/PSSA composite beads.

Composite beads	Thickness of pellet (cm)	Conductivity (S/cm)
	0.069	1.85×10^{-5}
PA / PSSA (VIII)	0.075	$2 \times 10^{-3} \times 1.2 \times 10^{-4}$

PA	0.088	$3 \times 10^{0} - 4 \times 10^{0}$

2.0μm

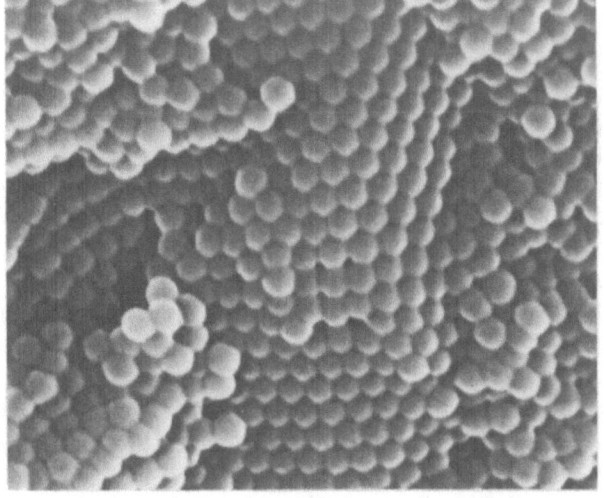

Fractured surface of a pellet made from IV before doping.

Figure 2

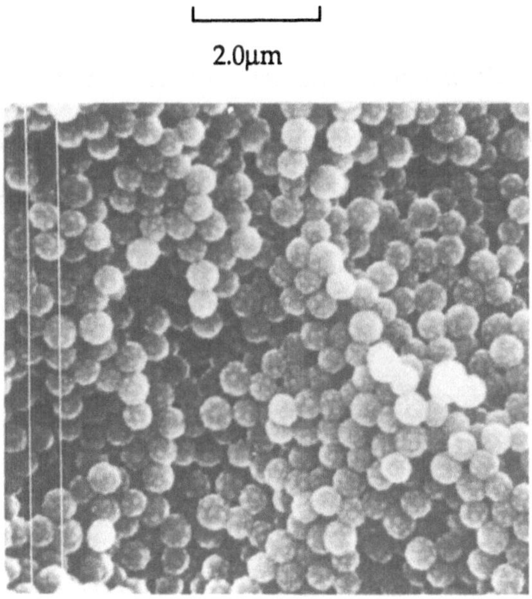

2.0μm

Fractured surface of a pellet made from VIII.

Figure 3

Acknowledgement: This work was supported by the ONR through grant number (N00014-88k-0302). We appreciate the help of Mr. Jack Worrel for the EDXS measurements at U. S. C. and Drs. Alan Gotcher and James Tse for the XPS measurements done at the Avery International Research Center.

References

1. a) H. Naarman and N. Theophilou, *Synth. Met.*, 22:1 (1987).
   b) J. Tsukamoto, *Japanese J. of Appl. Phys.*, 29:1 (1990).
2. S. Pekker and A. Janossy, Handbook of Conducting Polymers Vol 1, Terje A. Skotheim, ed., Marcel Dekker, Inc., New York (1986).
3. F.P. Regas, *Polymer.*, 25:249 (1984).
4. G.E. Wnek, J.C.W. Chien, F.E. Karasz and C.P. Lillya, *Polymer*, 20:1441 (1979).
5. D.R. Gagnon, F.E. Capistran, F.E. Karasz and R.W. Lenz, *Polymer Bulletin*, 12: 293 (1984).
6. J.M. Machado, J.B. Schlenoff and F.E. Karasz, *Macromolecules*, 22:1964 (1989).
7. S. Antoun, F. E. Karasz and R. W. Lenz, *J. Polym. Sci.*, Part A, *Polym. Chem.*, 26: 1809 (1988).
8. A. G. Green and A. E. Woodhead, *J. Chem. Soc.*, 2388 (1910).
9. R. de Surville, M. Jozefowicz, L. T. Yu, J. Perichon and R. Buvet, *Electrochim.Acta*, 13 (1968).
10. J. C. Chiang and A. G. MacDiarmid, *Synth. Met.*, 13:193 (1986).
11. A. G. MacDiarmid, J. C. Chiang and A. F. Richter, *Synth. Met.*, 13:285 (1987).
12. E.M. Genies, A.A. Syed and C. Tsintavis, *Mol. Cryst. Liq. Cryst.*, 121:181 (1985).
13. D. Zou, K. Gandhi, M. Park, L. Sun, D. Kriz, Y. Lee, G. Y. Kim, J.J. Aklonis and R. Salovey, *J. Polym. Sci. Polym. Chem.*, 28:1909 (1990).
14. R.W. Lenz, C. C. Han, J. Stenger-Smith and F.E. Karasz, *J. Polym. Sci., Part A, Polym. Chem. Ed.*, 26:3241 (1988).
15. D.R. Gagnon, J.D. Capistran, F.E. Karasz, R.W. Lenz and S. Antoun, *Polymer.*, 28: 567 (1987).
16. S. Antoun, F.E. Karasz and R.W. Lenz, *J. Polym. Sci., Part A, Polym. Chem. Ed.*, 26:1809 (1988).
17. Y. Cao, A. Andreatta, A. Heeger and P. Smith, *Polymer*, 31:2305 (1989).

TRANSPARENT ELECTRICALLY CONDUCTIVE COMPOSITES BY IN-SITU CRYSTALLIZATION OF (TSeT)$_2$Cl IN A POLYMER MATRIX

Ch. Kröhnke[*)], J. Finter, C. W. Mayer

Central Research Laboratories

J. Ansermet, H. Bleier, B. Hilti,
E. Minder, D. Neuschäfer

Materials Research
CIBA-GEIGY AG, CH-4002 Basle, Switzerland

[*)]present address: CIBA-GEIGY AG, Additives
Division, R-1038.5.09
CH-4002 Basle, Switzerland

ABSTRACT

A new chemical pathway was found which offers access to the thermostable, highly conductive tetraselenotetracene chloride radical cation salt (TSeT)$_2$Cl. This pathway allows the in-situ crystallization of (TSeT)$_2$Cl in a vitrifying polymer matrix. So highly conductive films (σ= 1-5 Scm^{-1}) with networks of (TSeT)$_2$Cl crystallites of a 1:1000 aspect ratio could be prepared at a fairly low percolation threshold of 0.4 - 1.6 % w/w. These films show metallic conductivity behavior in the 60 - 400 K range.

INTRODUCTION

Transparent electrically conductive materials are only accessible by composites of at least two phases, one transparent and one conducting phase. The design of volume conducting composites is limited by the laws of percolation. To achieve transparency very low loadings of conductive fillers are required. According to percolation theory, the shape of a filler determines the volume fraction to reach the percolation threshold. The lowest loadings (1-5 Vol.%) are possible with needles of high aspect ratios (length/diameter) > 100.

Besides conventional routes like filling polymers with high aspect ratio metal fibers or metallized glass fibers, an elegant route to such materials is the in-situ crystallization of a soluble organic electroconductors under non-equilibrium conditions in a vitrifying polymer matrix, a process called "reticulate doping".

This process is, however, limited to soluble organic CT-complex salts. Until now, only conductive TCNQ radical salts were successfully applied by this procedure. High conductivities are reported for some composites, thermal and hydrolytical stability of TCNQ crystals are, however, limited.

CRYSTALLIZATION OF TSeT-HALIDES IN A POLYMER MATRIX

Properties of (TSeT)$_2$-Halides

(TSeT)$_2$-halides combine high room temperature conductivity (σ = 2500 Ωcm^{-1}) with high thermal stability up to over 250 °C (see Fig. 1).They are, however, practically insoluble even in exotic organic solvents. Only the donor TSeT itself shows limited solubility in high boiling polar solvents. Until recently pure 2:1 halides could only be made by electrocrystallization in nitrobenzene in the presence of an organic ammonium halide as electrolyte. Chemical pathways like oxidation with Cl$_2$-gas or ferrous chloride lead to impure products mainly due to overoxidation of the donor.

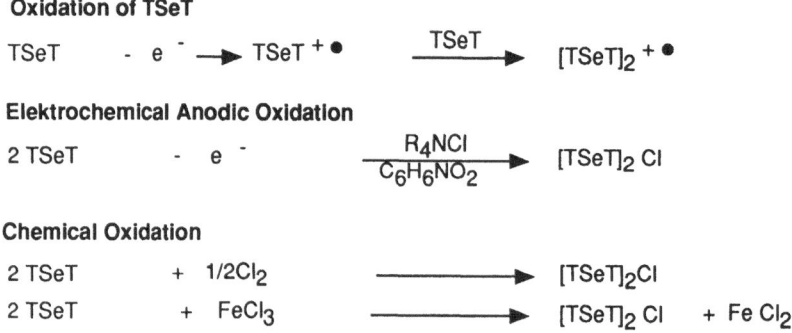

For crystallization of (TSeT)$_2$halides in a polymer film a selective chemical redox reaction had to be found. The thermally induced oxidation of TSeT with organic perhalides is a new synthesis pathway which offers access to the reaction and crystallization route. This reaction yields pure 2:1 complexes. The driving force is the insolubility of the formed CT-complex salt, the reaction proceeds during the increase of concentration due to solvent evaporation. Fig. 2 shows a typical experiment for the in-situ crystallization from a diluted solution. A solution of TSeT, polymer and solvent is mixed quickly with an excess of a

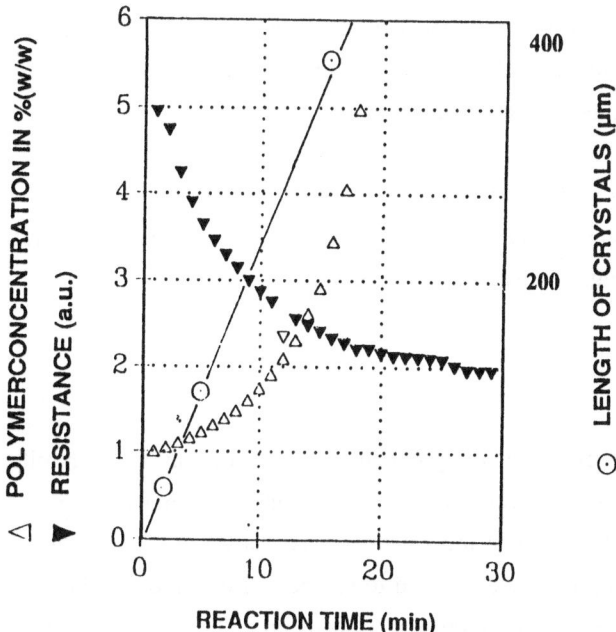

Fig. 1. Change of polymer concentration, length of needle crystals and resistance as function of time during reaction induced crystallization

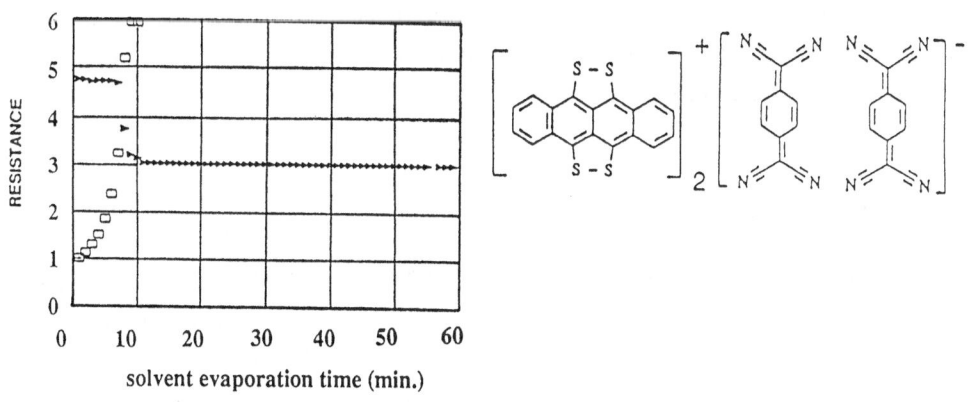

resistance (absolute units)

Fig. 2. Change of polymer concentration and resistance as function of time during reticulate doping of polycarbonate with TTT-TCNQ in o-dichlorobenzene at 150°C. ▶:change of resistance ☐ : change of concentration in w/w %

do

perchloroalkene and the solvent is allowed tom evaporate under isothermal conditions. Upon evaporation of the solvent the polymer concentration increases. Growth of the $(TSeT)_2Cl$ crystals starts however immidiately after addition of the halogenacceptor. The growth of these crystals is linear with respect to time indicating a homogeneous nucleation. The resistity of the reaction medium decreases continuously and reaches its minimum before the film solidifies.

The crystallization kinetics are different from reticulate doping of soluble CT-complexes where crystallization is very rapid and starts at the onset of supersaturation. An example for is shown in Fig. 2 for soluble TTT-TCNQ.

Morphology

The morphology of the resulting conductive network may be controlled mainly by the proper choice of solvent, concentrations and temperature. Besides that, the ratio of oxidant to donor influences the final properties, too. As shown in Fig. 3 a threefold molar excess of chlorination agent should be used to ensure full oxidation of the donor. As exspected the value of the percolation threshold depends upon the amount of CT-complex in a film. For the system TSeT, perchloropropene in poly(bisphenol-A carbonate) in chlorobenzene the conductivity of the CT-needle network is independent of CT-concentration beyond 0.8 %(w/w) loadings (Fig. 4). At a given CT-complex/polymer ratio, films with different conductivities and transparencies are accessible by control of the aspect ratio via choice of solvent and crystallization temperature. In many cases the needle crystals form hollow fibers. At least for the amorphous polymers we studied (polycarbonate, polystyrene, soluble polyimides, polysulfonethers, polyarylates) the morphology of the needle networks was not influenced of the type of polymer or viscosity.

Electrical Properties

The conductivities of the composite films depend upon the concentration of CT-complex and are typically around 1 mho cm^{-1} for 1-2 %(w/w) loadings. The best values were around 5 mho cm^{-1}. Surprisingly, composite films of $(TSeT)_2Cl$ in a polymer matrix show a metallic ohmic behavior. To the best of our knowledge this is the first example of a composite with metallic ohmic properties.As evident in Fig. 5 the properties of the macroscopic single crystal including the phase transition at 30 K are maintained in the network crystals. For such composite materials with insulating matrix transport of electrons is normally limited by tunneling processes over short insulating ranges. Therefore resitivity increases with decreasing temperature. For the observed metallic behaviour it is highly probable, that our needle networks exhibit intimate crystal-to-crystal

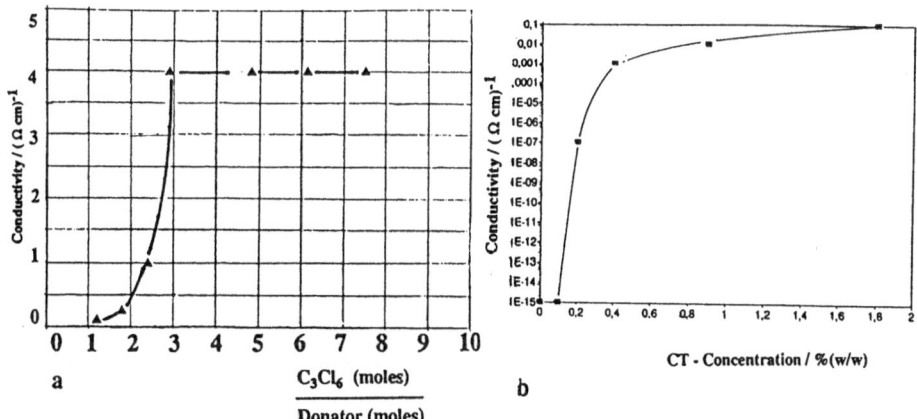

Fig. 3. a) Percolation threshold influenced by reaction conditions: Conductivity of a composite film as function of the molar excess of perchloropropene over TSeT-donor. b) Conductivity as function of TSeT$_2$Cl -content

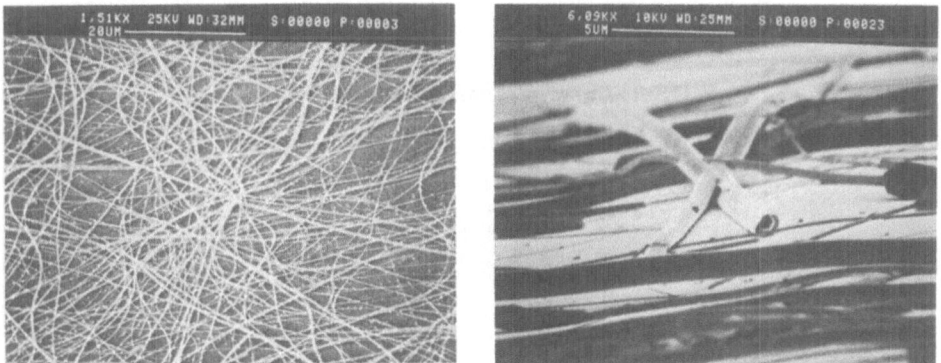

Fig. 4. Typical morphologies of (TSeT)$_2$Cl needle networks.

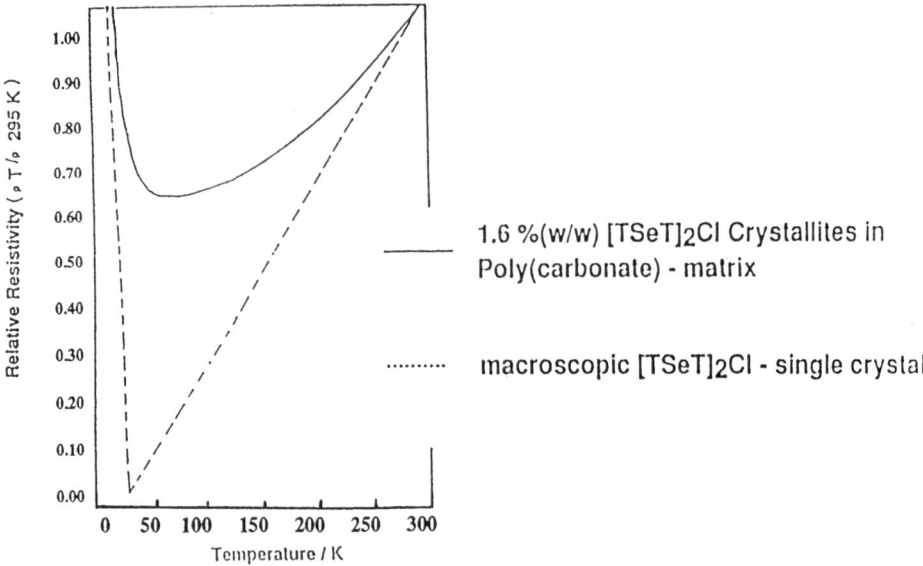

1.6 %(w/w) [TSeT]2Cl Crystallites in
Poly(carbonate) - matrix

.... macroscopic [TSeT]2Cl - single crystal

Fig. 5. Relative change of resistance of a (TSeT)$_2$Cl needle network in poly(carbonate) and of a macroscopic (TSeT)$_2$Cl single crystal

(TSeT)$_2$Cl-polycarbonate composite film

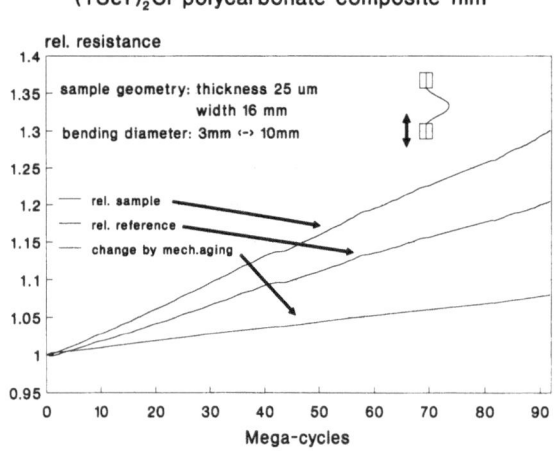

Fig. 6. Flexural stability: Change of relative resistance (R/R$_0$) as function of bending cycles.

contacts. To support this hypothesis first measurements of conductivity under elongation revealed a sudden breakdown of conductivity at elongations over 3% presumingly by mechanical distortion of the network.

Maximum current density is also an important figure for electrically conductive materials. Failure occurs at current densities around 10 A/cm^2.

In a two phase system of conductive needles embedded in a non conductive matrix the the conductivity should also depend upon the size of the sample with respect to the mesh size of the conductive needle network. When films were cut into slices of decreasing width, a minimum slice width of 100 μm was found./5/

Stability and Mechanical Properties

The polymer films show a moderate decrease in conductivity after prolonged exposure to a 85 °C, 85 % relative humidity atmosphere. Under this conditions the resistance of a condictive increases typically by 1 - 5 % in 1000 hours /3/. The embedded needles are also highly stable towards flexural deformation of the film. Upon continuous bending no failure of a film has been observed even after 16 million cycles (Fig. 6). The increase in resistance is mainly due to normal ageing.

References

1. M. Kryszewski, J. K. Jeszka, J. Ulanski and A. Tracz, Pure Appl. Chem. 56 (1984) 355-368

2. Eur. Pat. Appl. EP 285564 A1 (1988) to Ciba Geigy AG, Inv. J. Finter, B. Hilti, C. W. Mayer, E. Minder and J. Pfeifer

3. A. Tracz and M. Kryszewski, Makromol. Chem. Suppl. 15 (1989) 219 - 231

4. J. Finter, C. W. Mayer, J. Ansermet, B. Hilti, E. Minder, D. Neuschäfer, Proc. VII. Intl. Conf. Sci. Techn. Synth. Metals, (1990) to be published

5. D. Neuschäfer, J. Finter, D. Goltz, B. Hilti, H. Spahni, Proc. VII. Intl. Conf. Sci. Techn. Synth. Metals(1990), to be published

POLYMER ELECTROLYTES AND HYDROGELS FROM POLYETHYLENE GLYCOLS

CROSS-LINKED WITH A HYDROPHOBIC POLYISOCYANATE

Du Wei Xia, Arjen Sein and Johannes Smid

Polymer Research Institute, Chemistry Department
College of Environmental Science and Forestry
State University of New York
Syracuse, NY 13210, USA

INTRODUCTION

Solvent free polymer electrolytes have been extensively researched in recent years because of their potential as thin-film ion-conducting materials in high energy density batteries and electrochromic devices.[1-3] Our work in this area has been chiefly concerned with the morphology and conducting properties of salt-containing comb-like poly(methacrylate)s and polysiloxanes endowed with oligo-oxyethylene side chains.[4-5] Ion transport in such systems largely occurs by a free volume mechanism. This is facilitated when the polymeric component has a minimum of crystallinity and a low glass transition temperature. The presence of ion-chelating or polar additives can also have spectacular effects on the conductivity, since they can function as plasticizers or cause the dielectric constant of the material to increase. In this report the synthesis and conducting properties of lithium perchlorate and lithium triflate-containing networks of polyethylene glycols cross-linked with a hydrophobic tri- or tetrafunctional isocyanate are described. The cross-linking agents are well defined hydrosilylation products of α,α-dimethyl-meta-isopropenylbenzylisocyanate (m-TMI). With added propylene carbonate, the elastomeric solids have conductivities exceeding 10^{-3} S cm^{-1} at 25°C and reach values close to 10^{-2} S cm^{-1} at 90°C. The salt free networks when swollen in water exhibit a strong affinity for a variety of hydrophobic molecules. The latter apparently are bound to the large hydrophobic cross-linking sites. Preliminary results of the properties of these hydrogels are also discussed.

EXPERIMENTAL PROCEDURES

The triisocyanate T3TMI and the tetraisocyanate D4TMI, the structures of which are depicted in Scheme A, were synthesized in high purity and essentially quantitative yield by the bulk hydrosilylation of α,α-dimethyl-meta-isopropenylbenzylisocyanate (m-TMI from American Cyanamid) with methyltris(dimethylsiloxy)silane (T3H) and 2,4,6,8-tetramethylcyclotetrasiloxane (D4H), respectively.[6] Poly(ethylene glycol)s (PEG, MW 200-6,000) were dried azeotropically or freeze dried, and their hydroxyl content determined by a standard acetylation method. The o,o'-bis(2-aminopropyl) polyethylene glycols (Jeffamines, MW 900 and 2,000) were purchased from Fluka and the amine content carefully checked by titration with 0.2 N HCl and bromophenol blue as indicator.

Networks were made under a nitrogen atmosphere by reacting a 30% THF solution of T3TMI or D4TMI with the appropriate PEG (ratio NCO/OH=1) for

Contemporary Topics in Polymer Science, Vol. 7., Edited by
J.C. Salamone and J. Riffle, Plenum Press, New York, 1992 ⊀, 1992

229

T3TMI   D4TMI

Scheme A

about 24 hrs at 75°C, using as catalyst tetramethylethylene diamine (TMEDA) or dibutyldilauryltin. The curing process was completed at 85°C under vacuum. Polymer electrolytes were obtained by swelling the dry networks with a known quantity of lithium triflate or perchlorate dissolved in acetonitrile. A more facile method involves the reaction of T3TMI or D4TMI with one of the Jeffamines at 5°C in a solution of CH_2Cl_2/THF (4/1, v/v) containing a known amount of the lithium electrolyte. The cross-linking reaction is completed in about ten minutes when the mixture is allowed to warm up to room temperature. When used as hydrogel the CH_2Cl_2/THF was exchanged for dioxane, and the dioxane gradually replaced by water using water/dioxane mixtures of increasing water content.

Measurements. Conductivity measurements were carried out under a helium atmosphere over a temperature range of 25°C-90°C by means of a Hewlett Packard HP4192LF impedance analyzer. DSC scans of the polymer electrolytes and the salt free networks were obtained on a Perkin Elmer DSC-4 calorimeter at a heating rate of 20°C/min. Swelling of the networks were determined in water and dioxane at 25°C. The affinity of our hydrogels for organic species was studied with optical and fluorescent probes such as 8-anilinonaphthalene sulfonate (ANS), bromophenol blue (BPB) and picrate salts. More details of the measurements will be reported in a forthcoming publication.

RESULTS AND DISCUSSION

^{1}H NMR, chemical analysis and GPC data of T3TMI and D4TMI demonstrate that these polyfunctional aliphatic isocyanates are well defined products unlike many of the commercial multifunctional isocyanates such as the Desmodur's which are mixtures of compounds. The reaction of PEG with T3TMI or D4TMI in the presence of TMEDA or the tin compound proceeds smoothly. Reaction with the Jeffamines and the TMI cross-linkers is very rapid although slows down somewhat when $LiClO_4$ is present. No unreacted NCO groups can be detected in the networks by IR. DSC scans demonstrate that the cross-linking greatly reduces the crystallinity found in the higher molecular weight poly(ethylene glycol)s. The glass transition temperature of the network rapidly rises with decreasing PEG chain length. A linear relationship was found on plotting the inverse of T_g versus the concentration of cross-links expressed in terms of the molar concentration of D4TMI units.

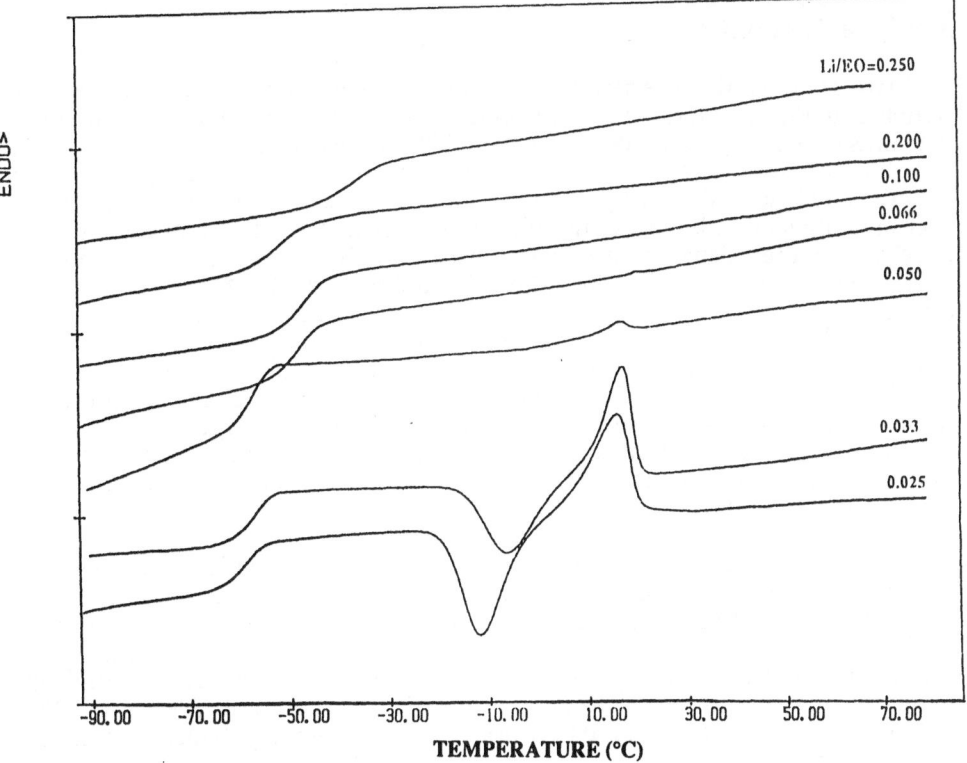

Figure 1. DSC scans showing the effect of LiClO$_4$ content on the crystallinity and T$_g$ of a D4TMI-PEG 2040 network containing 40 wt.% PC-NMA (4/1)

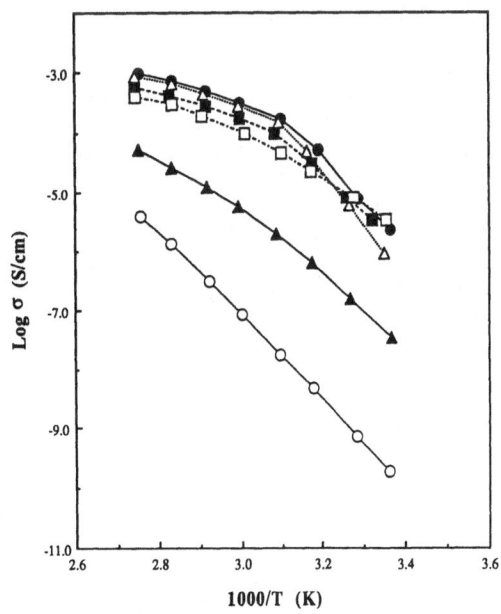

Figure 2. Temperature dependence of the conductivity of D4TMI-PEG networks containing LiClO$_4$ (EO/Li=15) for PEG of different molecular weight: MW of PEG=293 (○), 590 (■), 985 (▲), 2040 (□), 4240 (△), and 5750 (●).

A. Polymer Electrolytes

Lithium perchlorate and triflate, as well as other alkali salts,[7] easily solubilize in PEG networks, and their presence lowers the crystallinity that exists in networks for which PEG exceeds a MW of 1500. For example, a network of D4TMI and PEG 2000 which has a melt peak at 58°C turns completely amorphous when the ratio ethylene oxide units (EO) to Li reaches a value of 5. Addition of additives such as propylene carbonate or its mixtures with N-methylacetamide has a similar effect, and figure 1 demonstrates that in these gel polymer electrolytes crystallinity already has disappeared when the ratio EO/Li drops below 20. Although the loss of crystallinity facilitates ion transport, increasing the salt content also causes the T_g to rise. Earlier work on the comb poly(methacrylate)s and polysiloxanes with oligo-oxyethylene side chains showed a nearly linear increase in T_g with the ratio Li/EO.[4,5] Chelation of Li^+ with multiple EO binding sites causes the T_g to increase rapidly. As a result the conductivity passes through a rather sharp maximum as the salt content is increased in spite of a higher ion carrier content. The maximum occurs at a ratio EO/Li $\approx$ 15-20, and most of our experiments were performed at this ratio.

Figure 2 depicts temperature dependent conductivity plots for a series of $LiClO_4$-containing networks with increased PEG chain length. For networks with MW of PEG > 2000 a distinct break in the plot is found around 60°C, close to the PEO melting point. As expected, the conductivity rapidly increases with the EO unit content of the network, σ at 25°C being only 10^{-9} S cm^{-1} for the PEG 300 network and rising to 5×10^{-6} S cm^{-1} for the PEG 2000 network. A value of 10^{-3} S cm^{-1} is reached for the PEG 5750 network at 90°C where the polymer electrolyte is completely amorphous.

The profound effect of propylene carbonate (PC) on σ is illustrated in figure 3. For a $LiClO_4$-PEG 5750 network (EO/Li = 15) the σ at 25°C increases from 8×10^{-6} S cm^{-1} to 2×10^{-3} S cm^{-1} on adding 30 wt.% PC. AT 90°C the increase is from 2×10^{-3} S cm^{-1} to 9×10^{-3} S cm^{-1}. Similar observations are found when the electrolyte is lithium triflate although the conductivities are slightly below those for $LiClO_4$. In both cases the materials remain elastomeric solids. Even higher conductivities can be achieved with N-methylacetamide (NMA) as additive. The latter compound has a reported dielectric constant of 176 compared to 65 for PC although the viscosity of PC is lower. The latter parameter of course is important since a high viscosity lowers the ion mobilities. The best results were obtained with a PC/NMA mixture equal to 4 (v/v). When in this gel polymer electrolyte the $LiClO_4$ content is increased the usual maximum in σ at EO/Li $\approx$ 15-20 is not observed. As can be seen in figure 4, σ remains constant or even increases up to a ratio EO/Li = 5, then drops rather abruptly when more salt is added. The reason for this behavior may be that the T_g of this system does not appear to change when the ratio EO/Li changes from 15 to 5, but then increases when the ratio drops to 4 (see figure 2). Also, the much higher dielectric constant of the gel polymer electrolyte means that the ion carrier concentration rapidly increases with salt content since much more of the salt is dissociated.

It may be argued whether in the PC gel polymer electrolytes the lithium ions are still chelated to the ethylene oxide units of the PEG chains or whether they are at least in part solvated by PC. To gain more insight into this problem we measured ^{13}C NMR shifts in $LiClO_4$/TEGDME mixtures as a function of added PC. The results clearly show that even at 40 wt.% (60 mol.%) PC on the average only about 15% EO units in the solvation shell of Li^+ are replaced by PC.[8] This is consistent with the observation that in a 50/50 (wt.%) mixture of PC and TEGDME containing $LiClO_4$ (PC/Li = 2.2; EO/Li = 5.5) the carbonyl infrared absorption band is barely shifted, while in pure PC the lithium salt (PC/Li = 3) shifts the carbonyl band by 19 cm^{-1}. The same is found when PC is added to $LiClO_4$ dissolved in PEO, that is, no C=O shift is observed.[8] However, a significant shift is found with a random copolymer of ethylene oxide and propylene oxide is used. The latter monomer unit is a poor chelating agent and permits to PC to interact

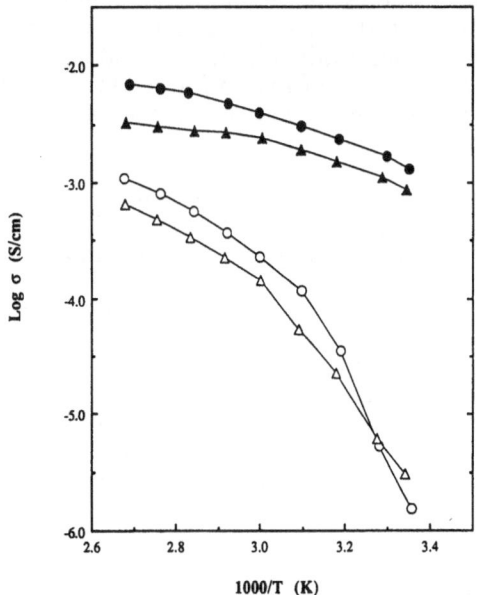

Figure 3. Temperature dependence of the conductivity for the D4TMI-PEG 5750 network containing $LiClO_4$ or $LiCF_3SO_3$ with or without PC. EO/Li = 15. $LiClO_4$ (○); $LiCF_3SO_3$ (△); $LiClO_4$ + 30 wt.% PC (●); $LiCF_3SO_3$ + 30 wt.% PC (▲).

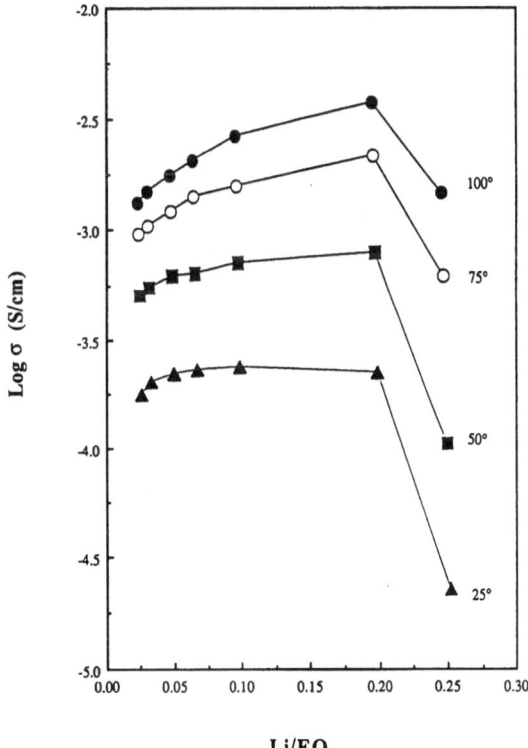

Figure 4. Dependence of conductivity on the $LiClO_4$ content (expressed as Li/EO) for a D4TMI-PEG 2040 network with 40 wt.% PC/NMA (4/1) at 25° (▲), 50° (■), 75° (○) and 100°C (●).

with $LiClO_4$. Competitive studies on the complexation of PC and linear polyethers to picrate salts in toluene[7] reveal that a ligand such as TEGDME has an affinity for LiPi nearly three orders of magnitude higher than that of PC. Hence, all available evidence points to the poor affinity of Li ions for PC relative to that of PEO chains. The main effect of PC is raising the dielectric constant of the polymer complex and increasing its free volume.

Dissociation constants for LiPi measured by optical spectroscopy in mixtures of TEGDME/PC at 25°C give values ranging from 1.5×10^{-6} M in pure TEGDME to 0.2 M^{-1} when 90 vol.% PC is added.[8] On increasing the temperature from 0° to 100°C the K_d for LiPi in TEGDME/20 wt.% PC was found to decrease by a factor five. In spite of the decrease in the free ion fraction, the conductivity at high temperature in the polymer electrolyte complexes is enhanced due to the decrease in viscosity and increase in free volume. It is important to point out that LiPi in TEGDME is a tight ion pair, which then dissociates into free ions on adding PC. At the salt concentrations used in polymer electrolytes (≈ 1 M), most of the current carrying species are probably triple ions, and the temperature dependence of triple ion equilibria may be quite different from that of ion pair dissociation equilibria.

B. Amphilitic Hydrogels

The D4TMI-PEG networks were found to swell in both aqueous and non-aqueous media. The extent of swelling in water for gels of different PEG chain length is depicted in figure 5. The water content at equilibrium rapidly increases with the content of EO units. The weight increase at 25°C for the D4TMI-PEG 5750 network amounts to a factor 6.6. Similar studies in dioxane yield for the same network a factor 8.2. The gels are all transparent, but for networks with PEG > 1500 transparency in water is only achieved when the organic solvent used in the synthesis is gradually replaced by water using dioxane-water mixtures of increasing water content. DSC studies of the water-swollen hydrogels reveal the presence of three forms of water (an example is shown in figure 6). This has also been reported for water-swollen linear PEO,[10] and points to the presence of water bound to EO units and around hydrophobic regions of the polymer.

In their behavior the D4TMI-PEG networks can be classified as amphilitic hydrogels. These are comprised of hydrophilic and hydrophobic polymer segments or domains which cause them to interact with both polar and apolar solvents. Such hydrogels often respond in dramatic fashion to environmental stimuli such as variations in temperature, pH, solvent composition and salt concentration. Discontinuous volume changes and complete gel collapse or dissolution can be the end result. In our systems we measured the change in cloud point (T_c) of the water swollen networks as a function of added salt. The behavior resembles that of linear polymers with EO units such as the comb poly(methacrylate)s with oligo-oxyethylene side chains.[12] For example, the T_c of the network D4TMI-PEG 900 decreases linearly with the molar concentration of added NaF, dropping from 82°C in water to 48°C in 0.8 M NaF. It increases to 98°C in an aqueous solution of 0.8 M NaSCN, a "salting in" effect typical for isocyanates.[12]

In our networks the hydrophobic domains are the D4TMI cross-links, the concentration of which in the swollen gels can easily be calculated. Whether these cross-links associate to larger hydrophobic domains, especially when the PEG chains are long, has not yet been established. We did observe from surface tension measurements such an aggregation in star polymers made from D4TMI and methoxypolyethylene glycols. One of the interesting properties of our hydrogels is their strong affinity towards a variety of organic molecules such as picrate salts, bromophenol blue (PBP), 8-anilinonaphthalene sulfonate (ANS), pyrene derivatives and tetraphenylboron anions. For example, a 10 ml 10^{-4} M solution of the blue dye BPB is decolorized completely on adding 50 mg D4TMI-PEG 2040 network, the gel turning dark blue. Preliminary studies with picrate salts yielded

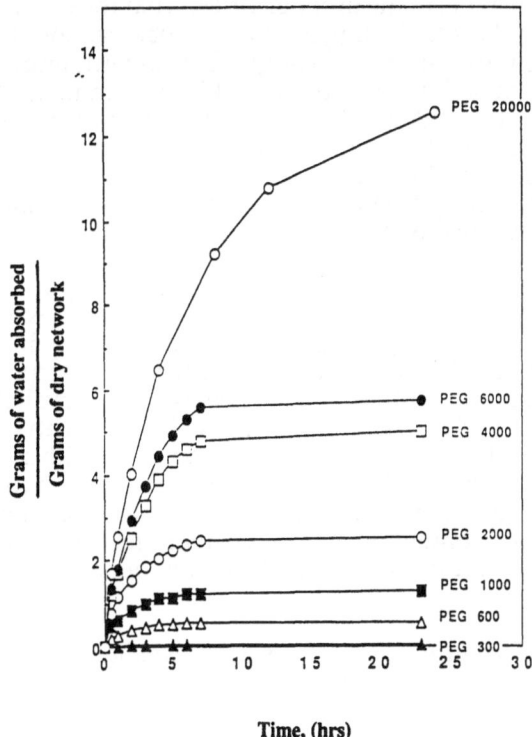

Figure 5. Equilibrium swelling of D4TMI-PEG networks in water as a function of the length of the PEG chain.

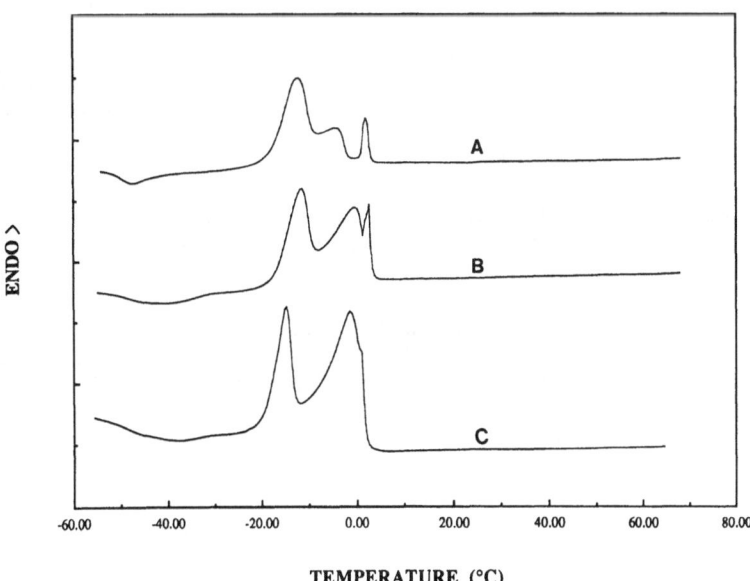

Figure 6. DSC tracings of water-containing networks of D4TMI-PEG 5750 networks as a function of water content. Ratio H_2O/EO units = 4 (A), 5 (B) and 6 (C).

intrinsic binding constants in the order of 10^5M. That the organic species bind to the hydrophobic D4TMI sites is suggested by the observation that in an aqueous ANS solution the hydrogel turns strongly fluorescent, implying that the gel entrapped fluorescent probe is in a hydrophobic environment. Addition of small quantities of NaBPh$_4$ to a yellow-colored gel with entrapped picrate anions rapidly turns colorless as the more hydrophobic BPh$_4^-$ anions replace the bound picrate anions. All the observed phenomena resemble those previously studied in our group on the binding of hydrophobic organic molecules to linear polysoap-like poly(crown ether)s in aqueous media.[13]

The binding of the organic molecules is not greatly influenced by high concentrations of added alkali ions. Conceivably, the alkali ions of the organic salts could bind to the PEG chains and augment the binding of hydrophobic anions to the gel. Apparently this is not the case. We intend to incorporate macrocyclic polyethers (crown ethers) into the gel to precisely accomplish this, that is, to regulate the binding of charged organic species by means of crown ether-complexable cations. This was effectively accomplished with soluble poly(crown ethers),[13] which bind cations in aqueous media, thereby changing the neutral polymer into a polycation capable to attract anionic molecules and to repel cationic species. Such a crown-containing amphilitic hydrogel could be used to regulate reactions catalyzed by the presence of hydrophobic domains as occurs in micellar catalysis.

Acknowledgment: The authors gratefully acknowledge the financial support of this research by the Polymers Program of the National Science Foundation, Grant No. DMR 8722245.

REFERENCES

1. M.B. Armand, Polymer electrolytes, *Ann. Rev. Mater. Sci.*, 16:245 (1986).
2. C.A. Vincent, Polymer electrolytes, *Chemistry in Britain*, 391 (1989).
3. D.F. Shriver and G.C. Farrington, Solid ionic conductors, *Chem. Eng. News.*, May 20:42 (1985).
4. D.W. Xia, D. Soltz and J. Smid, Conductivities of solid polymer electrolyte complexes of alkali salts with polymers of methoxypolyethyleneglycol methacrylates, *Solid State Ionics*, 14:221 (1984).
5. D. Fish, I.M. Khan, E Wu and J. Smid, Polymer electrolyte complexes of LiClO$_4$ and comb polymers of siloxane with oligo-oxyethylene side chains, *Brit. Polym. J.*, 20:281 (1988).
6. G.B. Zhou, R. Fragnito and J. Smid, Siloxanes with aliphatic isocyanate groups, *Polym. Bull.*, 22:85 (1989).
7. A. Killis, J.F. LeNest, H. Cheradame and A. Gandini, Ionic conductivity of polyether-polyurethane networks containing NaBPh$_4$, *Makromol. Chem.*, 183:2835 (1982).
8. D. Fish and J. Smid, to be published.
9. W.Y. Xu and J. Smid, results from this laboratory.
10. N.B. Graham, M. Zulfiqar, N.E. Nwachuku and A. Rashid, Interaction of poly(ethylene oxide) with solvents: Water-poly(ethylene glycol), *Polymer*, 30:528 (1989).
11. E.S. Matsuo and T. Tanaka, Kinetics of discontinuous volume-phase transition of gels, *J. Chem. Phys.*, 89:1695 (1988).
12. E. Nwankwo, D.W. Xia and J. Smid, Salt effects on cloudpoints and viscosities of polymethacrylates with pendant oligooxyethylene chains, *J. Polym. Sci., Phys. Edit.*, 26:581 (1988).
13. J. Smid, Binding of solutes to poly(vinylbenzocrown ether)s and poly(vinylbenzoglyme)s in aqueous media, *Pure Appl. Chem.*, 54:2129 (1982).

NONLINEAR OPTICAL PROPERTIES OF NEW DYE DOPED
PHOTOCROSSLINKABLE POLYMERS

S. Tripathy, L. Li[†], B. K. Mandal, J. Y. Lee and J. Kumar[†]

Departments of Chemistry and [†]Physics
University of Lowell
Lowell, Massachusetts 01854

INTRODUCTION

Polymeric materials are playing an increasingly important role in electronic and photonic applications[1-4]. This includes application in active devices utilizing optical effects attributable to the nonlinear polarization of the medium. A number of applications such as frequency mixing, second harmonic generation, optical bistability, optical parametric amplification and oscillation, electrooptic and all optical switching and modulation etc. have been proposed.

There is a separation of charges in a nonlinear optical (NLO) medium under the application of an electric field. For weak electric fields, the applied electric vector, E, induces a polarization, P, in the material which is linearly dependent on the electric field. $P_{linear} = \chi^{(1)} E$, where $\chi^{(1)}$ is the susceptibility tensor. The net polarization in the medium under the influence of several fields is a linear superposition of the effects of the same fields acting independently. At high intensities the optical fields can interact with each other through the higher order terms in the polarization vector. P can now be expanded as $\chi^{(1)} E + \chi^{(2)} E^* E + \chi^{(3)} E^* E^* E + ...$ The nonlinear susceptibilities, $\chi^{(2)}$, $\chi^{(3)}$, etc., of the medium represent the magnitudes of the higher order interactions with the applied fields. These properties are dictated by features at the molecular level; including electronic and molecular structure and the nature of the overall molecular packing.

Extensive studies have been made on poled NLO guest-host systems[5-7]. The limited solubility of the NLO species in a host matrix and the thermal relaxation of the induced nonlinear optical activity of the poled polymeric films are major disadvantages. Side chain polymers have attracted attention since a large number of NLO molecules may be covalently attached to the polymer chain and the problem of phase segregation of the NLO component is alleviated[8]. In this article, we present the second and third order NLO properties of a new class of guest-host system which possess excellent doping features and photoreactive characteristics.

Contemporary Topics in Polymer Science, Vol. 7., Edited by
J.C. Salamone and J. Riffle, Plenum Press, New York, 1992 ·k, 1992

237

MATERIALS REQUIREMENTS

For second order NLO properties, a noncentrosymmetric organization is essential. This is not necessary for third order NLO properties. In addition to large optical nonlinearity of the material, the ease of processing is also important from the standpoint of device fabrication. Strong electron donor and acceptor groups in a π-conjugated structure has been used to promote intramolecular charge transfer, which is important for large second order hyperpolarizability. These molecular systems can be incorporated into a host polymer for effective processing. Noncentrosymmetric alignment of the NLO species in the guest-host systems can be introduced by electric field induced poling[9]. Third order nonlinear susceptibility of guest-host systems incorporating the conjugated molecules is also of current interest because of the possibility of making low loss guided wave devices from these materials. The origin of large third order nonlinearity lies in the extended π-conjugation effect[3].

Design of nonlinear optical polymers based on the guest-host approach calls for molecular structures that, in addition to large molecular hyperpolarizability, will also be compatible with the host polymer. Any phase segregation of the NLO species will adversely effect the optical quality of the polymeric film. Transparency at the wavelength of operation is also desired. For second order application the NLO species are poled, resulting in a net polarization for the medium. In the poling process, a noncentrosymmetric organization is produced. In addition, the alignment yields the largest second order nonlinear coefficient in the direction of the poling field. Largest component of the molecular hyperpolarizability is usually along the molecular axis.

MATERIALS AND METHODS

Within the scheme of a guest-host system, suitable combinations of guest and host molecules have led to higher NLO concentrations and more stable NLO response. Copolymers such as styrene acrylonitrile (SAN) have been found to be better hosts compared to conventional glassy homopolymers such as polystyrene (PS) or polymethyl methacrylate (PMMA)[10]. Higher concentrations of polar dyes could be incorporated prior to phase segregation and the poled organization possessed greater temporal and thermal stability (Figure 1).

In our laboratory, a new route to achieve a stable crosslinked NLO polymer system by photochemical reaction has been developed. This class of guest-host systems has been designed from first principal to be compatible leading to extremely high levels of dye loading without phase segregation. Large third and second order nonlinear susceptibilities have been measured for the guest-host unpoled and poled crosslinked systems, respectively. Approaches to fabrication of nonlinear optical devices have also been developed which takes advantage of the excellent processability and photocrosslinking behavior of the polymers[11].

A donor-acceptor substituted azo dye was selected as the basic NLO species. These dyes have been reported to possess some of the largest molecular hyperpolarizabilities[12]. The NLO dye was functionalized at two sites using photoreactive chromophores.

Cinnamoyl (C_6H_5-CH=CH-CO-) and styryl-acryloyl (C_6H_5-CH=CH-CH=CH-CO-) groups were selected as the desired photoreactive groups. The phenyl rings may be further suitably derivatized to modify the spectral features, photosensitivity and photoselectivity. An example of this type of NLO molecules is 2,2'-bis(cinnamoyloxy) -4-diethylamino-4'-nitroazobenzene (CNNB-R) (Figure 2). Clearly, many other variations in structure may be introduced to carry out spectral tuning, optimization of the NLO properties, conformational features and the overall stability of the molecule[13].

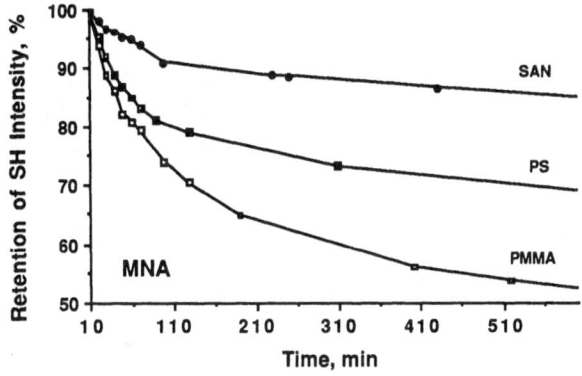

Figure 1. Temporal behavior of 2-methyl-4-nitroaniline (MNA) molecules in homo- and copolymers.

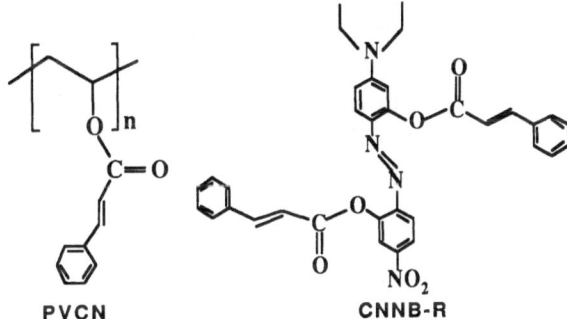

Figure 2. Structures of new photocrosslinkable guest-host system.

The desired host polymer should contain a large number of the same photoreactive group. It should have excellent film forming properties and preferably be glassy with a moderately high T_g to facilitate processing at not too high a temperature. Its overall compatibility with the guests should be helped by the presence of similar functional groups, but is not assured.

The simplest and commercially available polymer polyvinylcinnamate (PVCN) was chosen as a candidate host polymer (Figure 2). This polymer is synthesized by esterification of polyvinyl alcohol with cinnamoyl chloride[14]. A PVCN with a large degree of derivatization (>70%) is appropriate for effective solid state photoreaction process. Compatibility of this polymer among other parameters will be dominated by the interaction of the photoreactive groups with similar groups in the guest. The polymer backbone is relatively flexible and the guest dye has an articulated structure.

For third order nonlinear application, however, it is the concentration of the NLO dye that is of paramount importance, not the orientational aspects. The third order response is typically dependent on the extent and nature of electronic conjugation. Polydiacetylene, for example, shows very high $\chi^{(3)}$ because of the delocalized electronic structure along the π-conjugated polymer chains[15]. The large dye molecules posses quite reasonable molecular hyperpolarizability. A high concentration of these dyes without a concomittant deterioration of linear optical properties can lead to useful bulk third order NLO coefficients.

Processing

Thin films may be prepared by spin coating from a solution of PVCN and CNNB-R in toluene/chlorobenzene (1:3 v/v) mixture. 100 parts of PVCN and 0, 10, 20, 30, 40, 50 parts of CNNB-R by weight were made into a saturated solution. The solutions were filtered through a 1 μm membrane filter and then spun on glass or quartz plates at 3000 rpm, resulting in film of thickness from 0.5 - 1 μm as measured by ellipsometry. Films were dried at 60° - 70° C in a vacuum oven overnight.

Linear Spectroscopy

Thin films spin coated on quartz plates are used for UV-visible spectroscopy. The absorbance from the films at the characteristic wavelength at which the dye absorbs scales linearly with dye concentration, establishing a homogeneous distribution. UV-visible spectrum further establishes the desired wavelength for most effective photocrosslinking[16]. A small concentration (<1%) of conventional sensitizer such as coumarin or thiazoline dye can also be used to effectively shift the crosslinking radiation to longer wavelengths. Practically, any wavelength light from UV to mid visible is adequate for crosslinking in conjunction with an appropriate sensitizer.

The optical constant such as refractive index, and the thickness of a film are determined by ellipsometry. For polymer thin films the reflection ellipsometry technique is commonly used[17]. This technique is especially useful in the wavelength regions where the materials are strongly absorbing so that the transmission measurements are precluded.

The thickness and the refractive index of the polymer film can also be measured more precisely by optical waveguiding techniques. The planar wave guide technique in which the light is confined in one direction only is widely used for this purpose[18]. Table 1 summarizes some of the linear optical properties of photoreactive polymer films.

Poling and Photocrosslinking

Corona poling technique in the wire to plane geometry as well as parallel electrode poling techniques have been used to pole these classes of polymers in film form. Poling is carried out prior to crosslinking to align the NLO units and possibly poise the photoreactive groups for photo-induced solid state reaction.

The poling temperature was set about 70° C, which is 10° C below T_g of the uncrosslinked guest-host system. UV irradiation from a mercury lamp at 254 nm.

producing an intensity of 2 mW/cm^2 on the film surface was used for crosslinking. The optimum exposure time was established from the UV-visible spectra of the doped polymer films.

Photocrosslinking may be carried out by exposure to UV radiation. Crosslinks are formed by 2 + 2 photodimerization between an excited cinnamoyl group with another in the ground state. It is expected that a sufficient number of these pairs will belong to the same or different molecules to form a crosslinked network. The intermolecular crosslinking reaction between the photosensitive chromophores is represented in Figure 3.

Second Order Properties

The second order nonlinear susceptibility of the polymer has been measured by comparing the second harmonic intensity from the polymer films with that of Y-cut crystal quartz. The stability of the second harmonic signal with time after poling and crosslinking have been observed with PVCN doped with 20 % by weight of CNNB-R (Figure 4). The results clearly indicate that the poled and crosslinked polymer is quite stable in its second harmonic signal, while the polymer film which has not been crosslinked during the poling process shows substantial decay in SHG signal with time. The same UV cured polymer shows no decay in SHG intensity upon heating at 80°-85°C for long periods of time. Linear electrooptic coefficient of doped poled and crosslinked films have been measured. Second order coefficient, d_{33}, of up to 30 pm/V have been measured for these guest-host systems. The results of some of these measurements are summarized in Table 1.

Thermal Properties

Thermal analysis techniques were used to investigate thermal stability and relaxation behavior in the crosslinked and uncrosslinked samples. To measure T_g and ΔC_p at T_g upon photocrosslinking, a DSC set up has been used. DSC curves of PVCN and PVCN/20 % CNNB-R after different periods of crosslinking radiation are shown in Figure 5. As seen in Figure 6, UV-radiation does not lead to a large increase in T_g. UV exposure results in to extensive crosslinking and increase in the beam penetration depth. In the thin films unreacted sample material is decreased. As the material is completely crosslinked the medium loses the glass rubber transition.

Third-order nonlinear optical properties

Third-order nonlinear optical properties (namely, quadratic electrooptic coefficient) of this polymeric system at different dye concentrations have been measured using a modified Michelson interferometer[19] at 633nm. For this measurement, the polymer film was spin-coated from the filtered solution on an indium-tin-oxide (ITO) coated glass plate where the ITO layer was used as an electrode. A gold layer of 500Å thickness was thermally evaporated onto the polymer film to form the second electrode. The experimental set-up of this modified Michelson interferometer is schematically shown in (Figure 7).

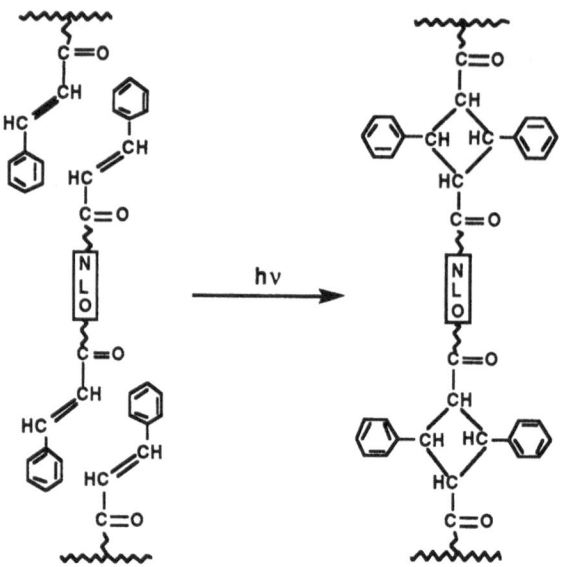

Figure 3. One of the photocrosslinking mechanisms in guest-host system.

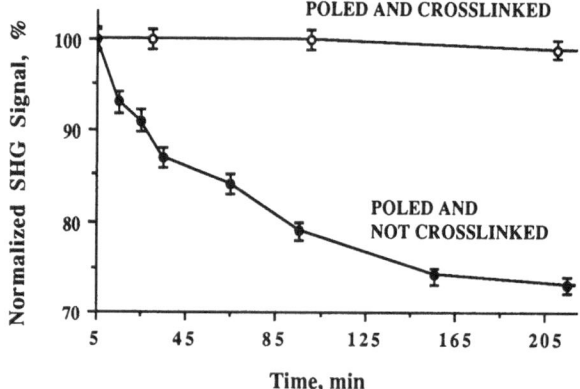

Figure 4. Stable NLO behavior of photocrosslinkable guest-host system.

Table 1. Properties of photocrosslinkable guest-host system.

PVCN/CNNB-R (%)	(10%)	(20%)
Thickness (μm)	0.5	0.5
Abs. max (nm)	520	520
T_g (°C)[a]	84.3	81.0
Refractive index		
λ (μm)		
0.532	1.632	1.634
0.632	1.677	1.685
1.000	1.613	1.625
d_{33} (pm/V)		
1.063μm	11.5	21.5
1.540μm	3.7	5.1

[a]Obtained from DSC (DuPont 2910 differential scanning calorimeter), 10°C/min (midpoint). T_g of PVCN is 88.1°C.

242

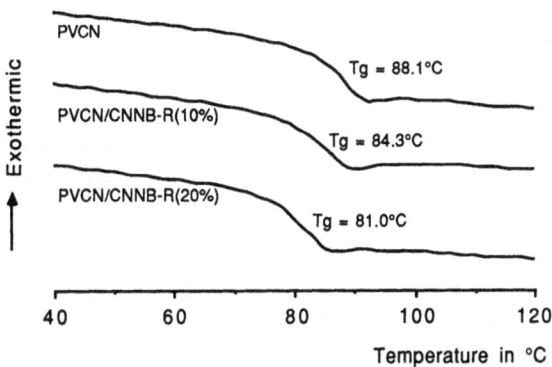

Figure 5. DSC curves of PVCN and PVCN doped with CNNB-R.

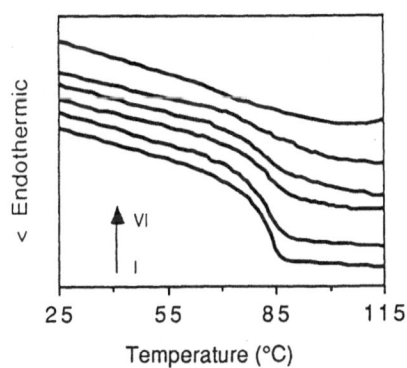

Figure 6. DSC curves of PVCN films doped with 10% CNNB-R. Crosslinking was performed by UV radiation at wavelength 254 nm; I=0, II=1, III=3, IV=5, V=10, VI=30 min respectively.

A HeNe laser beam is split into a sample beam and a reference beam. The sample beam propagates via the ITO coated glass and the polymer film, and is then reflected back by the gold electrode. An AC-modulating field (about 5 kHz) was applied across the polymer film. A detector was employed to measure the modulations at two half-intensity points of a selected interference fringe at 2ω, noted as $I_{2\omega}^+$ and $I_{2\omega}^-$. Then the quadratic electrooptic (E-O) coefficient, S_{1133}, as a measure of the third order optical nonlinearity is given by:

$$S_{1133} = (1/\pi n^3) [(I_{2\omega}^+ - I_{2\omega}^-)/(I_{max} - I_{min})] (\lambda d / V_m^2)$$

where n is the refractive index, λ is the laser wavelength, d is the thickness of the film, V_m is the modulating voltage, and I_{max} and I_{min} are the maximum and minimum intensities of the selected fringe, respectively.

The measured quadratic E-O coefficient as a function of dye concentration is shown in Figure 8. An approximately linear relationship exists between them, which indicates no aggregation in this dye doped polymer system. The quadratic E-O coefficients in this system are large compared with those of our former systems[20] based on azomethine dyes and those reported recently by Kuzyk et al[21] at the same dye concentrations.

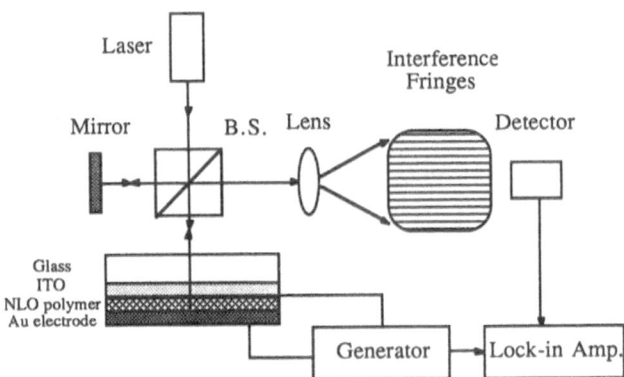

Figure 7.   Experimental set-up for the measurement of quadratic electrooptic coefficient.

In addition to the quadratic E-O coefficient measurement, the linear optical properties of those films were measured using an ellipsometer and a UV-visible spectrophotometer. The measured refractive index (at 633 nm) and normalized peak absorbance of those films are also approximately linear with the dye concentration (up to 33 wt%), which are shown in Figures 9a and 9b, respectively. These results also show that no phase segregation occurred in this dye doped polymer system for dye concentration up to 33 wt%.

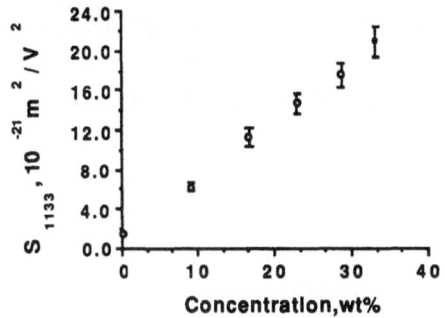

Figure 8. Effect of quadratic E-O coefficient with
the concentration on the NLO dye.

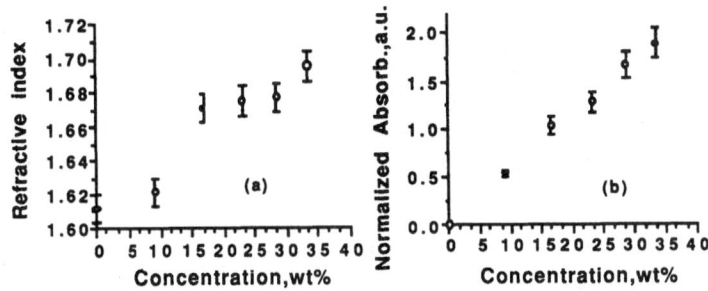

Figure 9. Linear optical properties of dye doped polymer film. (a) Change in
refractive index with increasing dye concentration. (b) Change in
normalized absorbance with increasing dye concentration.

ACKNOWLEDGEMENT

Partial support from the Office of Naval Research is greatfully acknowledged.

REFERENCES

1. "Nonlinear Optical Properties of Organic and Polymeric Materials", D. J. Williams, ed., ACS Symp. Ser. No. 233, ACS, Washington, D.C. (1983).
2. H. Gibbs, "Optical Bistability", Academic Press, New York (1987).
3. "Nonlinear Optical and Electroactive Polymers", P. N. Prasad and D. R. Ulrich, eds., Plenum, New York (1988).
4. "Nonlinear Optical Properties of Organic Materials III", G. Khanarian, ed., SPIE, San Diego, California (1990).
5. K. D. Singer, J. E. Sohn and S. J. Lalama, Appl. Phys. Lett., 49(5), 248 (1986)
6. S. Matsumoto, K. Kubodera, T. Kurihara and T. Kaino, Appl. Phys. Lett., 51, 1 (1987).

7.   H. L. Hampsch, J. Yang, G. K. Wong and J. M. Torkelson, Macromolecules, 23, 3640 (1990).

8.   K. D. Singer, M. G. Kuzyk, W. R. Holland, J. E. Sohn, S. J. Lalama, R. B. Comizzoli, H. E. Katz and M. L. Schiling, Appl. Phys. Lett., 52, 1800 (1988).

9.   M. A. Mortazavi, A. Knoesen, S. T. Kowel, B. G. Higgins and A. Dienes, J. Opt. Soc. Am., B (6), 733 (1989).

10. B. K. Mandal, Y. M. Chen, R. J. Jeng, T. Takahashi, J. C. Huang, J. Kumar and S. Tripathy, Eur. Polym. J., May (1991).

11. P. Miller, J. Y. Lee, R. J. Jeng, W. H. Kim, B. K. Mandal, S. Tripathy and J. Kumar, submitted to Appl. Phys. A.

12. K. D. Singer, M. G. Kuzyk and J. E. Sohn, J. Opt. Soc. Am. B (4), 968 (1987).

13. B. K. Mandal, Y. M. Chen, J. Y. Lee, J. Kumar and S. Tripathy, Appl. Phys. Lett. 58 (22) 3 June (1991).

14. A. Reiser, "Photoreactive Polymers", John Wiley & Sons, New York (1989).

15. P. A. Chollet, F. Kajzor and J. Messier, "Nonlinear Optics of Organics and Semiconductors", T. Kobayashi, ed., Springer-Verlag, Berlin (1989).

16. B. K. Mandal, J. Y. Lee, X. F. Zhu, Y. M. Chen, E. Prakeenavincha, J. Kumar and S. Tripathy, Synth. Met. 43 (1-2) 2803, June 7 (1991).

17. R. M. A. Azzam and N. M. Bashara, "Ellipsometry and Polarized Light", North-Holland, New York (1986).

18. R. Ulrich and R. Tonge, Appl. Opts., 12, 2901 (1973).

19. X. F. Zhu, Y. Chen, B. K. Mandal, R. J. Jeng, J. Kumar and S. Tripathy, Opt. Commun. (submitted).

20. J. Kumar, A. K. Jain, M. Cazeca, J. Ahn, R. S. Kumar and S. K. Tripathy, SPIE, 1147, 177 (1989).

21. M. G. Kuzyk, J. E. Sohn and C. W. Dirk, J. Opt. Soc. Am. B (6), 842, (1990).

MOLECULAR ENGINEERING OF SIDE CHAIN LIQUID CRYSTALLINE POLYMERS

Virgil Percec and Dimitris Tomazos

Department of Macromolecular Science
Case Western Reserve University
Cleveland, Ohio 44106

SOME GENERAL CONSIDERATIONS ON SIDE CHAIN LIQUID CRYSTALLINE POLYMERS

The field of side chain liquid crystalline polymers was recently reviewed.[1] Therefore, this paper will discuss only recent progress made on their molecular engineering mainly by living polymerization reactions. Most of the present discussion will be made on side chain liquid crystalline polymers with mesogenic groups normally attached to the polymeric backbone.[2] Figure 1 outlines the concept of side chain liquid crystalline polymers. It has been theoretically predicted[3] that the conformation of the polymer backbone should get distorted in the liquid crystalline phase. Both small-angle neutron scattering (SANS) experiments[4-8] and X-ray scattering experiments,[9-11] have shown that the statistical random-coil conformation of the polymer backbone is slightly distorted in the nematic phase and highly distorted in the smectic phase.

Let us first consider very briefly the influence of various parameters (i.e., nature of flexible spacer and its length, nature and flexibility of the polymer backbone and its degree of polymerization) on the phase behavior of a side chain liquid crystalline polymer. According to some thermodynamic schemes which were described elsewhere,[12,13] the increase of the degree of polymerization decreases the entropy of the system and therefore, if the monomeric structural unit exhibits a virtual or monotropic mesophase, the resulting polymer should most probably exhibit a monotropic or enantiotropic mesophase. Alternatively, if the monomeric structural unit displays an enantiotropic mesophase, the polymer should display an enantiotropic mesophase which is broader. It is also possible that the structural unit of the polymer exhibits more than one virtual mesophase and therefore, at high molecular weights the polymer will increase the number of its mesophases. All these effects were observed with various polymer systems.[2]

The length of the flexible spacer determines the nature of the mesophase. Long spacers favor smectic phases while short spacers favor nematic phases. This effect is similar to that observed in low molar mass liquid crystals.

At constant molecular weight the rigidity of the polymer backbone determines the thermodynamic stability of the mesophase. According to the thermodynamic schemes described previously[12,13] the isotropization temperature of the polymer with more rigid backbone should be higher. However, experimentally this situation is reversed. The highest isotropization transition temperature is observed for polymers with more flexible backbones. This conclusion is based on systematic investigations performed with two mesogenic groups which are constitutional isomers i.e., 4-methoxy-4'-hydroxy-α-methylstilbene (4-MHMS)[14] and 4-hydroxy-4'-methoxy-α-methylstilbene (4'-MHMS)[15] and polymethacrylate, polyacrylate, polysiloxane and polyphosphazene backbones[14-16] (Scheme 1). This dependence can be explained by assuming that a more flexible backbone uses less energy to

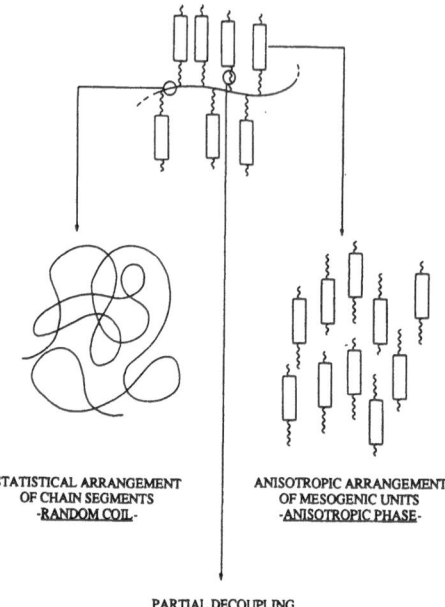

Figure 1. Schematic representation of side chain liquid crystalline polymers showing the necessity of decoupling the mesogenic groups and the polymer backbone through flexible spacers.

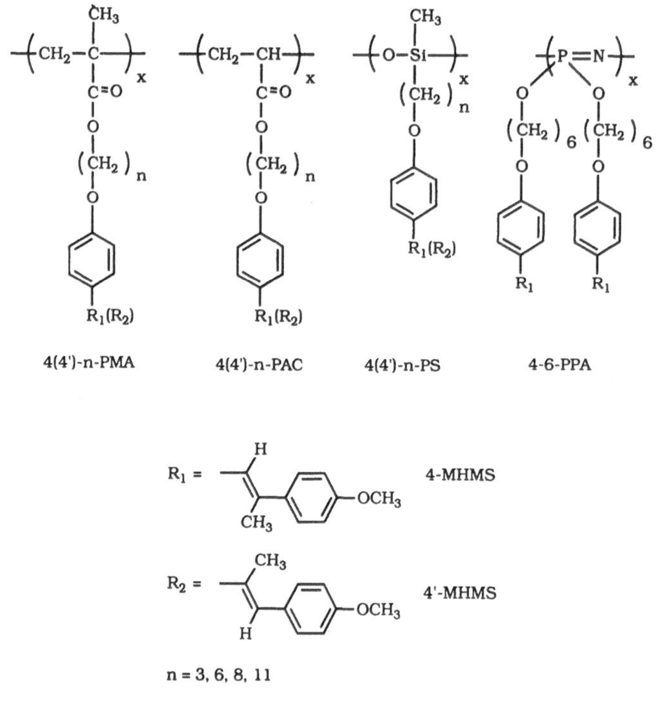

4(4')-n-PMA 4(4')-n-PAC 4(4')-n-PS 4-6-PPA

R₁ = 4-MHMS

R₂ = 4'-MHMS

n = 3, 6, 8, 11

Scheme 1. Polymethacrylates, polyacrylates, polysiloxanes and polyphosphazene based on 4-MHMS and 4'-MHMS constitutional isomeric mesogenic groups.

get distorted and therefore, generates a more decoupled polymer system. In fact the more flexible backbones do not generate only higher isotropization temperatures but also a higher ability towards crystallization. However, contrary to all expectations the entropy change of isotropization is higher for those polymers which are based on more rigid backbones and therefore, exhibit lower isotropization temperatures (Figure 2a,b).[17] This contradiction between the values of the entropy change and the isotropization temperatures can be accounted for by a different mechanism of distortion of different polymer backbones as outlined in Figure 3. That is, while a rigid backbone gets more extended and therefore, in the smectic phase it can cross the smectic layer, in the case of a flexible backbone it gets squeezed between the smectic layers. The higher configurational entropy of the flexible backbone versus that of the rigid backbone in the smectic phase can account for the difference between the entropy change of isotropization from Figure 2. At shorter spacer lengths, there is not much difference between the contribution of various backbone flexibilities since most probably, in order to generate a mesophase they should get extended. Therefore the entropy change of isotropization is less dependent of backbone flexibility (Figure 2).

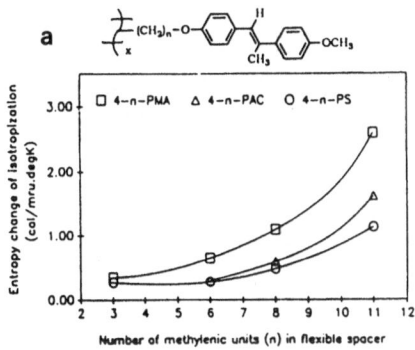 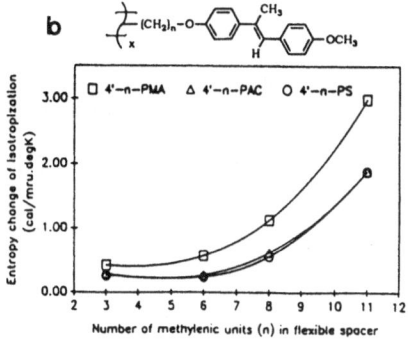

Figure 2. a) The dependence between the entropy change of isotropization (ΔS_i) determined from the cooling DSC scans, the nature of the polymer backbone and the number of methylenic units (n) in the flexible spacer for the series of polymers based on 4-MHMS isomer.

b) The dependence between the entropy change of isotropization (ΔS_i) determined from the cooling DSC scans, the nature of the polymer backbone and the number of methylenic units (n) in the flexible spacer for the series of polymers based on 4'-MHMS isomer.

Based on this discussion it is quite obvious that copolymers containing structural units with and without mesogenic groups and flexible backbone display a microphase separated morphology in their smectic phase (Figure 4).[2,9,11] Therefore, the highest degree of decoupling is expected for copolymers containing mesogenic and nonmesogenic structural units and highly flexible backbones, i.e., microphase separated systems. In this last case, when the monomeric structural unit of the polymer exhibits a virtual mesophase, the high molecular weight polymer might also display only a virtual or a monotropic mesophase. The transformation of a virtual and/or monotropic mesophase of the homopolymers into an enantiotropic mesophase can be most conveniently accomplished by making copolymers based on two monomers which are constitutional isomers, such as monomers based on 4-MHMS and 4'-MHMS.[18,19] Since the structural units of the homopolymers based on 4-MHMS and 4'-MHMS are isomorphic within their liquid crystalline phase, but not within their crystalline phase, the crystalline melting transition decreases while the mesophase exhibits a continuous almost linear dependence on composition. As a consequence, the virtual or monotropic mesophase of the homopolymer becomes enantiotropic.[18,19] Finally, the molecular weight at which the isotropization temperature becomes independent of molecular weight should be, and indeed is, dependent on the flexibility of the polymer backbone. For example, the isotropization temperature of polysiloxanes[20,21] containing mesogenic side groups is molecular weight dependent up to much higher molecular weights than that of polymethacrylates containing mesogenic side groups.[2]

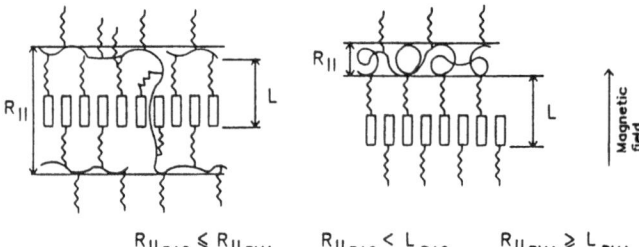

a Theoretical (M. Warner)

b Experimental (Saclay Group)

$$R_{|| PAC} \lessgtr R_{|| PMA} \qquad R_{|| PAC} < L_{PAC} \qquad R_{|| PMA} \gtrless L_{PMA}$$

Figure 3. a) Schematic representation of the theoretical distortion of the statistical random-coil conformation of the polymer backbone in the nematic and smectic phases;
b) Two possible modes of distortion of the random-coil conformation of a rigid (left) and a flexible (right) polymer backbone. $R_{/\!/}$ and $R_{\perp}$ refer to the radius of gyration parallel and perpendicular to the magnetic field, respectively.

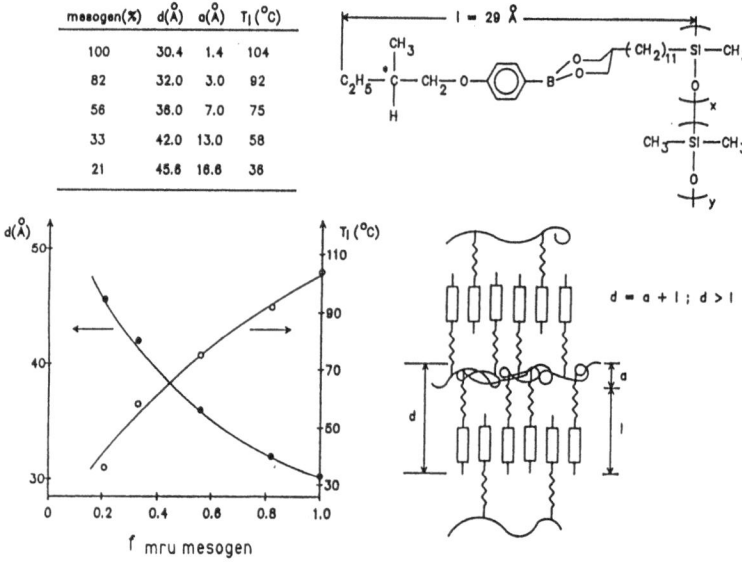

mesogen(%)	$d(\overset{\circ}{A})$	$a(\overset{\circ}{A})$	$T_i(^\circ C)$
100	30.4	1.4	104
82	32.0	3.0	92
56	38.0	7.0	75
33	42.0	13.0	58
21	45.6	16.6	36

$d = a + l \; ; \; d > l$

Figure 4. Microphase separated morphology of smectic copolymers.

Based on this discussion and on the thermodynamic discussion described previously,[12,13] we can easily consider that the "polymer effect" can provide via its molecular weight and backbone flexibility the same effect. In an oversimplified way it can be considered that it provides an overall change in the entropy of the system. Through this change, it can transform, in a reversible way, a virtual mesophase into a monotropic and subsequently into an enantiotropic one. In addition, the kinetic factors provided by the glass transition and crystallization should always be considered. For example, the formation of a mesophase located in the close proximity of a glass transition temperature becomes kinetically controlled or even can be kinetically prohibited.

MOLECULAR ENGINEERING OF LIQUID CRYSTALLINE POLYMERS BY LIVING POLYMERIZATION

General Considerations

Several polymerization methods were investigated in order to develop living polymerization procedures for the preparation of side chain liquid crystalline polymers with well defined molecular weight and narrow molecular weight distribution. They include cationic polymerization of mesogenic vinyl ethers,[22,23] cationic ring opening polymerization of mesogenic cyclic imino ethers,[24] group transfer polymerization of mesogenic methacrylates,[25-28] and polymerization of methacrylates with methylaluminium porphyrin catalysts.[29] Cationic polymerization has been proved to be the most successful since it can be used to polymerize under living conditions mesogenic vinyl ethers containing a large variety of functional groups.[30-51] Scheme 2 provides some representative examples of mesogenic vinyl ethers which could be polymerized by a living mechanism with our preferred initiating system (i.e., CF_3SO_3H, $(CH_3)_2S$, CH_2Cl_2, 0 °C).[52] As we can observe from Scheme 2, vinyl ethers containing nucleophilic groups such as methoxybiphenyl,[44] electron-withdrawing groups such as cyanobiphenyl,[30-35,44,45] nitrobiphenyl and cyanophenylbenzoate,[44] double bonds like in 4-alkoxy-α-methylstilbene,[46] double bonds and cyano groups like in 4-cyano-4'-α-cyanostilbene,[34] aliphatic aromatic esters,[36] acidic protons and perfluorinated groups,[44,48] oligooxyethylene and aromatic ester groups,[49] crown

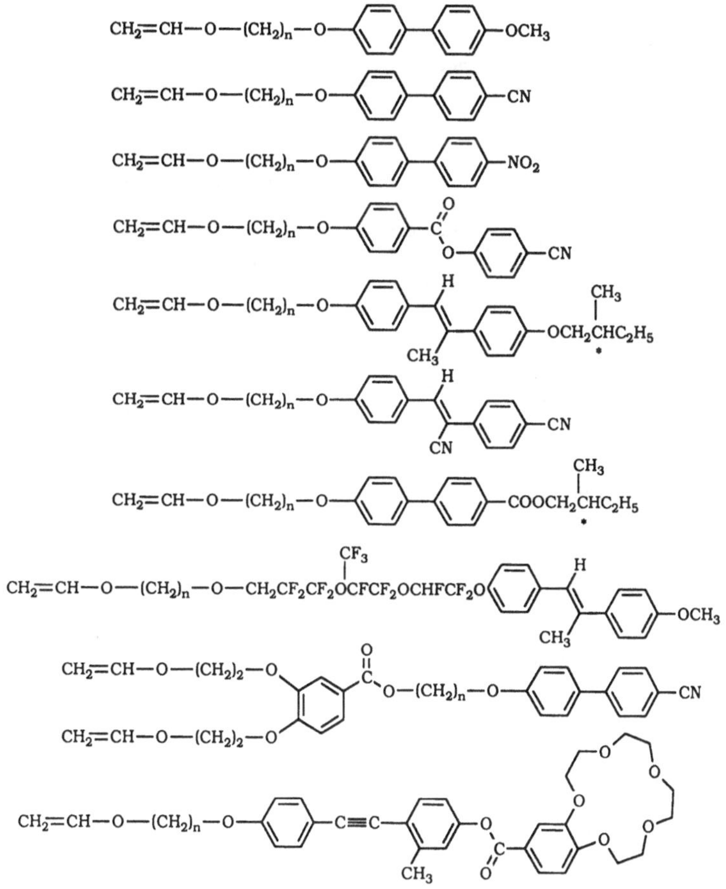

Scheme 2. Representative examples of mesogenic vinyl ethers which can be polymerized by living cationic mechanism.

ethers and triple bonds,[50] all can be polymerized by a living cationic mechanism. In addition, cationic polymerization of any of these monomers can be performed in melt phase either in liquid crystalline phase or in isotropic phase by using thermal[53], or photo cationic initiators.[54,55] When the polymerization is performed in liquid crystalline phase with alligned films of liquid crystalline monomers, perfectly alligned single crystal liquid crystalline polymer films are obtained.[54,55] In the following two subchapters we will discuss two topics. The first one refers to the influence of molecular weight on the phase transitions of poly{ω-[(4-cyano-4'-biphenylyl)oxy]alkyl vinyl ether}s with alkyl groups containing from four to eleven methylene units. In the second one we will demonstrate the molecular engineering of phase transitions of side chain liquid crystalline polymers by azeotropic living copolymerization experiments.

<u>Influence of Molecular Weight on the Phase Transitions of Poly{ω-[(4-cyano-4'-biphenylyl)oxy]alkyl vinyl ether}s</u>

Scheme 3 outlines the general method used for the synthesis of ω-[(4-cyano-4'-biphenylyl)oxy]alkyl vinyl ethers (<u>6-n</u>) and of the model compound for the polymer with degree of polymerization of one i.e., ω-[(4-cyano-4'-biphenylyl)oxy]alkyl ethyl ethers (<u>8-n</u>). We will use over the entire discussion the same short notations as in the original publications. The synthesis and characterization of poly(<u>6-n</u>) and (<u>8-n</u>) with n = 2, 3, 4,[31] 5, 7,[32] 6, 8,[45] 9, 10,[33] and 11[30] will be briefly discussed. Details are available in the original publications. All polymers have polydispersities of about 1.10. Scheme 4 outlines the polymerization mechanism and the structure of the resulted polymers. This structure was confirmed by 300 MHz 1-D and 2-D ^{1}H-NMR spectroscopy.[56]

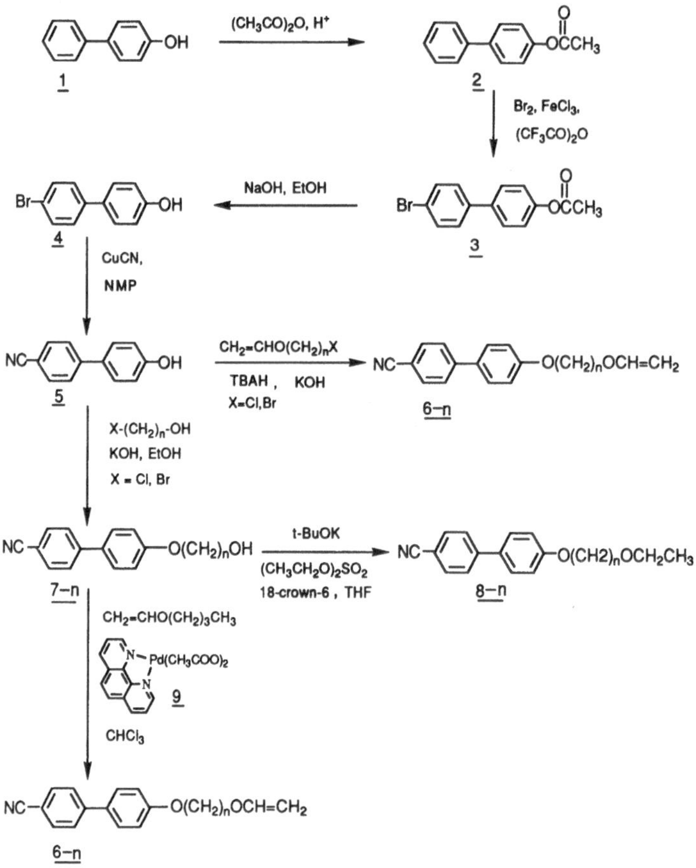

Scheme 3. Synthesis of ω-[(4-cyano-4'-biphenylyl)oxy]alkyl vinyl ethers (<u>6-n</u>).

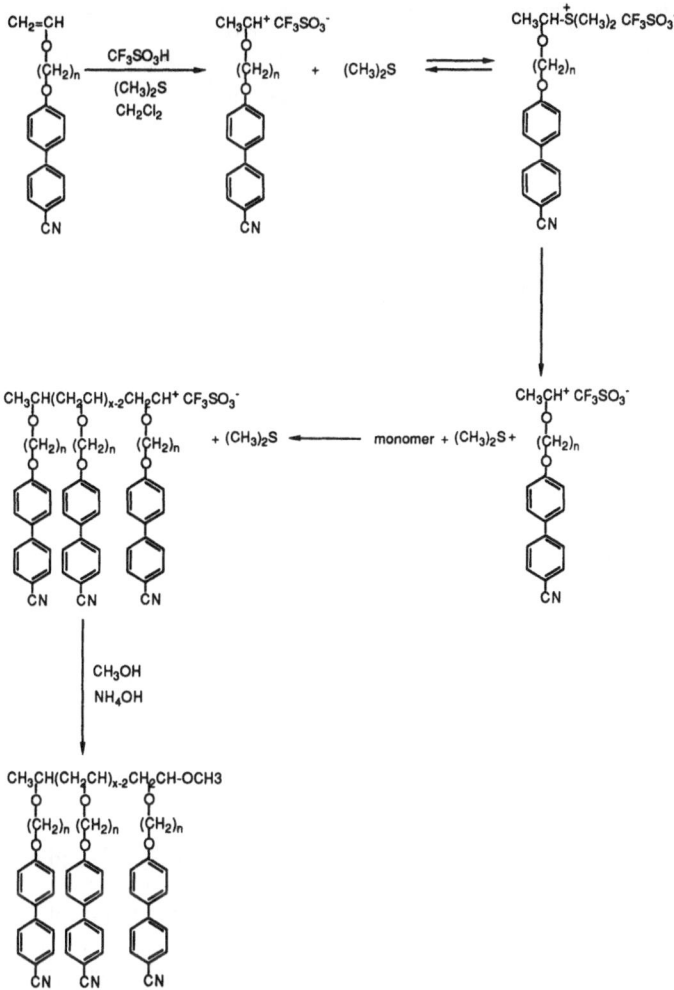

Scheme 4. Mechanism of living cationic polymerization of ω-[(4-cyano-4'-biphenylyl)oxy]alkyl vinyl ethers (6-n).

All data were classified according to their similarities. Figure 5 presents the dependence of phase transition temperatures of poly(6-n) with n = 3, 4, 7 and 9 as a function of molecular weight. These data were collected from second heating scans. The data for 8-n are not plotted . 8-3 is crystalline, 8-4 and 8-7 exhibit a monotropic nematic mesophase while 8-9 monotropic nematic and smectic mesophases. As we can observe from Figure 5 by increasing the molecular weight all four polymers show a broadening of the thermal stability of their mesophase. The mesophase of 8-7 and 8-9 changes from nematic to s_A by increasing the degree of polymerization from one to about 3.

Figure 6 presents similar data for poly(6-2), poly(6-6) and poly(6-8). In all cases the nature of the mesophase is molecular weight dependent. Poly(6-2) has a nematic mesophase only at degrees of polymerization lower than 5. 8-2 is only crystalline. At degrees of polymerization higher than 5, poly(6-2) is only glassy. This is because its glass transition temperature becomes higher than the isotropization temperature and therefore, the mesophase is kinetically prohibited. 8-6 exhibits an enantiotropic nematic mesophase. At low degrees of polymerization poly(6-6) and poly(6-8) exhibit nematic and s_A mesophases. Due to the difference between the slope of the dependences of the nematic phase transition temperature on molecular weight and of the s_A phase transition temperature on molecular weight, above a certain molecular weight the nematic phase disappears. Both poly(6-6) and poly(6-8) show a second smectic mesophase (s_X, i.e. unassigned). Qualitatively, this behavior is in agreement with the influence of molecular weight on phase transitions predicted by thermodynamics.[12,13] Quantitative predictions of these phase diagrams require more theoretical research.

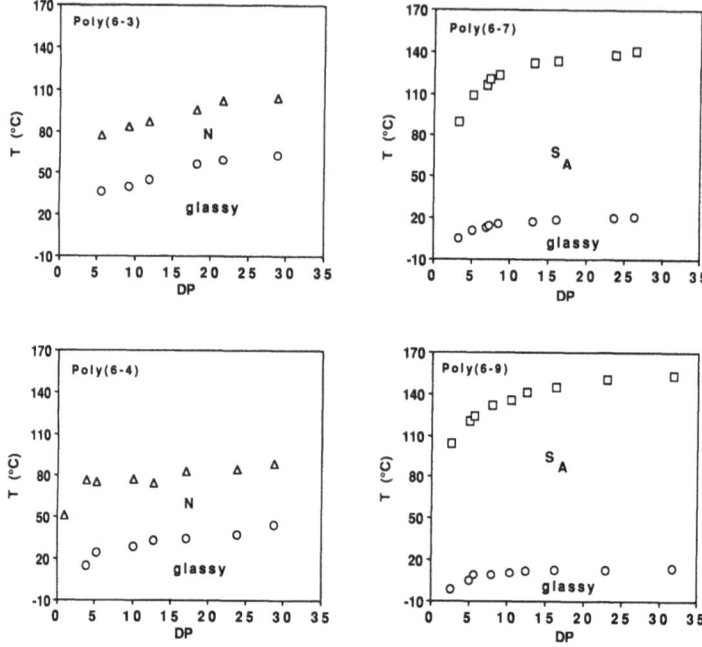

Figure 5. The influence of molecular weight on the phase behavior of poly(6-3), poly(6-4), poly(6-7) and poly(6-9) (determined from second DSC heating scans).

Finally, Figure 7 presents the behavior of poly(6-5), poly(6-10) and poly(6-11). 8-5 shows an monotropic nematic phase, 8-10 a monotropic s_A phase while 8-11 an enantiotropic s_A mesophase. Poly(6-5) exhibits above a degree of polymerization of 10 the unusual sequence isotropic-nematic-s_{Ad}-n_{re}-glassy.[57] This will be discussed in more detail in a subsequent subchapter. At high molecular weights poly(6-10) and poly(6-11) exhibit s_A and s_X phases.

As a general observation we can mention that polymers with short spacers (n = 2, 3, 4) and medium length spacers containing an odd number of methylene units (n = 7, 9) do not generate polymorphism at different molecular weights. Polymers with medium length and an even number of methylene units (n = 6, 8), as well as polymers with long length with both even and odd numbers of methylenic units (n = 10, 11) generate a rich polymorphism which is molecular weight dependent. The borderline polymer is poly(6-5) which is the only one displaying n and s_A mesophases over a broad range of molecular weights and therefore, also generates the reentrant nematic mesophase.[57]

Molecular Engineering of Liquid Crystalline Phases by Living Cationic Copolymerization

In order to tailor make mesophases of side chain liquid crystalline copolymers we first need to synthesize copolymers with constant molecular weight and controllable composition. Copolymer composition is conversion dependent in all statistic copolymerizations. The only exception is provided by azeotropic copolymerizations in which the copolymer composition is identical to the monomer feed at any conversion.[58] This situation is provided by monomers with $r_1 = r_2 = 1$. Since the reactivity of the polymerizable vinyl ether groups is not spacer length dependent, all 6-n monomers have the same reactivity. Therefore, all 6-n pairs of monomers lead to azeotropic copolymerizations, and when the copolymerization is performed under living conditions they lead to copolymers with controllable molecular weight. The azeotropic copolymerization of various pairs of 6-n monomers is outlined in Scheme 5. We will discuss selected examples of copolymers prepared from monomer pairs which give rise to homopolymers exhibiting nematic and nematic, s_A and s_A, nematic and s_A, and glassy and s_A phases as their highest temperature mesophases.

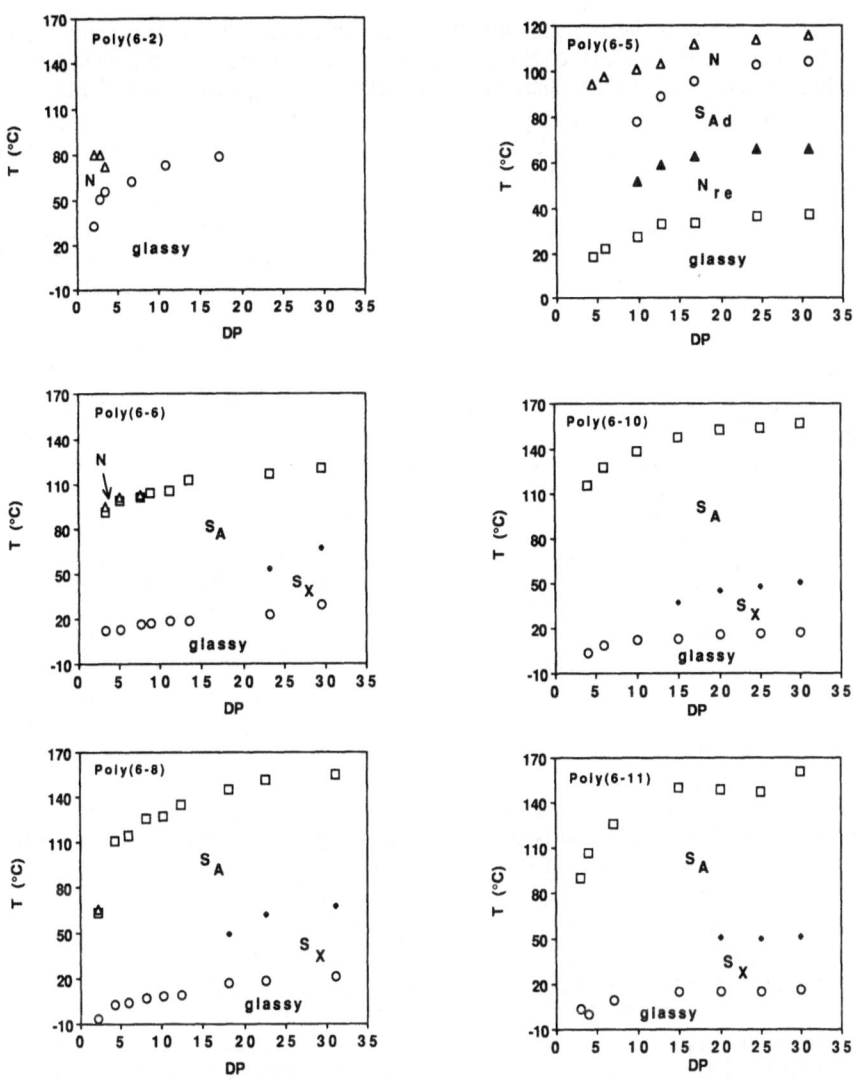

Figure 6. The influence of molecular weight on the phase behavior of poly(6-2), poly(6-6) and poly(6-8) (determined from second DSC heating scans).

Figure 7. The influence of molecular weight on the phase behavior of poly(6-5), poly(6-10) and poly(6-11) (determined from second DSC heating scans).

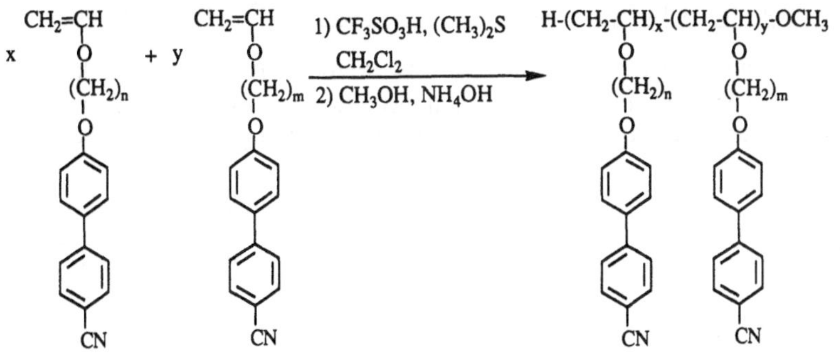

Scheme 5. Azeotropic copolymerization of ω-[(4-cyano-4'-biphenylyl)oxy]alkyl vinyl ether (6-n) monomer pairs.

Figure 8 presents the dependence of phase transition temperatures obtained from second DSC heating scans (a,d), cooling scans (b, e) and the enthalpy changes associated with the highest temperature mesophase of copolymers poly[(6-3)-co-(6-5)]X/Y and poly[(6-6)-co-(6-11)]X/Y. The degrees of polymerization of all copolymers are equal to 20.[59] Copolymers poly[(6-3)-co-(6-5)]X/Y are based on a monomer pair which gives rise to two homopolymers displaying an enantiotropic nematic mesophase as their highest temperature mesophase. As we can observe from Figure 8a,b,c the nematic-isotropic transition temperature and its associated enthalpy change show linear dependences of composition. This means that the structural units derived from poly(6-3) and poly(6-5) are isomorphic into their nematic mesophase. However, the same two structural units are isomorphic within the s_A mesophase exhibited by poly(6-5) only over a very narrow range of compositions. The linear dependence of the isotropization temperature is predictable by the Schroeder-Van Laar equations.[60] The same discussion is valid for the copolymer system poly[(6-6)-co-(6-11)]X/Y except that the isotropization temperature of these copolymers exhibit an upward curvature. This upward curvature is also predicted by the Schroeder-Van Laar equations[60] and is due to the more dissimilar enthalpy changes associated with the isotropization temperatures of the two homopolymers.

Figure 9 presents the phase diagrams of copolymers poly[(6-3)-co-(6-11)]X/Y[57,59] and poly[(6-5)-co-(6-11)]X/Y.[57] Both sets of copolymers have degrees of polymerization of 20. Both pairs of copolymers are based on monomers which give rise to homopolymers exhibiting nematic and s_A as their highest temperature mesophases. However, poly(6-5) displays a nematic and a s_A mesophase, while poly(6-3) only a nematic mesophase. Both sets of copolymers display continuous dependences of their highest temperature mesophase with a triple point at a certain composition. This triple point generates over a very narrow range of compositions copolymers exhibiting the sequence isotropic-nematic-s_A-n_{re}. Again the shape of the dependences of the phase transition temperature on composition obeys the Schroeder-Van Laar equations.[60]

Figure 10 presents two sets of phase diagrams obtained from monomer pairs giving rise to homopolymers which exhibit isotropic and s_A mesophases as their highest temperature mesophases, i.e., poly[(6-2)-co-(6-8)]X/Y with degree of polymerization of 10,[57,61] and poly[(6-2)-co-(6-11)]X/Y with degree of polymerization of 15.[57,62] Both sets of copolymers display a similar phase diagram. Over a certain range of compositions the two structural units are isomorphic within the s_A phase, after which follows a triple point. After this triple point the two structural units are isomorphic within a newly generated nematic mesophase. Both copolymers generate within a certain range of compositions on the left side of the triple point the sequence isotropic-nematic-s_A-n_{re}.[57] Again the shape of the dependence of the highest temperature mesophase on composition is predictable by the Schroeder-Van Laar equations. This means that the stuctural units of all binary copolymers based on an identical mesogenic unit but different spacer lengths behave as an ideal solution. This behavior allows the engineering of mesomorphic phase transition temperatures and of their thermodynamic parameters in a straight forward manner by living azeotropic copolymerizations. The same behavior was demonstrated for monomer pairs which both give rise to homopolymers exhibiting a chiral smectic C mesophase.[36,63]

<u>Side Chain Liquid Crystalline Polymers Exhibiting a Reentrant Nematic Mesophase</u>

The reentrant nematic phase (n_{re}) was discovered in 1975 in low molar mass liquid crystals.[64] Since then it has received substantial theoretical and experimental interest.[65-73]

The first side chain liquid crystalline polymers exhibiting a n_{re} phase were reported in 1986.[74,75] Some other examples of polymers exhibiting the sequence isotropic-nematic-s_{Ad}-n_{re} were reported in the meantime.[57,76,77-80] All these polymers are based on mesogenic units containing a cyano group, five or six atoms in the flexible spacer and a polyacrylate or polyvinyl ether backbone. The replacement of these quite flexible backbones with a more rigid one like polymethacrylate does not allow the formation of the n_{re} phase. As discussed in the previous subchapter a n_{re} mesophase can be generated by copolymerization of two monomers which lead to homopolymers with nematic or isotropic and s_A as their highest temperature mesophases, since these copolymers exhibit a triple point on their phase diagrams.[57] According to our experimental results any polymer which exhibits the sequence

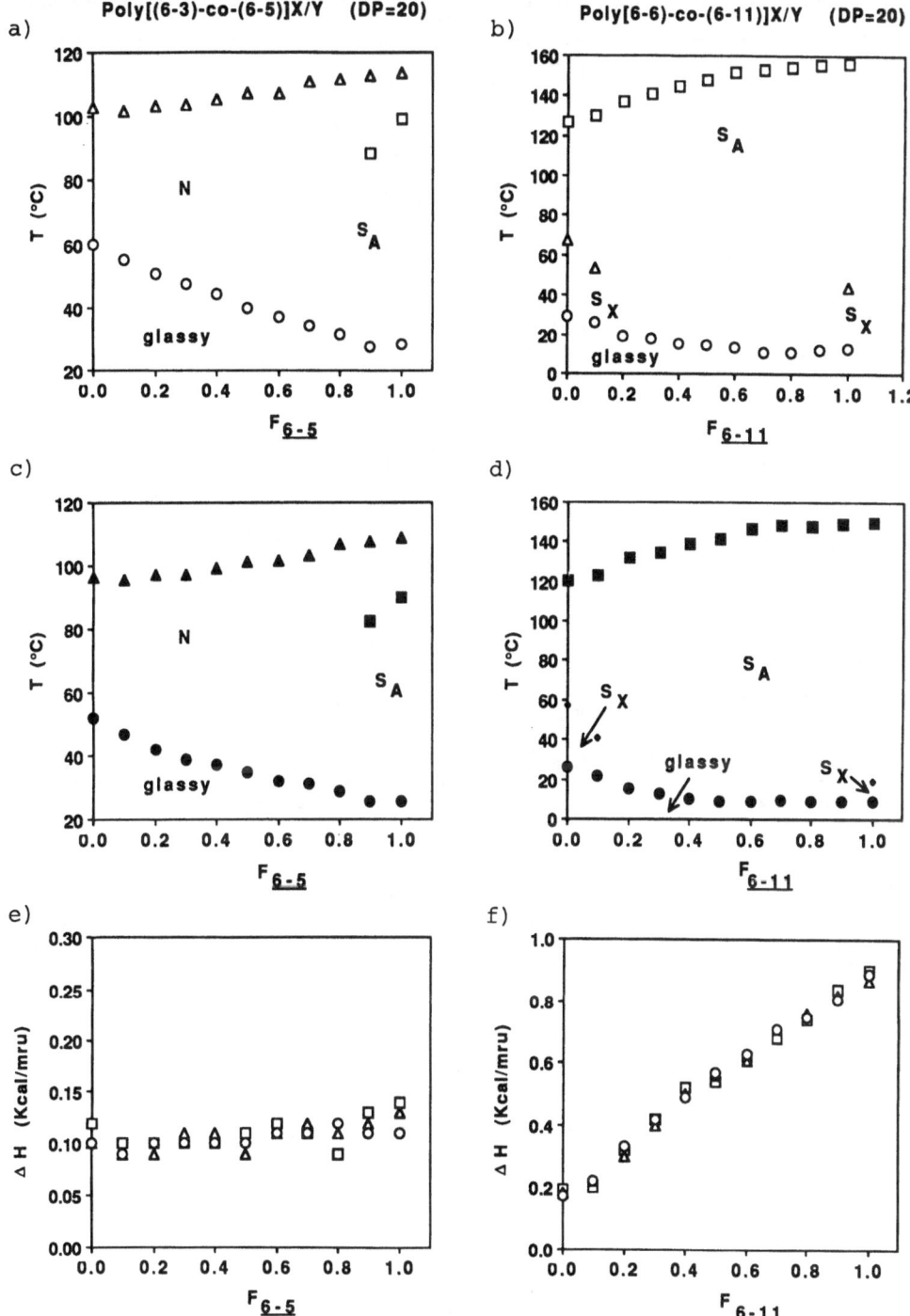

Figure 8. The dependence of phase transition temperatures obtained from second DSC heating scan (a,d), cooling scan (b,e), and the enthalpy changes associated with their highest temerature mesophase of copolymers poly[(6-3)-co-(6-5)]X/Y and poly[(6-6)-co-(6-11)]X/Y (all with degrees of polymerization equal to 20).

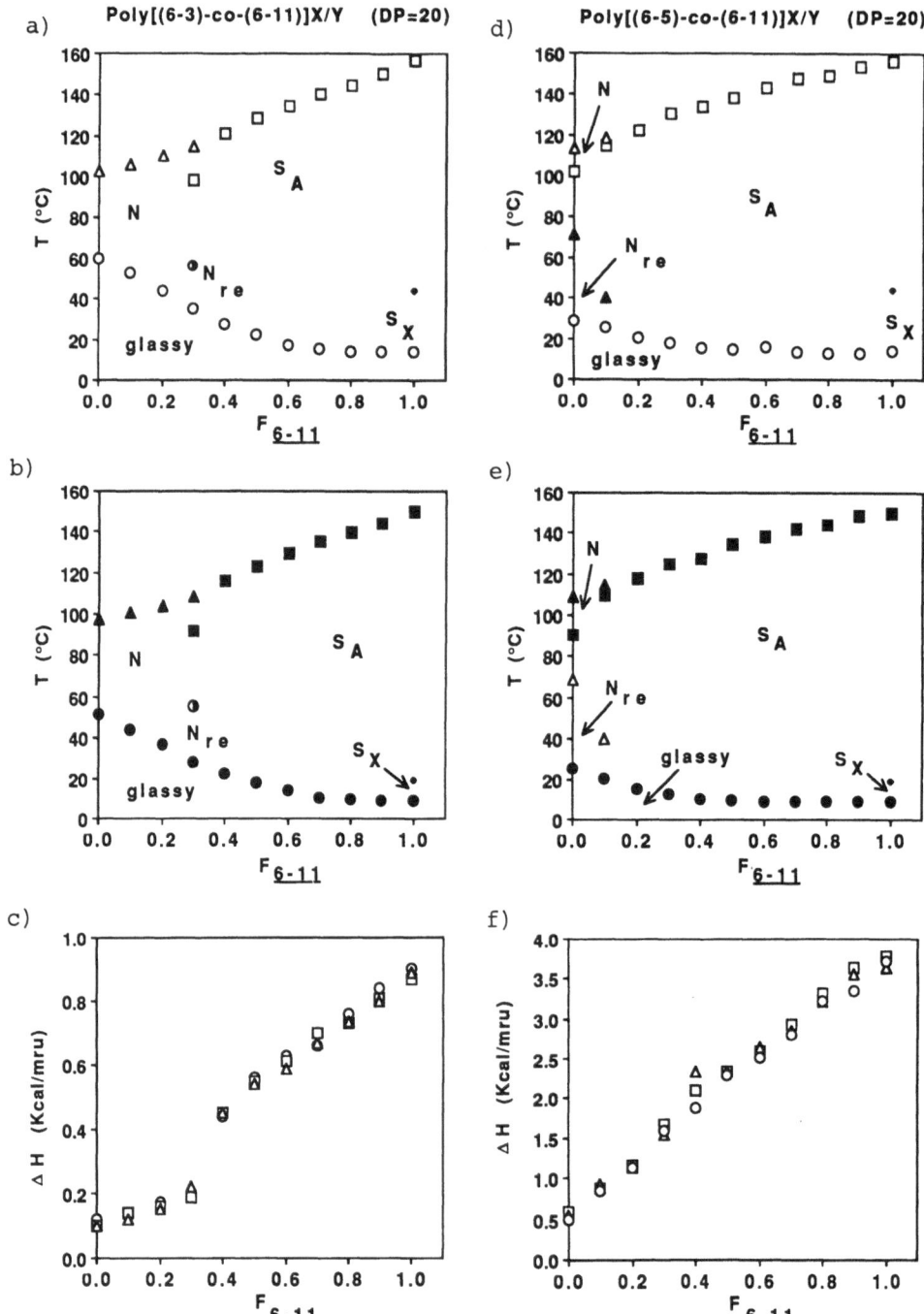

Figure 9. The dependence of phase transition temperatures obtained from second DSC heating scan (a,d), cooling scan (b,e), and the enthalpy changes associated with their highest temerature mesophase of copolymers poly[(6-3)-co-(6-11)]X/Y and poly[(6-5)-co-(6-11)]X/Y (all with degrees of polymerization equal to 20).

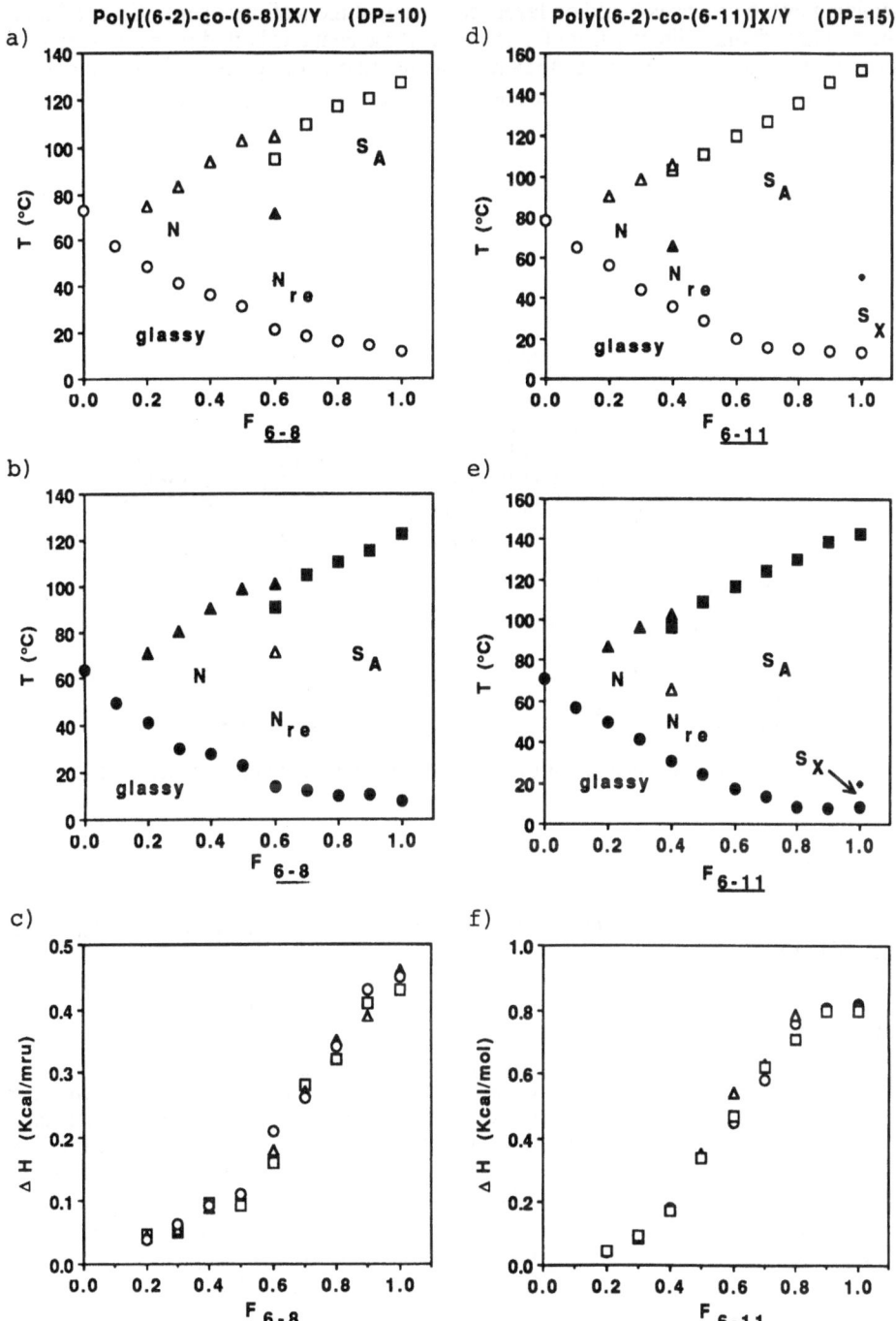

Figure 10. The dependence of phase transition temperatures obtained from second DSC heating scan (a,d), cooling scan (b,e), and the enthalpy changes associated with their highest temerature mesophase of copolymers poly[(6-2)-co-(6-8)]X/Y and poly[(6-2)-co-(6-11)]X/Y (with degrees of polymerization equal to 10 and 15, respectively).

isotropic-nematic-s_A should also display a n_{re} phase. The most probable mechanism for the generation of a n_{re} phase is outlined in Figure 11.[73] The most stable s_A phase of mesogens containing cyano groups is based on layers containing dimers of mesogens. On cooling, the nematic phase formed directly from the isotropic phase contains both dimeric mesogens and monomeric mesogens and so does the first s_A phase. In order to go from the less ordered s_A phase to the s_A phase based on dimeric mesogens, a n_{re} phase is required (Figure 11).[73]

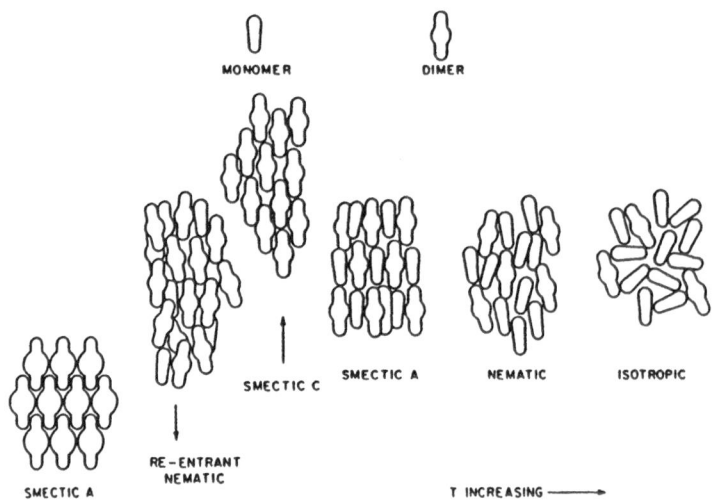

Figure 11. The mechanism of formation of the re-entrant nematic mesophase.

LIQUID CRYSTALLINE POLYMERS CONTAINING CROWN ETHERS AND POLYPODANTS

Mesomorphic host-guest systems of low molecular weight and polymer liquid crystals containing macroheterocyclic ligands and polypodants provide a novel approach to self-assembled systems which combine selective recognition with external regulation.[81-84] Three basic architectures can be considered for liquid crystalline polymers containing crown ethers (Figure 12): main chain liquid crystalline polymers containing crown ethers in the main chain of the polymer and side chain liquid crystalline polymers containing crown ethers either in the mesogenic group or in the main chain. Alternatively, the same series of polymers with polypodants instead of crown ethers can be considered.

Main chain polyamides and polyethers containing crown ethers were reported.[85,86] A variety of side chain liquid crystalline polymers containing crown ether groups at one end of the mesogenic unit were designed.[87-93] Side chain liquid crystalline polymers containing crown ethers in the main chain were synthesized by living cationic cyclopolymerization and cocyclopolymerization of 1,2-bis(2-ethenyloxyethoxy)benzene derivatives containing mesogenic side groups.[94,95] Polymers containing crown ethers in the side groups dissolve ion-pairs and behave as copolymers containing two different mesogenic groups, i.e., complexed and uncomplexed. Their behavior is similar to that of copolymers derived from two different mesogenic groups. Therefore, their phase behavior is directed by molecular recognition.[96] The use of oligooxyethylenic spacers in main chain,[97] and side chain[98-100] liquid crystalline polymers leads to liquid crystalline polypodants. Both main chain[97] and side chain[100] liquid crystalline polypodants dissolve large amounts of alkali metal salts, and the resulting liquid crystalline polyelectrolytes are ionic conductors.[101]

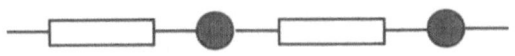

II. Side Chain Liquid Crystalline Polymers

A. Crown ether ligand as part of the mesogenic unit

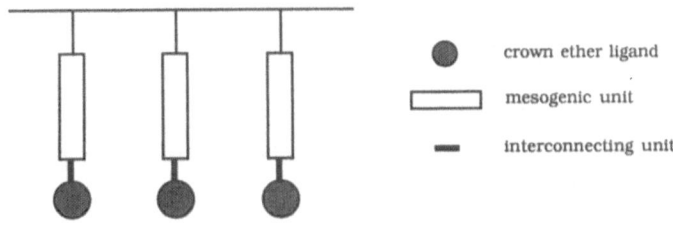

crown ether ligand

mesogenic unit

interconnecting unit

B. Crown ether ligand as part of the polymer backbone

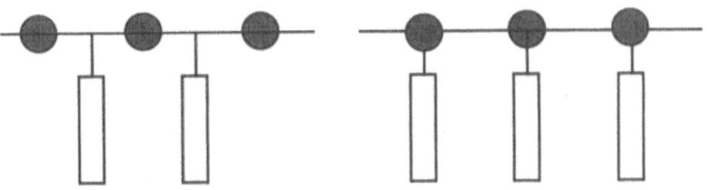

Figure 12. The molecular architecture of liquid crystalline polymers containing crown ether ligands.

MOLECULAR RECOGNITION DIRECTED SELF-ASSEMBLY OF SUPRAMOLECULAR LIQUID CRYSTALLINE POLYMERS

The molecular recognition of complementary components leads to systems able to self-assembly or self-organize i.e., systems capable to generate spontaneously a well defined supramolecular architecture from their components under a well-defined set of conditions.[81,82] Although self-assembly is a well recognized process in biological systems,[102,103] the general concept of self-assembly of synthetic molecules by molecular recognition of complementary components, received a revived interest only after it was integrated by Lehn in the new field of supramolecular chemistry.[81,82,104,105] Several examples in which molecular recognition induces the association of complementary nonmesomorphic components into a low molar mass or polymeric supramolecular liquid crystal are described below. The principles of formation of a mesogenic supramolecule from two complementary components is outlined in Scheme 6. The particular example used by Lehn et al.[106] to generate a supramolecular mesogenic group which exhibits a hexagonal columnar mesophase is by formation of an array of three parallel hydrogen bonds between groups of uracil and 2,6-diaminopyridine type as those depicted in Scheme 6. The transplant of the same concept to the generation of a supramolecular liquid crystalline polymer is outlined in Scheme 7.[107] The complementary moieties used TU_2 and TP_2 are uracil (U) and 2,6-diacylamino-pyridine (P) groups connected through tartaric acid esters (T). The tartaric acid (T) unit provides in addition, the opportunity to investigate the effect of changes in chirality on the species formed. Thus the components LP_2, LU_2, DP_2, MP_2 and MU_2 are derived from L(+), D(-) and meso (M) tartaric acid respectively. Although all monomers

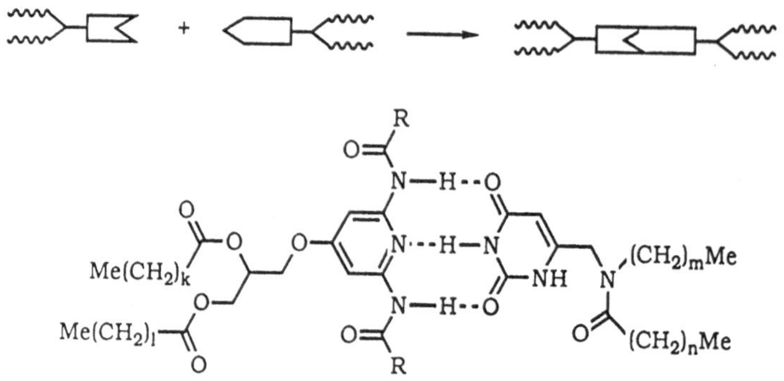

Scheme 6. Formation of a mesogenic supramolecule from two complementary components.

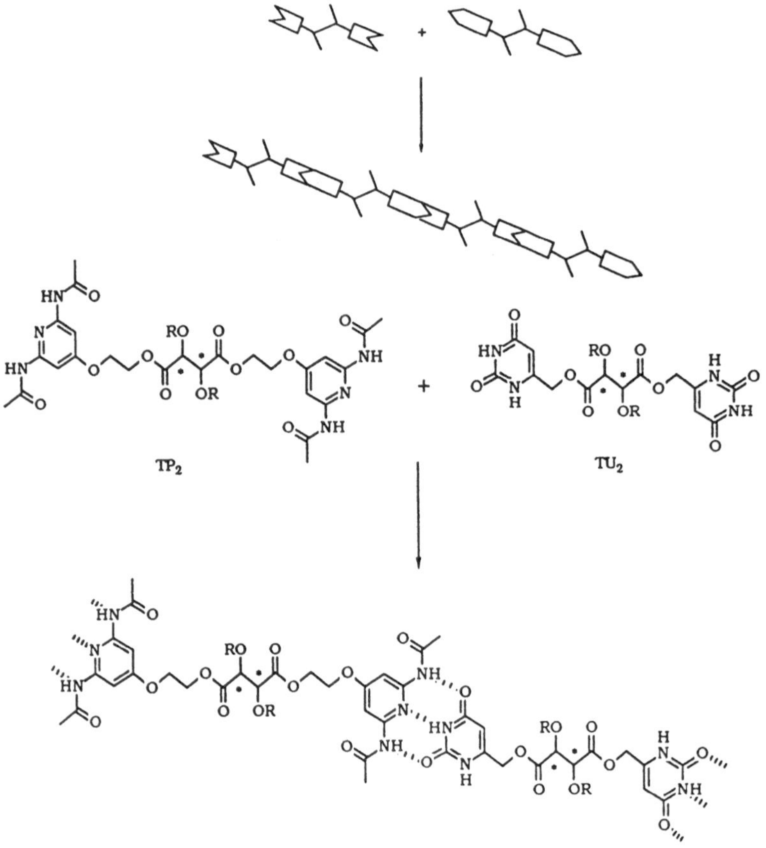

$(TP_2 \cdot TU_2)_n$

Scheme 7. Generation of a supramolecular liquid crystalline polymer.

(LP$_2$, LU$_2$, DP$_2$, MP$_2$ and MU$_2$) are only crystalline, the corresponding supramolecular "polymers" obtained through hydrogen bonding (LP$_2$ + LU$_2$, DP$_2$ + LU$_2$ and MP$_2$ and MU$_2$) exhibit hexagonal columnar mesophases. These hexagonal columnar mesophases are generated from cylindrical helical suprastructures.[107]

An additional example of supramolecular liquid crystalline polymer obtained through the hydrogen bonding of nonmesomorphic monomers was recently reported.[108] Examples in which a mesophase was generated through dimerization of carboxylic acid derivatives via hydrogen bonding were available in the classic literature on liquid crystals and were extensively reviewed.[109,110] New and interesting examples on the generation of nonsymmetrical liquid crystalline dimers,[111] twin dimer[112] and side chain liquid crystalline polymers[113] by specific hydrogen bonding "reactions" continue to be reported (Scheme 8).

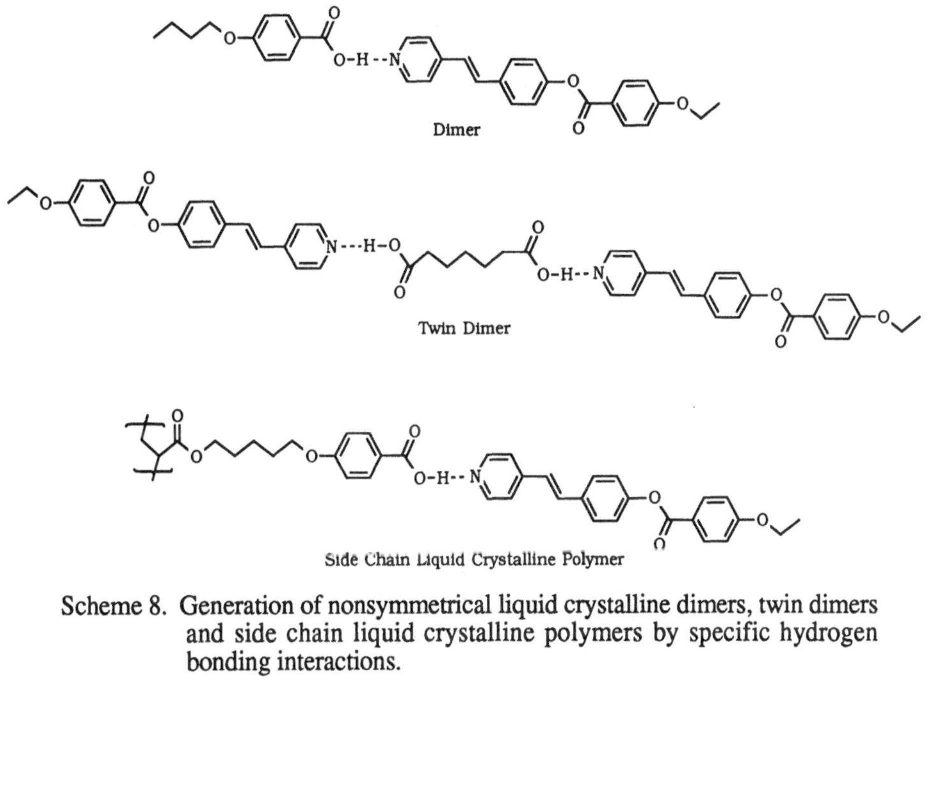

Dimer

Twin Dimer

Side Chain Liquid Crystalline Polymer

Scheme 8. Generation of nonsymmetrical liquid crystalline dimers, twin dimers and side chain liquid crystalline polymers by specific hydrogen bonding interactions.

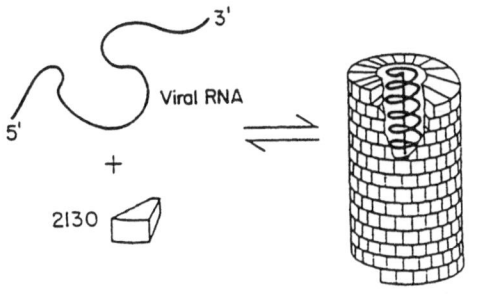

Tobacco Mosaic Virus

Figure 13. Self-assembly of tobacco mosaic virus (TMV). The protein subunits define the shape of the helix and the RNA defines the helix length. All information for assembly is contained within the component parts.

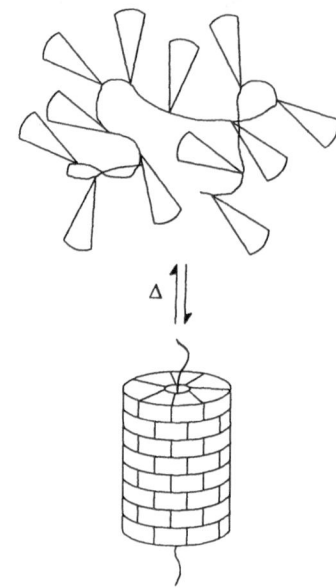

Figure 14. The self-organization of a randomly coiled flexible polymer
containing tapered side groups into a rigid rod-like columnar
structure.

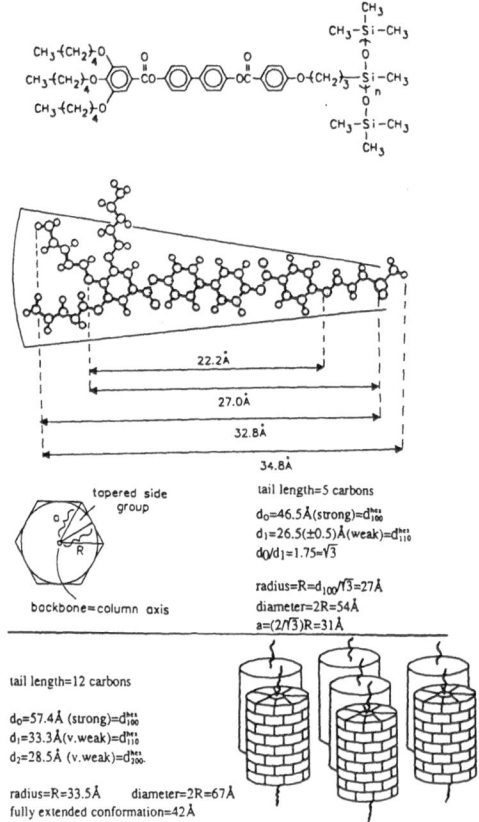

Figure 15. A representative structure of a flexible polymer containing tapered
side-groups and its self-assembling into a columnar structure
which exhibits a hexagonal columnar mesophase.

Recently, a new approach to molecular recognition directed self-assembly of a liquid crystalline supramolecular structure by a mechanism which resembles that of self-assembly of tobacco mosaic virus (TMV) was reported.[114] The self-assembly mechanism of TMV is outlined in Figure 13.[103] The synthetic approach can be summarized as follows. A flexible polymer backbone containing tapered side groups self-organizes the side groups into a column which surrounds the polymer backbone (Figure 14). These polymers exhibit thermotropic hexagonal columnar mesophases (Figure 15). Although the number of chains penetrating through the center of the column is not yet known and requires further research (Figure 15), it seems that this self-assembling system is complementary to those elaborated by Lehn et al.[106,107] In the model elaborated by Lehn et al.[106,107] the complementary pairs are self-organized through hydrogen bonding type interactions (endo-recognition), while in the last case[103,114] only the shape of tapered side groups is responsible for the generation of a polymeric column (exo-recognition).

ACKNOWLEDGMENTS

Financial support by the National Science Foundation and the Office of Naval Research is gratefully acknowledged.

REFERENCES

1. C. B. McArdle, "Side Chain Liquid Crystal Polymers", Blackie, Glasgow (1989).
2. V. Percec and C. Pugh, in: "Side Chain Liquid Crystal Polymers", ed. C. B. McArdle, Chapman and Hall, New York, (1989), p. 30; V. Percec and D. Tomazos, Molecular Engineering of Liquid Crystalline Polymers, in "Comprehensive Polymer Science", Supplement 1, Sir G. Allen and J. C. Bevington Eds., Pergamon Press, Oxford, (1992), in press
3. M. Warner, in reference 1, p. 1
4. C. Nöel, in "Side Chain Liquid Crystal Polymers", ed. C. B. McArdle, Chapman and Hall, New York, (1989), p. 159
5. (a) C. Nöel, Makromol. Chem., Macromol. Symp., 22:95 (1988); (b) P. Davidson, L. Noirez, J. P. Cotton and P. Keller, Liq. Cryst., 10:111 (1991) and references cited therein; (b) G. Pepy, J. P. Cotton, F. Hardouin, P. Keller, M. Lambert, F. Moussa, L. Noirez, A. Lapp and C. Strazielle, Makromol. Chem., Macromol. Symp., 15:251 (1988) and references cited therein
6. L. Noirez, J. P. Cotton, F. Hardouin, P. Keller, F. Moussa, G. Pepy and C. Strazielle, Macromolecules, 21:2889 (1988)
7. P. Davidson, L. Noirez, J. P. Cotton and P. Keller, Liq. Cryst., 10:111 (1991)
8. F. Hardouin, S. Mery, M. F. Achard, L. Noirez and P. Keller, J. Phys. II, 1:511 (1991)
9. V. Percec, B. Hahn, M. Ebert and J. H. Wendorff, Macromolecules, 23:2092 (1990)
10. H. Mattoussi, R. Ober, M. Veyssie and H. Finkelmann, Europhys. Lett., 2:233 (1986)
11. F. Kuschel, A. Madicke, S. Diele, H. Utschik, B. Hisgen and H. Ringsdorf, Polym. Bull., 23:373 (1990)
12. V. Percec and A. Keller, Macromolecules, 23:4347 (1990)
13. A. Keller, G. Ungar and V. Percec, in "Advances in Liquid Crystalline Polymers", ed. R. A. Weiss and C. K. Ober, ACS Symp. Ser 435, Am. Chem. Soc., Washington DC, (1990)
14. V. Percec and D. Tomazos, J. Polym. Sci. Polym. Chem. Ed., 27:999 (1989)
15. V. Percec and D. Tomazos, Macromolecules, 22:2062 (1989)
16. V. Percec, D. Tomazos and R. A. Willingham, Polym. Bull., 22:199 (1989)
17. V. Percec and D. Tomazos, Polymer, 31:1658 (1990)
18. V. Percec and D. Tomazos, Macromolecules, 22:1512 (1989)
19. V. Percec and D. Tomazos, Polymer, 30:2124 (1989)
20. R. D. Richards, W. D. Hawthorne, J. S. Hill, M. S. White, D. Lacey, J. A. Semiyen, G. W. Gray and T. C. Kendrick, J. Chem. Soc. Chem. Comm., 95 (1990)
21. G. W. Gray, in reference 1 p. 106
22. J. M. Rodriguez-Parada and V. Percec, J. Polym. Sci. Polym. Chem. Ed., 24:1363 (1986)

23. V. Percec and D. Tomazos, Polym. Bull., 18:239 (1987)
24. J. M. Rodriguez-Parada and V. Percec, J. Polym. Sci. Polym. Chem. Ed., 25:2269 (1987)
25. V. Percec, D. Tomazos and C. Pugh, Macromolecules, 22:3259 (1989)
26. C. Pugh and V. Percec, Polym. Prepr., Am. Chem. Soc., Div. Polym. Chem., 26:303 (1985)
27. W. Kreuder, O. W. Webster and H. Ringsdorf, Makromol. Chem. Rapid Commun., 7:5 (1986)
28. M. Hefft and J. Springer, Makromol. Chem. Rapid Commun., 11:397 (1990)
29. T. Kodaira and K. Mori, Makromol. Chem. Rapid Commun., 11:645 (1990)
30. V. Percec, M. Lee and H. Jonsson, J. Polym. Sci., Polym. Chem. Ed., 29:327 (1991)
31. V. Percec and M. Lee, J. Macromol. Sci.-Chem., A28:651 (1991)
32. V. Percec, M. Lee and C. Ackerman, Polymer, 33:703 (1992)
33. V. Percec and M. Lee, Macromolecules, 24:2780 (1991)
34. V. Percec, A. S. Gomez and M. Lee, J. Polym. Sci., Polym. Chem. Ed., 29:1615 (1991)
35. V. Percec, C. S. Wang and M. Lee, Polym. Bull., 26:15 (1991)
36. V. Percec, Q. Zheng and M. Lee, J. Mater. Chem., 1:611 (1991)
37. T. Sagane and R. W. Lenz, Polymer, 30:2269 (1989)
38. V. Percec and M. Lee, Macromolecules, 24:1017 (1991)
39. T. Sagane and R. W. Lenz, Polym. J., 20:923 (1988)
40. T. Sagane and R. W. Lenz, Macromolecules, 22:3763 (1989)
41. V. Heroguez, A. Deffieux and M. Fontanille, Makromol. Chem. Macromol. Symp., 32:199 (1990)
42. V. Heroguez, M. Schappacher, E. Papon and A. Deffieux, Polym. Bull., 25:307 (1991)
43. E. Papon, A. Deffieux, F. Hardouin and M. F. Achard, Liq. Cryst., in press
44. H. Jonsson, V. Percec and A.Hult, Polym. Bull., 25:115 (1991)
45. V. Percec and M. Lee, Macromolecules, 24:1017 (1991)
46. V. Percec, C. S. Wang and M. Lee, Polym. Bull., 26:15 (1991)
47. V. Percec, Q. Zheng and M. Lee, J. Mater. Chem, 1:611 (1991)
48. R. Rodenhouse, V. Percec and A. E. Feiring, J. Polym. Sci. Polym. Lett. Ed., 28:345 (1990)
49. R. Rodenhouse and V. Percec, Adv. Mater., 3:101 (1991)
50. R. Rodenhouse and V. Percec, Polym. Bull., 25:47 (1991)
51. S. G. Kostromin, N. D. Cuong, E. S. Garina and V. P. Shibaev, Mol. Cryst. Liq. Cryst., 193:177 (1990)
52. C. G. Cho, B. A. Feit and O. W. Webster, Macromolecules, 23:1918 (1990)
53. H. Jonsson, P. E. Sundell, V. Percec, U. W. Gedde and A. Hult, Polym. Bull., 25:649 (1991)
54. H. Jonsson, H. Andersson, P. E. Sundell, U. W. Gedde and A. Hult, Polym. Bull., 25:641 (1991)
55. H. Jonsson, V. Percec, U. W. Gedde and A. Hult, Makromol. Chem. Macromol. Symp., 54/55:83 (1992)
56. V. Percec, M. Lee, P. L. Rinaldi and V. E. Litman, J. Polym. Sci. Polym. Chem. Ed., 30:1213 (1992)
57. V. Percec and M. Lee, J. Mater. Chem., 1:1007 (1991)
58. D. A. Tirrell, in "Encyclopedia of Polymer Science and Engineering", ed. H. F. Mark, N. M. Bikales, C. G. Overberger and G. Menges, 2nd Ed., Wiley, New York, (1986), Vol. 4, p. 192
59. V. Percec and M. Lee, Macromolecules, 24:4963 (1991)
60. T. Schroeder, Z. Phys. Chem., 11:449 (1893); J. J. Van Laar, Z. Phys. Chem., 63:216 (1908); G. R. Van Hecke, J. Phys. Chem., 83:2344 (1979); M. F. Achard, M. Mauzac, M. Richard, M. Sigaud and F. Hardouin, Eur. Polym. J., 25:593 (1989)
61. V. Percec and M. Lee, Polymer, 32:2862 (1991)
62. V. Percec and M. Lee, Polym. Bull., 25:131 (1991)
63. V. Percec, Q. Zheng and M. Lee, J. Mater. Chem., 1:1015 (1991)
64. P. E. Cladis, Phys. Rev. Lett., 35:48 (1975)
65. P. E. Cladis, R. K. Bogardus, W. B. Daniels and G. N. Taylor, Phys. Rev. Lett., 39:720 (1977)

266

66. D. Guillon, P. E. Cladis and J. Stamatoff, Phys. Rev. Lett., 41:1598 (1978)
67. P. E. Cladis, R. K. Bogardus, and D. Aadsen, Phys. Rev. Ser. A, 18:2292 (1978)
68. N. H. Tinh, J. Chim. Phys., 1983, 80:83 (1983)
69. J. W. Goodby, T. M. Leslie, P. E. Cladis and P. L. Finn, in "Liquid Crystals and Ordered Fluids", ed. A. C. Griffin and J. F. Johnson, Plenum, New York, p. 203 (1984)
70. G. Sigaud, N. H. Tinh, F. Hardouin and H. Gasparoux, Mol. Cryst. Liq. Cryst., 69:81 (1981)
71. F. Hardouin, A. M. Levelut, M. F. Achard and G. Sigaud, J. Chim. Phys., 80:53 (1983)
72. F. Hardouin, Physica A., 140:359 (1986)
73. P. E. Cladis, Mol. Cryst. Liq. Cryst., 165:85 (1988)
74. P. Le Barny, J. C. Dubois, C. Friedrich and C. Noel, Polym. Bull., 15:341 (1986)
75. T. I. Gubina, S. G. Kostromin, R. V. Talrose, V. P. Shibaev and N. A. Plate, Vysokomol. Soed. Ser. B, 28:394 (1986)
76. V. Shibaev, Mol. Cryst. Liq. Cryst., 155:189 (1988)
77. N. Lacoudre, A. Le Borgne, N. Spassky, J. P. Vairon, C. L. Jun, C. Friedrich and C. Noel, Makromol. Chem. Macromol. Symp., 24:271 (1989)
78. T. I. Gubina, S. Kise, S. G. Kostromin, R. V. Talrose, V. P. Shibaev and N. A. Plate, Liq. Cryst., 4:197 (1989)
79. S. G. Kostromin, V. P. Shibaev and S. Diele, Makromol. Chem., 191:2521 (1990)
80. C. Legrand, A. Le Borgne, C. Bunel, N. Lacoudre, P. Le Barny, N. Spassky and J. P. Vairon, Makromol. Chem., 191:2979 (1990)
81. J. M. Lehn, Angew. Chem. Int. Ed. Engl., 27:89 (1988)
82. J. M. Lehn, Angew. Chem. Int. Ed. Engl., 29:1304 (1990)
83. D. J. Cram, Angew. Chem. Int. Ed. Engl., 27:1009 (1988)
84. C. J. Pedersen, Angew. Chem. Int. Ed. Engl., 27:1021 (1988)
85. G. Cowie and H. H. Wu, Br. Polym. J., 20:515 (1988)
86. V. Percec and R. Rodenhouse, Macromolecules, 22:2043 (1989)
87. R. Rodenhouse and V. Percec, Polym. Bull., 25:47 (1991)
88. V. Percec and R. Rodenhouse, Macromolecules, 22:4408 (1989)
89. G. Ungar, V. Percec and R. Rodenhouse, Macromolecules, 24:1996 (1991)
90. V. Percec and R. Rodenhouse, J. Polym. Sci., Polym. Chem. Ed., 29:15 (1991)
91. J. S. Wen, G. H. Hsiue and C. S. Hsu, Makromol. Chem., Rapid Commun., 11:151 (1990)
92. R. Rodenhouse and V. Percec, Makromol. Chem., 192:1873 (1991)
93. G. H. Hsiue, J .S. Wen and C. S. Hsu, Makromol. Chem., 192:2243 (1991)
94. R. Rodenhouse, V. Percec and A. E. Feiring, J. Polym. Sci. Polym. Chem. Ed., 28:345 (1990)
95. R. Rodenhouse and V. Percec, Adv. Mater., 3:101 (1991)
96. V. Percec and G. Johansson, Macromolecules, submitted
97. T. D. Shaffer and V. Percec, J. Polym. Sci., Polym. Chem. Ed., 25:2755 (1987)
98. J. M. Rodriguez-Parada and V. Percec, J. Polym. Sci., Polym. Chem. Ed., 24:1363 (1986)
99. C. J. Hsieh, C. S. Hsu, G. H. Hsiue and V. Percec, J. Polym. Sci., Polym. Chem. Ed., 28:425 (1990)
100. V. Percec and D. Tomazos, to be published
101. C. J. Hsieh, G. H. Hsiue and C. S. Hsu, Makromol. Chem., 191:2195 (1990)
102. For a discussion of self-organization in biological systems see: M. Eigen and L. DeMaeyer, Naturwissenschaften, 53:50 (1966); M. Eigen, Naturwissenschaften, 58:465 (1971)
103. For a review on the self-assembly of tobacco mosaic virus (TMV) which represents the best understood self-organized biological system see: A. Klug, Angew. Chem. Int. Ed. Engl., 22:565 (1983)
104. For general reviews on self-assembly see: J. S. Lindsey, New J. Chem., 15:153 (1991); D. Philp and J. F. Stoddart, Synlett., 445 (1991); G. M. Whitesides, J. P. Mathias and C. T. Seto, Science, 254:1312 (1991); for other representative contributions in this field see: P. L. Anelli, N. Spencer and J. F. Stoddart, J. Am.

Chem. Soc., 113:5131 (1991) and references cited therein; C. T. Seto and G. M. Whitesides, J. Am. Chem. Soc., 113:712 (1991); C. T. Seto and G. M. Whitesides, J. Am. Chem. Soc., 112:6409 (1990); J. Rebek, Jr., Angew. Chem. Int. Ed. Engl., 29:245 (1990)

105. F. Vogtle, "Supramolekulare Chemie", B. G. Teubner, Stuttgart, (1989)

106. M. J. Brienne, J. Gabard, J. M. Lehn and I. Stibor, J. Chem. Soc. Chem. Commun., 1868 (1989)

107. C. Fouquey, J. M. Lehn and A. M. Levelut, Adv. Mater., 2:254 (1990)

108. R. Fornasier, M. Tornatore and L. L. Chapoy, Liq. Cryst., 8:787 (1990)

109. a) G. W. Gray, "Molecular Structure and the Properties of Liquid Crystals", Academic Press, London and New York, (1962); b) G. W. Gray, in "Liquid Crystals and Plastic Crystals", ed. G. W. Gray and P. A. Winsor, Ellis Harwood Ltd, Chichester, (1974), p. 125; c) G. W. Gray, in "The Molecular Physics of Liquid Crystals", ed. G. R. Luckhurst and G. W. Gray, Academic Press, London, (1979), p. 14; d) G. W. Gray, in "Polymer Liquid Crystals", ed. A. Ciferri, W. R. Krigbaum and R. B. Meyer, Academic Press, New York, (1982), p. 5.

110. R. Eidenschink, Angew. Chem. Int. Ed. Engl. Adv. Mater., 28:1424 (1989)

111. T. Kato and J. M. J. Frechet, J. Am. Chem. Soc., 111:8533 (1989)

112. T. Kato, A. Fujishima and J. M. J. Frechet, Chem. Lett., 919 (1990)

113. T. Kato and J. M. J. Frechet, Macromolecules, 22:3819 (1989)

114. V. Percec, J. Heck and G. Ungar, Macromolecules, 24:4957 (1991)

NLO DEVICE STRUCTURES FROM POLYDIACETYLENES

Gregory L. Baker

Bellcore
331 Newman Springs Road
Red Bank, NJ 07701-7040

INTRODUCTION

Communications, the processing and transmission of information, is evolving from technologies based on electronics, to technologies that rely on optical methods. Optical fibers now carry the bulk of long haul telephone communications, and optical fiber is expected to replace copper as the medium of choice for short haul communications within the next decade. Optics may also play a role in signal processing. Carrying out some processing steps now reserved for electronics by optical methods may result in dramatically increased switching speeds and simplified switch architectures. Optics also seems better suited for parallel processing schemes.

Most all-optical signal processing schemes are based on nonlinear optical phenomena. These rely on the observation that the refractive indices of materials are intensity dependent as shown in equation 1

$$n = n_0 + n_2 I \tag{1}$$

where n, n_0, and n_2 are the refractive index at intensity I, the zero-intensity refractive index, and the intensity dependent refractive index coefficient, respectively. The nonlinear index coefficients for most materials are small, and there has been much interest in discovering new materials with large optical nonlinearities.[1] To date, the conjugated polymers have the largest nonresonant optical nonlinearities yet measured,[2-4] and not surprisingly, there have been several attempts to demonstrate NLO device concepts in conjugated polymers.[5-8]

Just as important as the nonlinear refractive index is a collection of physical properties that must be compatible with the high-intensity optical fields that are expected in NLO devices. Among these are optical clarity, high optical damage thresholds, and thermal stability. Candidate materials must also be processable, so appropriate structures can be fabricated. In this contribution, I review some of these materials requirements by using our work on poly(5,7-dodecadiyne-1,12-diolbis(n-butoxycarbonylmethylurethane)) (poly(4BCMU)) as a prototype material.

WAVEGUIDE DEVICES

Although many all-optical devices have been proposed, I will concentrate on the nonlinear directional coupler. This device has been studied extensively by theory and simulation,[9] and has been realized for the case of dual-core optical fibers.[10] The basic device (Figure 1) consists of two parallel channels of an NLO material, spaced so that light propagating through one channel is coupled to the second through the overlap of the light's evanescent

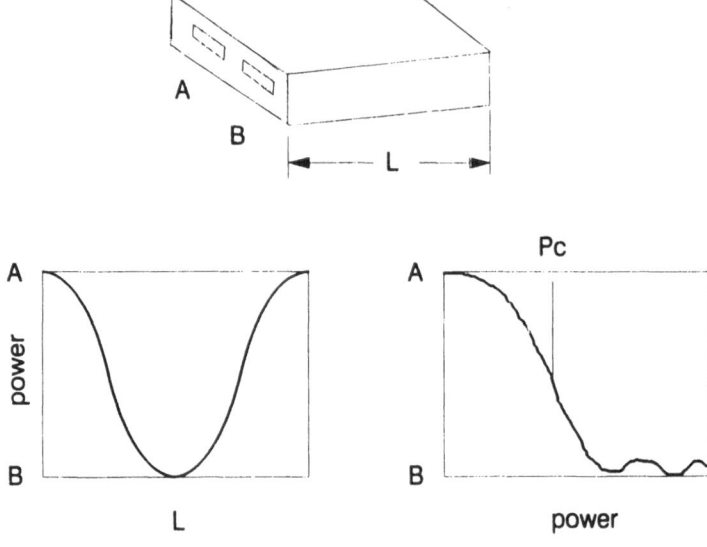

Figure 1. (top) A linear directional coupler. Power injected into guide A is transferred periodically from guide A to guide B over the length of the guides (lower left). (lower right) At the critical power for switching, p_c, light normally emanating from guide A instead exits guide B, making the device a *non*linear directional coupler.

field. In such a device, power is transferred periodically from one guide to the other, with complete transfer occurring over a distance of L_c, the critical coupling length. Choosing an appropriate guide length allows light injected into one guide to appear at the output of either guide. By making use of the intensity dependence of the refractive index (Eq. 1), such a device can be made to work as a switch. At high intensities, the refractive index (and hence the coupling of the two guides) is altered, and the relative intensities at the output of the guides is changed. For an idealized pair of guides having a length L_c, the switching behavior will appear as shown schematically in the lower right of Figure 1. The nonlinear directional coupler is an elementary building block for all-optical switching.

Polydiacetylenes suitable for NLO experiments come in two acceptable forms: single crystals and low-loss isotropic films. Mixed morphologies containing distinct domains generally are too lossy because of light scattering from the domains. In our work we focussed on isotropic films because of the processability offered by soluble polymers. We found that poly(4BCMU) can be deposited from cyclopentanone solutions by spin-casting techniques[11,12] to give optically clear films as thick as 1.5 μm with no signs of radial birefringence. The films are not truly isotropic since waveguiding experiments[11] show that the refractive index normal to the substrate was 1.53, while that parallel to the substrate was 1.60. This well-known phenomenon has been observed for other polymers,[13] and is attributed to the thinning of the polymer during drying. The birefringence observed for poly(4BCMU) is much larger than for most polymers because of the large contribution to the refractive index by the π electrons of the polymer backbone.

Waveguiding experiments carried out at 1.06μm on poly(4BCMU) films [11,14] yield estimated losses of about 1 dB/cm for the TM mode (light polarized normal to the substrate) and 5 dB/cm for the TE mode (polarization parallel to the substrate). Recent Photothermal Deflection Spectroscopy results[15] gave similar values for the optical losses, and suggested that NLO experiments could be carried out in such films from 0.7 to 1.3 μm. These losses are within the acceptable range for polymer-based devices, but exceed those of glasses by more than 10^3. Losses greater than about 5 dB/cm cause severe heating problems and introduce a significant thermal nonlinearity to the device. The source of optical losses in the near-IR were extensively examined in the context of polymeric optical fibers.[16] Most losses were attributed to a combination of Raleigh scattering, residual linear absorption, and absorption at vibrational overtones. The latter is the dominant loss mechanism for $\lambda > 1.3~\mu$m, but can be reduced by deuteriation or fluorination. Scattering losses can be minimized by careful exclusion of dust and other contaminants.[17]

PATTERNING TECHNIQUES

We investigated four methods for generating waveguide structures in thin films of polydiacetylenes. The first, direct optical lithography, makes use of the sensitivity of most conjugated materials to photooxidation. Poly(4BCMU) degrades slowly (days to weeks) by photooxidation when exposed to a normal laboratory environment, converting the normally acetone-insoluble polymer to a degraded polymer with some acetone solubility. Wegner et al.[18] exploited photooxidation and described the use of soluble polydiacetylenes as photoresist materials for the fabrication of electronic circuits. By intentionally exposing selected areas of the polydiacetylene film, the exposed regions could be degraded and removed, leaving behind unchanged material. Because of the high absorption coefficients of polydiacetylenes and their low sensitivity to visible light, only thin films ($< 0.4~\mu$m) could be exposed and developed, layers too thin for practical waveguide formation.

Optical Lithography: Rochford et al.[19] observed similar results with laser sources operating at 442 and 514 nm. They found that for long exposure times, the optical absorption spectrum of poly(4BCMU) decreased uniformly in intensity, an effect they attributed to photooxidation. Patterns could be written into films, but the exposure times (hours) were excessive.

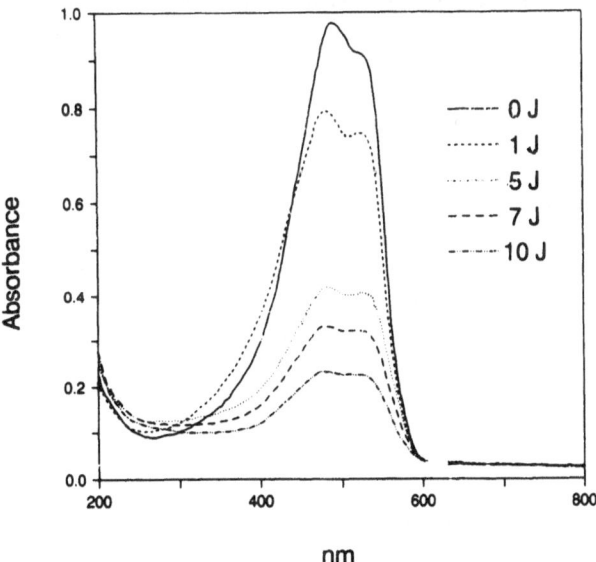

Figure 2. Change in the visible absorption spectrum for thin polydiacetylene films exposed to deep-UV radiation. The doses indicated in the figure are in mJ/cm^2, and are about 10^3 smaller that for exposures with 514 nm light.

At the same time,[20,21] we were exploring the deep-UV ($\lambda < 300$nm) patterning of poly(4BCMU) films. For us, shorter wavelength radiation was attractive since the absorption spectrum of poly(4BCMU) has a minimum near 270 nm, and thus the deeper penetration of light into the polymer film would enable thicker structures to be defined. Also, we felt that the higher energy of deep-UV photons would make degradation more likely and reduce the exposure times needed to define structures.

Our initial results confirmed our expectations. Deep-UV degraded films behaved to longer wavelength radiation, except that the exposure times needed to define structures were

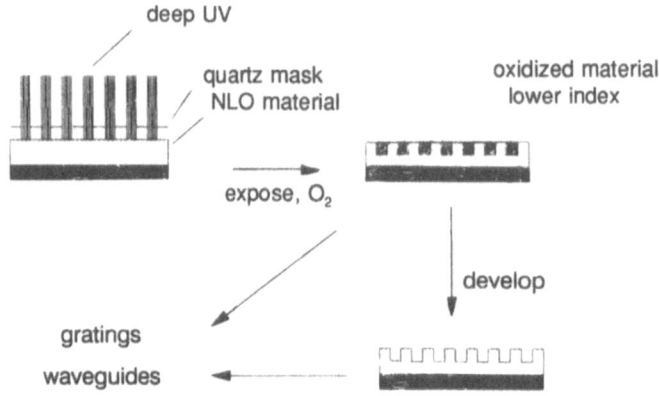

Figure 3. Grating in poly(4BCMU) formed by exposure to deep-UV through a quartz mask (contact printing).

reduced. Because the exposed polymer has a lower refractive index than the unchanged polymer,[19] structures such as gratings (Figure 3) are readily visible immediately after the polymer is exposed. As outlined in Figure 4, such structures can be used as-generated or alternatively, the exposed regions can be removed using 2-butanol as a developer. By

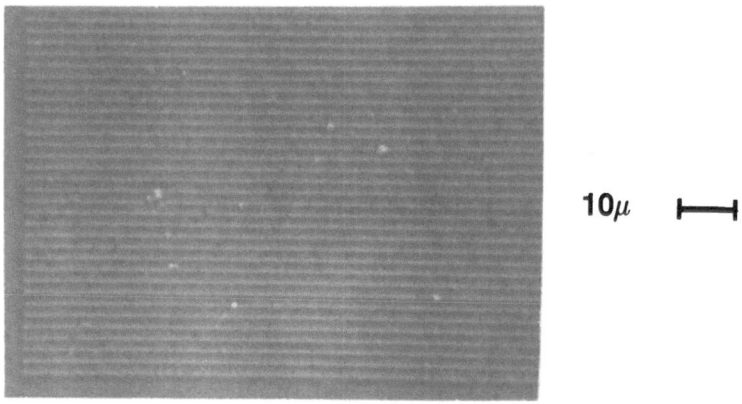

Figure 4. Scheme for forming optical structures in poly(4BCMU) by deep-UV optical lithography.

carrying out exposures in the presence of oxygen and in vacuum, we confirmed the degradation chemistry as photooxidation. Characterization of the exposed polymer by GPC showed large decreases in average molecular weights and a bimodal molecular weight distribution. The bimodality results from shorter chains absorbing more strongly at shorter wavelengths, and thus having a higher probability for chain scission than long chains. In addition, because the low-energy edge of the poly(4BCMU) absorption spectrum is dominated by an exciton of finite size, the absorption spectrum is largely independent of the chain length until the chain becomes comparable to the size of the exciton.

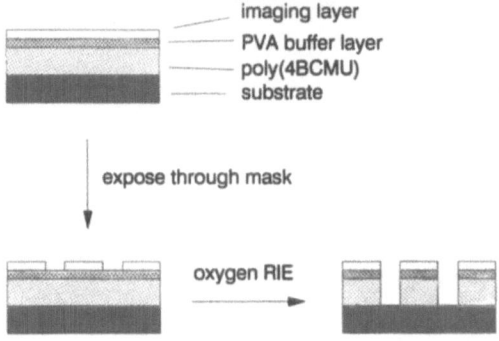

Figure 5. Bilayer lithographic route to waveguide structures in poly(4BCMU).

Bilayer lithography: Bilayer lithography is a second route to waveguide formation. In this approach, an etch mask is formed on the surface of the NLO polymer, and then the pattern is transferred though the polymer to the substrate in an oxygen reactive ion etching (RIE) step. To form an effective etch mask, imaging polymers must have a high silicon content (>10%), so that they develop an SiO_2 skin during etching that inhibits further etching. Regions where the imaging layer is removed are unprotected and are rapidly etched. Nearly vertical wall profiles can be obtained for micron-thick films, but the walls are often rough and may contribute to scattering losses.

For poly(4BCMU), a silicon-containing polyacetylene resist (brominated poly(1-trimethylsilyl-1-propyne))[22,23] worked well as the imaging layer because of its sensitivity to deep-UV radiation, and its silicon content (>20%) confers high oxygen plasma resistance. In order to deposit the resist on poly(4BCMU) without damaging the optical quality of the film, a buffer layer of poly(vinyl alcohol) was spun on the poly(4BCMU) before coating with the resist. Exposure with 10 mJ/cm^2 of deep-UV and development with n-butanol yielded the etch mask, and then the pattern was transferred through the poly(4BCMU) layer to the substrate. The advantage of the bilayer approach is its versatility since in principle it is applicable to any NLO polymer that does not form a refractory oxide in oxygen plasmas.

Epitaxial Alignment: Waveguide definition by techniques that do not rely on patterning the NLO material are particularly attractive since the NLO polymer is unaffected by the chemistry of the patterning process. We explored two such schemes. The first, epitaxial alignment of polymers,[24] exploits the surface forces acting between a grooved (or buffed) polymer substrate and a thin film of deposited polymer. On buffed surfaces, low molar mass liquid crystals align along the buffing direction, and we found that poly(4BCMU) behaves similarly providing a route to waveguide structures. As outlined in Figure 6, poly(butylene terephthalate) and nylons were rubbed with a polyester cloth, and then poly(4BCMU) was spin-coated onto the buffed layer. By heating the polymers near the melting transition for poly(4BCMU) (140-145°C), the poly(4BCMU) aligned along the buffing direction. We quantified the effect by observing the film between crossed polarizers. (Figure 7) When heated above 110°C, the temperature where the hydrogen bonds between the poly(4BCMU) side chains melts, the polymer became mobile and began to align with the surface. The alignment increased when cooled, and improved further when the heating-cooling cycle was repeated. When heated above T_m, the polymer disordered and the birefringence was lost.

Because the refractive index of the poly(4BCMU) film is changed, this phenomenon can be used to form waveguide structures by including a patterning step. Shown is Figure 8 are the results of a modified version of the previous experiment, where following buffing, portions of the polymer were removed lithographically. Since poly(4BCMU) aligns only on anisotropic surfaces, the polymer deposited on the buffed regions became ordered during the heat treatment, while in other regions there was no alignment. Seen under cross polarizers, the

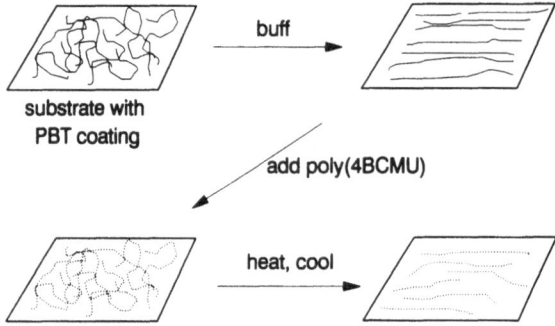

Figure 6. Scheme for the epitaxial alignment of Poly(4BCMU) on buffed PBT substartes.

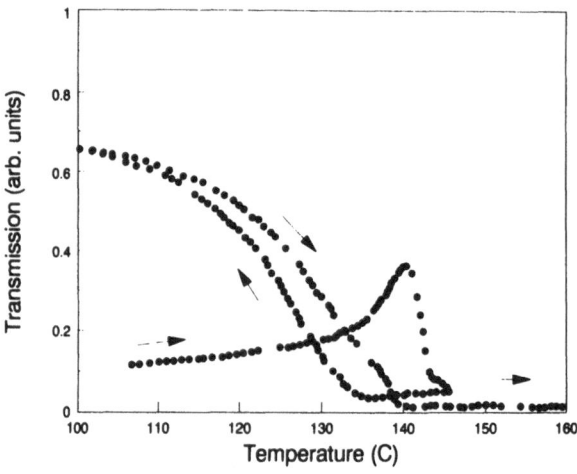

Figure 7. The evolution of birefringence for a poly(4BCMU) film deposited on buffed poly(butylene terephthalate). Heating rate: 5 °C/min.

Figure 8. Epitaxially aligned film of poly(4BCMU) as seen under
crossed polarizers. The light and dark regions correspond
to aligned and unaligned regions, respectively. The width
of the narrowest feature is ≈0.1 mm.

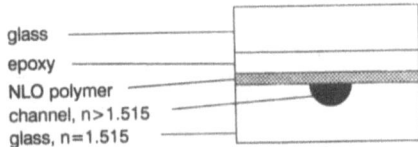

glass
epoxy
NLO polymer
channel, n>1.515
glass, n=1.515

Figure 9. Composite channel waveguide cross-section.

preferential alignment is manifested as a mosaic of dark and light areas corresponding to the unaligned and aligned regions respectively. The magnitude of the birefringence measured in the nonabsorbing region (620 nm) of the poly(4BCMU) spectrum is ≈0.14, large enough to define waveguides and gratings.

Composite Channel Waveguides. The fourth method for waveguide definition is the formation of composite channel waveguides.[25,26] Here the patterning is confined to a glass substrate, and the polymer is simply spin-coated on the patterned substrate to produce the polymer waveguide structures (Figure 9). The substrate is prepared by masking a glass slide with aluminum in a negative tone image of the waveguide pattern. When immersed in a KNO_3 melt, potassium ions are exchanged for the sodium ions of the glass, and the refractive index in the exchanged regions increases. The aluminum is then removed, the polymer added, and the ends of the substrate are polished.

The key feature of the composite guide is that the polymer layer is intentionally kept too thin to support guiding in the polymer alone, but that modes that can propagate in both the glass and the polymer can be supported. By having high index channels in the glass, the modes are confined to well-defined regions of the composite structure. A numerical simulation for a symmetric glass-polymer-glass structure (Figure 10), shows that most of the

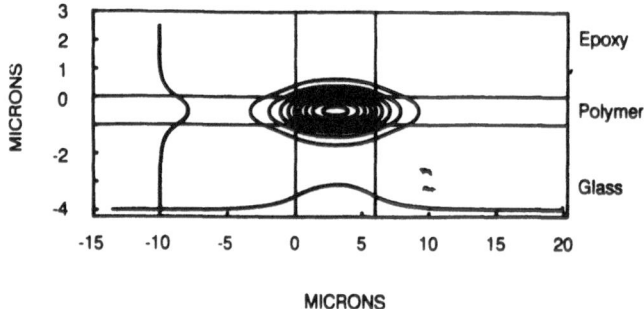

Figure 10. Numerical simulation of the intensity distribution in composite channel waveguide structures.

light in such a structure is confined to the polymer layer, with the high index region in the glass defining the channel. Most of our nonlinear optical measurements have been performed on such structures since they are robust structures that can be polished to give reasonable coupling efficiencies. The composite guides also are extremely versatile since the identical substrate can be reused for many NLO polymer samples.

SUMMARY

Waveguide devices structures can be made using a variety of methods, including optical lithography, bilayer lithography, epitaxial alignment and the use of composite channel waveguides. These varied techniques allow the the patterning technique to be chosen to match the chemical or physical properties of the NLO polymer.

REFERENCES

1. See for example: "Nonlinear Optical Properties of Organic Molecules and Crystals" Vol 1,2; D.S. Chemla and J. Zyss eds.; Academic Press, New York, NY; 1987.

2. F. Kajzar, S. Etemad, G.L. Baker, J. Messier *Solid State Commun.* **1987,** *63,* 1113.

3. G.L. Baker, S. Etemad, F. Kajzar *Proc. SPIE* **1987,** *824,* 102.

4. W.-S. Fan, S. Benson, J.M.J. Madey, S. Etemad, G.L. Baker, F. Kajzar *Phys. Rev. Lett* **1989,** *62,* 1492.

5. P.D. Townsend, J.L. Jackel, G.L. Baker, J.A. Shelburne III, S. Etemad *Appl. Phys. Lett.* **1989,** *55,* 1829.

6. M. Thakur, D.M. Krol *Appl. Phys. Lett.* **1990,** *56,* 1213.

7. K. Rochford, R. Zanoni, G.I. Stegeman, W. Krug, E. Miao, M.W. Beranek *Appl. Phys. Lett.* **1991,** *58,* 13.

8. J. Valera, A. Darzi, A.C. Walker, W. Krug, E. Miao, M. Derstine *Electron. Lett.* **1990,** *26,* 222.

9. For a review of NLO device structures see: G.I. Stegeman, E.M. Wright *Opt. Quant. Elect.* **1990,** *22.* 95.

10. S.R. Friberg, Y. Silberberg, M.K. Oliver, M.J. Andejco, M.A. Saifi P.W. Smith *Appl. Phys. Lett.* **1987,** *51,* 1135.

11. P.D. Townsend, G.L. Baker, N.E. Schlotter, C.F. Klausner, S. Etemad *Appl. Phys. Lett.* **1988,** *53,* 1782.

12. P.D. Townsend, G.L. Baker, N.E. Schlotter S. Etemad *Synth. Met.* **1989,** *28,* D633.

13. J.D. Swalen, M. Tacke, R. Santo, J. Fischer *Opt. Commun.* **1976,** *18,* 387.

14. W. Krug, E. Miao, M. Derstine, J. Valera *J. Opt. Soc. B* **1989,** *6,* 726.

15. M. Sinclair, C.H. Seager, D. McBranch, A.J. Heeger, G.L. Baker *Proc. Mat. Res. Soc.* **1991,** (in press).

16. Tanaka, Sawada, Takoshima, Wakatsuki *Fiber Integ. Opt.* **1988** *7,* 139.

17. T. Kaino, K. Jinguji, S. Nara *Appl. Phys. Lett.* **1982,** *41,* 802.

18. G. Wegner, R.J. Leyrer, M.A. Muller German Patent DE-OS 3346716.

19. K.B. Rochford, R. Zanoni, Q. Gong, G.I. Stegeman *Appl. Phys. Lett.* **1989,** *55,* 1161.

20. P.D. Townsend, G.L. Baker, J.L. Jackel, J.A. Shelburne III, and S. Etemad *Proc. SPIE* **1990,** *1147,* 256.

21. G.L. Baker, C.F. Klausner, J.A. Shelburne III, N.E. Schlotter, J.L. Jackel *Synth. Met.* **1989,** *28,* D639; G.L. Baker, C.F. Klausner US 4,824,522 **1989.**

22. A.S. Gozdz, G.L. Baker, C. Klausner, M.J. Bowden *Proc. SPIE* **1987,** *771,* 18; G.L. Baker, M.J. Bowden, A.S. Gozdz, C.F. Klausner US 4,863,834 **1989.**

23. G.L. Baker, C.F. Klausner, A.S. Gozdz, J.A. Shelburne III T.N. Bowmer *ACS Adv. Chem. Ser.* **1990,** *224,* 663.

24. J.S. Patel, S.-D. Lee, G.L. Baker, J.A. Shelburne III *Appl. Phys. Lett.* **1990,** *56,* 131.

25. J.L. Jackel, N.E. Schlotter, P.D. Townsend, G.L. Baker, S. Etemad *Proc. SPIE* **1988,** *971,* 239.

26. N.E. Schlotter, J.L. Jackel, P.D. Townsend, G.L. Baker *Appl. Phys. Lett.* **1990,** *56,* 13; G.L. Baker, J.L. Jackel, N.E. Schlotter US 4,834,480 **1989.**

RECENT SYNTHETIC DEVELOPMENTS IN MAINCHAIN CHROMOPHORIC

NONLINEAR OPTICAL POLYMERS

John D. Stenger-Smith,* J. W. Fischer, R. A. Henry,
J. M. Hoover and G. A. Lindsay

Chemistry Division, Research Department
Naval Weapons Center, China Lake, CA 93555

INTRODUCTION

When the chromophores (dipoles) of a polymer are oriented electrically (electric field poling) and mobility is frozen below the glass transition temperature of the polymer, these materials can be used as frequency doublers (second harmonic generation (SHG)), and materials for waveguides and electro-optical devices.[1]

Materials with good second-order nonlinear optical (NLO) properties include inorganic, semi-organic and organic crystals. Another type of material is polymer (host) and chromophore (guest) mixtures which can be oriented in an electric field to show good second-order nonlinear optics but the second-order optical properties are usually not very stable.[2] A class of materials that shows great promise is sidechain chromophoric polymers, which have the chromophore chemically bonded to the polymer chain .[3-6] These systems which have been oriented show good transmittance (almost no scattering), stability and second-order nonlinear optics. The second-order NLO properties of these sidechain polymers at or near their glass transition temperatures degrades rapidly.[7]

Progressing from guest-host systems (chromophores dissolved in a polymer matrix) to sidechain polymer systems (chromophores chemically attached to a polymer backbone), one arrives at another class of materials where the dipole is actually part of the polymer backbone, i.e., mainchain polymer systems. In fact, there are some reports on mainchain chromophoric copolymers in which the chromophores are attached head to tail.[8] Figure 1 summarizes the four types of polarized organic materials.

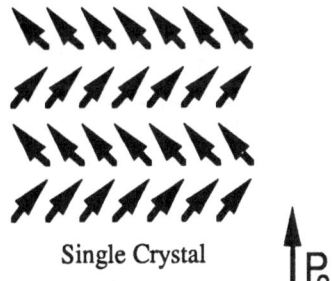

Single Crystal

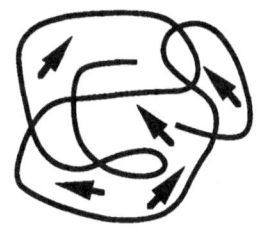

Guest-Host System
(dye-polymer)

P_0

Contemporary Topics in Polymer Science, Vol 7., Edited by
J.C. Salamone and J. Riffle, Plenum Press, New York, 1992

 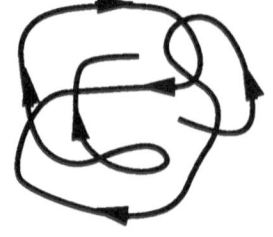

Side-Chain Chromophore
Polymer

Main-Chain Chromophore
Polymer

Figure 1. Types of polarized organic materials.

Mainchain and sidechain chromophoric polymers with second-order NLO properties can be best described by the following models:

Mainchain Chromophore   Sidechain Chromophore

$$-\left(S\text{-}A\text{-}C\text{-}D\right)_{x}-$$

$$\left(\;\right)_x \quad \left(\;\right)_x$$

S	S
D	A
C	C
A	D

Where:
 D = electron donating group (1 or more)
 A = electron withdrawing group (1 or more)
 C = connector that is conjugated with A and D and may contain 1 or more substituents
 X = number of repeat units or degree of polymerization, varies from about 10 to >1000
 S = a flexible spacer group of any length and composition

One unique feature of mainchain chromophoric polymers is that the dipoles or chromophores are directly bonded to one another. This means that it may be possible to get enhanced second-order NLO properties, and because the chromophores are in the polymer mainchain, it should be much more difficult for the chromophores to relax, which should increase the stability of the second-order NLO properties. Another advantage is that every polymer repeat unit has a chromophore, therefore the chromophore density is high (25×10^{20} chromophores/cm^3) in the case of poly((4-N-ethylene-N-ethylamino)-α-cyanocinnamate).[9]

In this report, the synthesis and characterization of three new mainchain chromophoric NLO polymers shown below are presented.

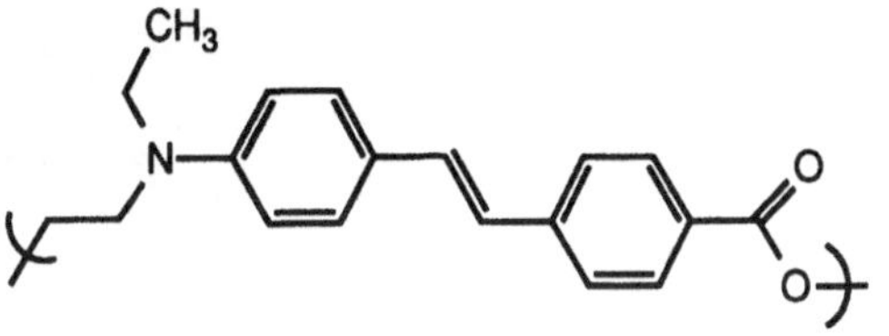

1

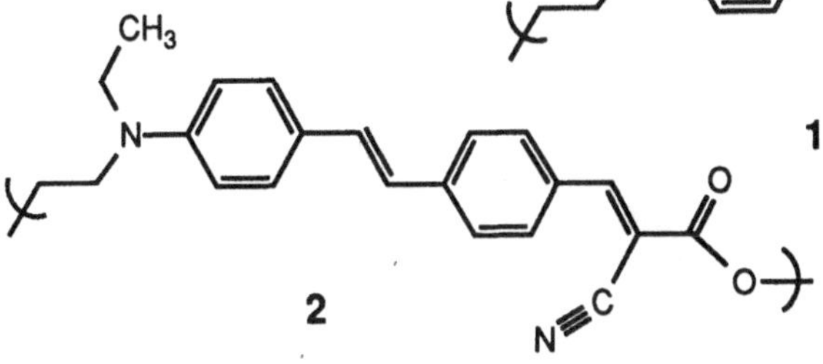

2

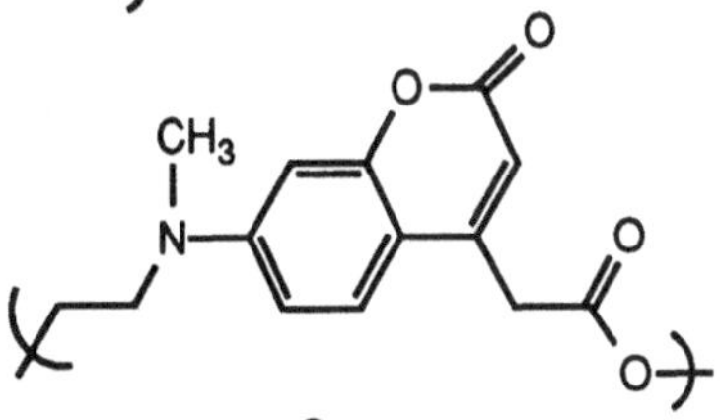

3

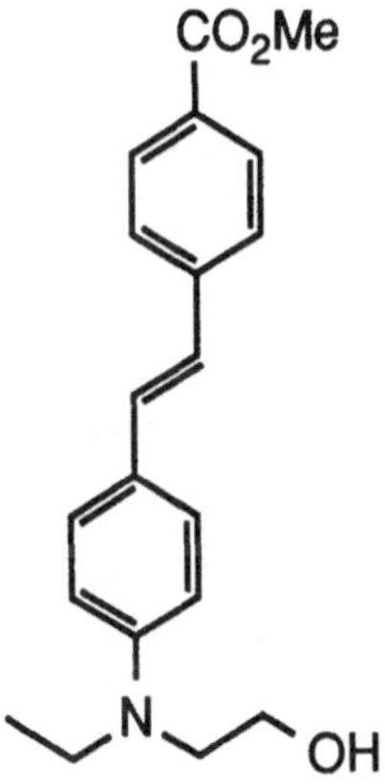

RESULTS AND DISCUSSION

Preparation of Monomer 1

4-Carbomethoxy benzyl triphenylphosphonium chloride (7.6 g, 17 mmol) was added in one portion to a solution of lithium diisopropyl amide (17 mmol) in tetrahydrofuran (150 mL) at 0°C. The resulting orange suspension was stirred for 15 minutes at 20°C under a nitrogen atmosphere. 4-(N-ethyl-N-2-*t*-butyldimethyl siloxy ethyl)-benzaldehyde (5.2 g, 17 mmol) dissolved in tetrahydrofuran (10 mL) was added in one portion and the resulting orange suspension was refluxed with stirring for 18 hours under a nitrogen atmosphere. The clear dark-orange solution was then cooled to ambient temperature and the solvent was removed under reduced pressure to yield an orange-yellow solid which was purified on a silica gel chromatography column eluting with 20% ethyl acetate in hexane. A yellow oil was collected and recrystallized from ethyl acetate-hexane. The yield of this solid was 1.65 g (22%).

The O-silylated dye (1.45 g, 3.3 mmol) was dissolved in a mixture of acetic acid (25 mL) and water (5 mL) and stirred at ambient temperature for 3 days. Volatiles were removed under reduced pressure and the yellow residue was recrystallized from ethyl acetate-hexane to give the desired dye in 73% yield (0.78 g).

Preparation of Monomer 2

4-Carbomethoxy-4'-[N-ethyl-N-2-*t*-butyldimethylsiloxy ethyl]stilbene was prepared as described in synthesis of monomer 1.

Lithium aluminum hydride (76 mg, 2 mmol) was added in portions to a stirring solution of the above silylated methyl ester (0.38 g, 0.9 mmol) in anhydrous ethyl ether (25 mL) at ambient temperature. This suspension was brought to reflux and stirred under nitrogen for 16 hours, cooled to ambient temperature, and carefully quenched with saturated Rochells' salt (5 mL). The ether layer was separated, dried over magnesium sulfate, and volatiles removed under reduced pressure to yield trans-4-(N-ethyl-N-2-*t*-butyldimethylsiloxy ethyl)-4'-hydroxymethylene stilbene in 54% yield (0.20 g).

Dimethyl sulfoxide (0.21 mL, 3.0 mmol) was added to oxalyl chloride (0.13 mL, 1.5 mmol) in anhydrous methylene chloride (20 mL) stirring under nitrogen at -40°C. After 3 minutes, the above alcohol (0.56 g, 1.4 mmol), dissolved in methylene chloride (5 mL), was added to this solution and stirred for 15 minutes. Triethyl amine (0.95 mL, 6.8 mmol) was then added and the orange solution is warmed to ambient temperatuare and stirred for approximately 40 minutes. The solution was then poured into water (10 mL), the layers were separated, and the aqueous layer was further extracted with methylene chloride (25 mL). Combining the methylene chloride layers, washing with saturated sodium chloride (25 mL), drying over sodium sulfate, and removal of volatiles under reduced pressure, afforded 0.60 g of trans-4-(N-ethyl-N-2-*t*-butyldimethylsiloxy ethyl)-4'-formyl-stilbene. This represents a slightly greater than quantitative yield. The unknown trace amount of impurity was found to have no affect on the subsequent steps and was not removed. Attempts to purify and store this compound led to decomposition. Once this aldehyde was made, it was used immediately.

The above aldehyde (1.3 g, 3.2 mmol), ethyl cyano acetate (0.35 mL, 3.3 mmol), glacial acetic acid (1.0 mL), and piperidene (0.5 mL) were mixed in toluene (30 mL) and stirred at reflux under nitrogen for 18 hours. This dark red solution is poured into water (50 mL) and extracted into ethyl ether (2 x 25 mL). Combination of the ether layers and washing with water (25 mL), saturated sodium chloride (25 mL), drying over sodium sulfate, and solvent removal gave a dark red oil which was purified on a silica gel chromatography column eluting with 20% ethyl acetate in hexane. The dark red solid from this chromatography was recrystallized from ethyl acetate-hexane to afford 0.65 g (40% yield of trans-4-(N-ethyl-N-2-*t*-butyldimethylsiloxy)-4'-(E-ethylene-2-carboxyethyl-2-cyano)-stilbene.

The above O-silylated dye (200 mg, 0.4 mmol) was dissolved in glacial acetic acid (10 mL) and water (1 mL) and stirred at ambient temperature for 2 days. All of the volatile material was removed under reduced pressure to afford the desired dye compound as a dark red solid which is purified by recrystallization from ethyl acetate-hexane. Final yield of the dye was 70 mg representing a 45% yield.

Preparation of Monomer 3

A melt of resorcinol (24 g) and 2-(methylamino)-ethanol was flushed with nitrogen and heated under an air condenser in an oil bath at 175-180°C for 6 hours. The temperature was then raised to 198-201°C and held for 1.5 hours. After the purple viscous mass had cooled, it was dissolved in 150 mL of boiling ethyl acetate. The solution was chilled overnight at 5°C after which the supernatant was decanted from the tar and evaporated. The yield of product of suitable use in the next synthetic step was about 30 g. Dimethyl acetonedicarboxylate (13.9 g), 13.6 g of the product from the previous step, 3.0 g of anhydrous zinc chloride, and 35 mL of methanol were refluxed for 23.5 hours. After cooling, the solution was poured over 60 g of ice, 20 mL of water, and 3 mL of concentrated hydrochloric acid. The purple oil which separated was extracted into methylene chloride. The methylene chloride solution was washed once with cold water, dried over sodium sulfate, filtered and evaporated (yield 13.6 g of semi-solid). After 2 days at room temperature, the solid was filtered from the liquid, washed twice with -20°C methanol, and dried. The yield was 1.34 g, m.p. 150-154°C. This material was then recrystallized from methanol to give yellow-orange plates, m.p. 158.5-160°C. The ^{1}H NMR spectrum for this compound was consistent with that expected for methyl 7-[N-methyl-N-(2-hydroxyethyl)amino]-coumarin-4-acetate. Elemental analysis for nitrogen: for $C_{15}H_{17}NO_5$, % N expected: 4.81; found: 4.93.

General Polymerization Procedure

The respective monomer (0.74 g) and 1 small drop of dibutyltindilaurate (0.01 g) were placed in a 15-mL round-bottom flask equipped with a stir bar and then the system was evacuated and purged with dry nitrogen (3 X). The system was then kept on dry nitrogen purge and lowered into a preheated 160°C oil bath and stirred. After 1 hour the material in the flask was very viscous (almost solid). The system was then evacuated and kept at 160°C for

Table 1. Summary of the Thermal Properties of the Polymers[a]

Polymer	Tg, °C	Tm, °C
1	110	215
1 (2nd heat)	120	. . .[b]
2	140	. . .
3	90	. . .

[a] Heating rate, 10°C/min.
[b] Melting point disappears presumably due to further polymerization, as evidenced by IR spectroscopy.

several hours. The system was then cooled to room temperature under reduced pressure, and the polymer removed for analysis. In the case of polymer 3, the polymer was dissolved in m-cresol and precipitated into methanol.

It was possible to draw highly oriented fibers of polymer 1 by heating the polymer to 235°C and drawing with a capillary tube or a pair of forceps.

Table 2. Summary of the Optical Properties of the Polymers/Monomers[a]

Polymer/Monomer (solvent)	λ_{max}	Cutoff[b]
1 (CH_2Cl_2)	375 nm	460 nm
2 ($CHCl_3$)	465 nm	590 nm
3 (DMF)	380 nm	440 nm

[a] UV/VIS spectra of a polymer and its respective monomer were almost identical.

[b] Essentially transparent beyond this point.

SUMMARY

Three new mainchain chromophoric polymers were synthesized and characterized. The physical properites of these polymers were compared to those of poly((4-N-ethylene-N-ethylamino)-α-cyanocinnamate). Films of the polymers can be made by either solution or melt casting of films. It was possible to obtain an oriented fiber of a polymer, namely poly(4'-[N-ethylene-N-(2-hydroxyethyl)amino] stilbene-4-formate, most likely because it has a melt transition temperature at around 210°C. Nonlinear optical studies of these polymers are in progress.

ACKNOWLEDGMENTS

This work was supported in part by a grant from the Office of Naval Research under contract N00014-90-WX24028 and by the Independent Research program at the Naval Weapons Center. The authors wish to thank D. G. Paull and P. Ashton for technical assistance. One of us (JDSS) would like to thank the Office of Naval Technology/American Society for Engineering Education for additional support.

REFERENCES

1. D. J. Williams, *Angew. Chem. Int. Ed. Engl.*, **23**, 690 (1984).
2. M. A. Mortazavi, A. Knoesen, S. T. Kowel, B. G. Higgins, and A. Dienes, *J. Opt. Soc. Am. B*, **6-4**:733 (1989).
3. (a) R. C. Hall, G. A. Lindsay, S. T. Kowel, L. M. Hayden, B. L. Anderson, B. G. Higgins, P. Stroeve, and M. P. Srinivasan, SPIE *Advances in Nonlinear Polymers and Inorganic Crystals, Liquid Crystals and Laser Media*, **824**:121 (1988).
   (b) R. C. Hall, G. A. Lindsay, S. T. Kowel, B. L. Anderson, B. G. Higgins, and P. Stroeve, *Mat. Res. Soc. Symp. Proc.*, **109**:351, (1988).
   (c) J. M. Hoover, G. A. Lindsay, S. T. Kowel, B. L. Anderson, B. G. Higgins, and P. Stroeve, *Thin Solid Films*, in press.
4. A. C. Griffin, A. M. Bhatti, and R. S. L. Hung, SPIE *Molecular and Pomeric Optoelectronic Materials: Fundamentals and Applications*, **682**:65, (1986).
5. K. D. Singer, W. R. Holland, M. G. Kuzyk, G. L. Wolk, H. E. Katz, M. L. Schilling, and P. A. Cahill, SPIE *Nonlinear Optical Properties of Organic Materials II*, **1147**:233 (1989).
6. F. R. Ore, L. M. Hayden, G. F. Sauter, P. L. Pasillas, J. M. Hoover, R. A. Henry, and G. A. Lindsay, SPIE *Nonlinear Optical Properties of Organic Materials II*, **1147**:26 (1989).
7. G. A. Lindsay, J. M. Hoover, A. Knoesen, M. Mortazavi, S. Kowel, *ACS Polymer Preprints* 255, April 1990.
8. G. D. Green, J. I. Weinschenk, III, J. E. Mulvaney, and H. K. Hall, *Macromol.* **20**:722, (1987).
9. (a) J. D. Stenger-Smith, J. W. Fischer, R. A. Henry, J. M. Hoover, and L. M. Hayden, *Makromol. Chem. Rapid. Commun.* **11**:141 (1990).
   (b) J. D. Stenger-Smith, J. W. Fischer, R. A. Henry, J. M. Hoover, and L. M. Hayden, *ACS Polymer Preprints*, 375, April 1990.

NEW FERROCENE COMPLEXES AND POLYMERS FOR NONLINEAR OPTICAL APPLICATIONS

Michael E. Wright and Edward G. Toplikar

Department of Chemistry and Biochemistry
Utah State University
Logan, Utah 84322-0300

INTRODUCTION

Ferrocene was the starting point for organometallic chemistry nearly four decades ago and is still today one of the most studied and versatile organometallic building blocks.[1] Ferrocene has been incorporated in polymeric systems to alter bulk properties of the material.[2] Ferrocene possesses excellent thermal and photochemical stability and can also protect polymeric systems from photodegradation.[3] In addition, the ferrocene building block has been used in conducting polymers[4] and in main chain liquid crystalline polyesters.[5]

Nonlinear Optical (NLO) materials is a relatively new area of chemistry and has caught the attention of both the polymer and organometallic chemist.[6] The use of ferrocene derivatives in NLO applications was first independently studied by two research groups.[7] As anticipated, the metal center in ferrocene was found to serve as an excellent electron-donor.[8] The crystalline compounds (see below) exhibited very high second harmonic generation (SHG) efficiencies relative to urea.[7]

M = Mo(NO)(L)Cl

Several key questions regarding the utilization of organometallic NLO systems remain unanswered. For instance, there is no example of an organometallic NLO system which has been poled and then shown to have SHG activity. Thus, it was not established that an organometallic NLO material would even survive the strong electric field (~5 KV) applied during the poling process. Other key questions have also not yet been addressed. For example, determining the ability (or lack) of polymeric-organometallic materials to retain alignment and can they survive the laser irradiation. The mission of our research effort is to answer these questions concerning polymeric organometallic NLO materials. This paper presents the initial stages of our research endeavor in polymeric-organometallic NLO materials.

Recent work from our laboratory developed new methodology for the preparation of novel *bis*-functionalized ferrocene complexes (Scheme I).[9] The synthetic strategy gave us the ability to functionalize the cyclopentadienyl rings in a sequential manner. The strategy proved to be invaluable for ferrocene monomer synthesis and permits an array of novel complexes to be prepared.

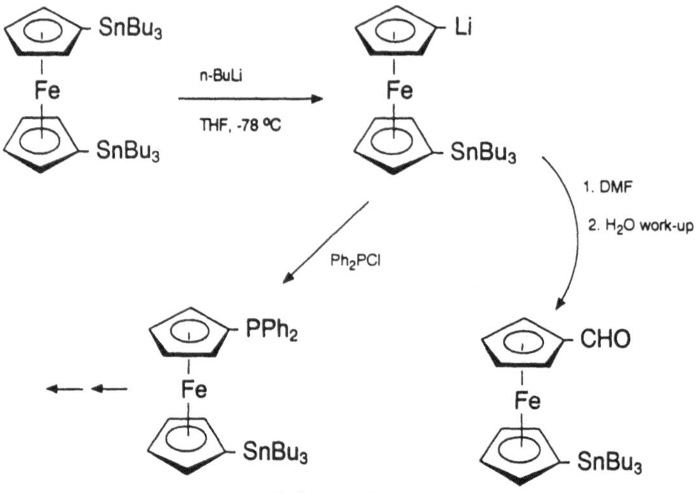

Scheme I

RESULTS AND DISCUSSION

Complex **4** was prepared in good yield through a sequence involving selective transmetalation of the tri-*n*-butylstannyl groups combined with established condensation chemistry of ferrocene carboxaldehydes (Scheme II).[10] Compound **4** crystallizes as long needles (some were 3 cm long!) but were unfortunately not suitable for a single-crystal X-ray diffraction study. The UV-Vis spectrum for compound **4** gave a λ_{max} of 515 nm ($\varepsilon = 2.24 \times 10^3$). The compound forms beautiful purple solutions which are air-stable for hours. In the solid state the material is air-stable indefinitely.

Compound **4** was suspended in poly(methyl methacrylate) (PMA) and then cast into a thin film. The "polymer solution" of **4** was then subjected to corona poling at elevated temperature.[11] The poling process afforded a polymer film which exhibited significant SHG (Scheme III). Using a primary frequency of 1064 nm the SHG band at 532 nm (*i.e.* green light) was visible to the naked eye. At this time there were no reference standards available for quantification of the data. After standing for several days the film showed little, if any SHG activity.

There are two very significant points which can be drawn from the results above. First, it was very clear that *the ferrocene "polymer solution" survived the poling process.* No apparent damage occurred to the polymer film (*e.g.* electrical arcing). Secondly, *the ferrocene systems responded to the poling process.* That is, the ferrocene compound has a significant permanent dipole to facilitate alignment.

Modification of **4** with methacryoyl chloride afforded the new monomer **5** in excellent yield (Scheme IV). Monomer **5** was copolymerized with methyl methacrylate (5/95 molar ratio, respectively) using the free-radical initiator AIBN. Copolymer **6** was purified by multiple precipitations into pentane and methanol. At this point we know the copolymer will cast good films but do not have any results on poling and SHG measurements.

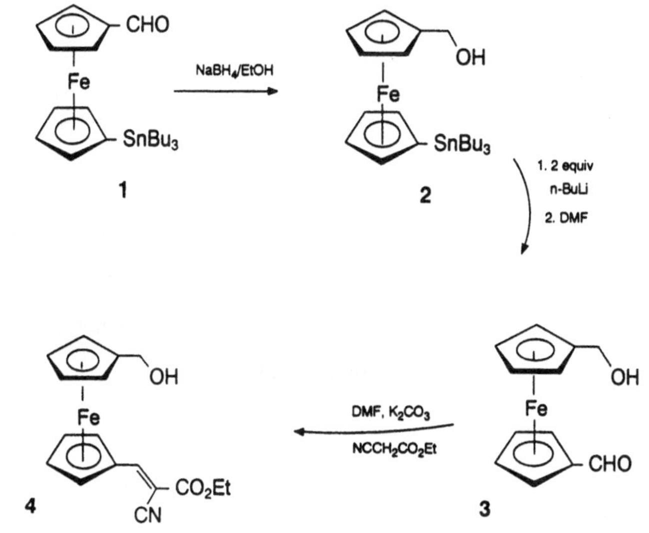

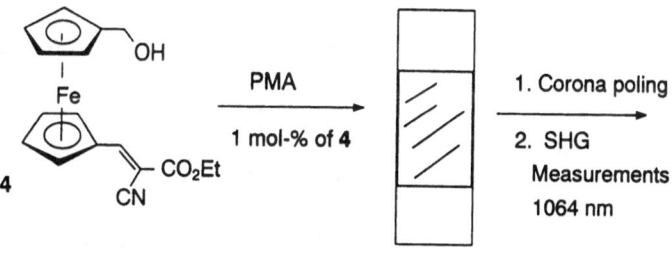

Scheme II

Scheme III

Polymer film supported
on a glass slide

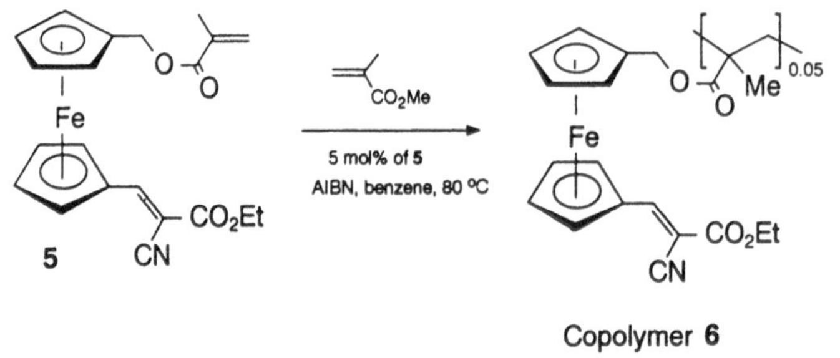

Scheme IV

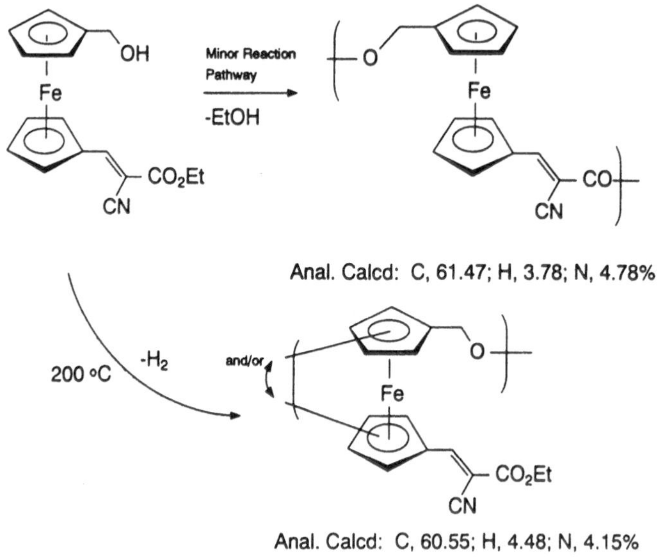

Anal. Calcd: C, 61.47; H, 3.78; N, 4.78%

Anal. Calcd: C, 60.55; H, 4.48; N, 4.15%

Scheme V

Another specific goal of our research program is to prepare homopolymers of NLO active ferrocene units. To accomplish this we carried out the homopolymerization of **4** (Scheme V). The polymerization was conducted under a nitrogen atmosphere as a melt at 200 °C. The polymer was isolated as a very tough and brittle glassy material. The polymeric material was collected and extracted with hot benzene for 24 h and then dried at reduced pressure for 24 h at 65 °C. Polymer **7** was assigned the structure illustrated in Scheme V based on spectroscopic, analytical data, and additional experiments discussed below. Polymer **7** was also prepared using the Lewis acid catalyst $Bu_2Sn(laurate)_2$ and found to afford a similar polymeric material. The latter conditions even though at 150 °C did promote elimination of HCN from the polymer as evidenced by a carbon-carbon triple bond stretch ($\nu_{C\equiv C}$ 2357 & 2341 cm^{-1}) in the infrared spectrum and a lowering of the nitrogen content.

Thermal gravimetric analysis (N$_2$, 10 °C/min ramp) of **7** exhibited a weight loss of ~8% by 300 °C which corresponds to ejection of HCN from the homopolymer. Continued heating of the sample to 700 °C resulted in a total weight loss of 32%. Subjecting a sample to differential scanning calorimeter (N$_2$, 10 °C/min ramp) showed the loss of HCN and the final mode of decomposition were exothermic events.

Ferrocenylalkanol derivatives have been homopolymerized using Lewis acid catalysis.[12] The mode of polymerization has been explained through formation of the α-ferrocenyl carbocation,[13] followed by a homo- or heteroannular Friedel-Crafts alkylation of a η^5-cyclopentadienyl ring of another ferrocene unit. We believe complex **4** deviates from this type of reaction pathway because of the very electron-withdrawing β-(α-cyanoacrylate) group. Depicting a logical resonance contributor to the structure of **4** illustrates why nucleophilic attack of the η^5-cyclopentadienyl ring is deemed reasonable (see below). Attack of the fulvene ring (*i.e.* heteroannular) represents conjugate addition of an alcohol to a very electron-poor olefin. The addition of alcohols to such olefins (*i.e.* good Michael exceptors) is in fact a well established organic reaction.[14] We have been unable to model this reaction and thus at this point we cannot prove beyond a shadow of a doubt our mechanism is indeed correct.

We have initiated a systematic X-ray diffraction study on ferrocene systems having electron-poor olefins attached to the cyclopentadienyl ring. In particular we are looking for structural evidence to support the fulvene resonance contribution to the ground state. To date, we have completed the single-crystal X-ray structure of $(\eta^5\text{-}C_5H_5)(\eta^5\text{-}C_5H_4CH=C(CN)CO_2Et)Fe$ (**8**).[10] A drawing of the structure is displayed in Figure I. The structure was refined to $R = 3.14\%$ with a GOF of 1.30. There are distinct bond length values which are consistent with a fulvene-like structure. However, at this point with only one structure in the series we are hesitant to put forth any strong conclusions. We are continuing to prepare analogues and accumulate more structural data.

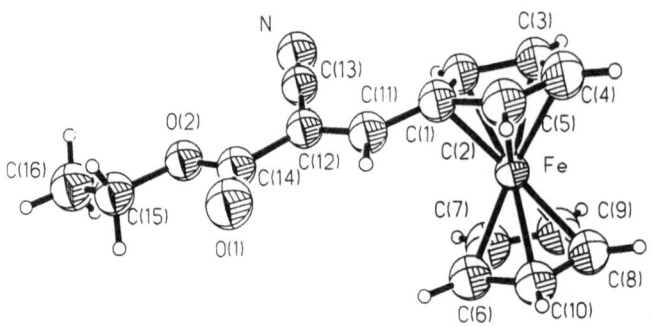

Figure I. Drawing of the single-crystal molecular structure of $(\eta^5\text{-}C_5H_5)(\eta^5\text{-}C_5H_4CH=C(CN)CO_2Et)Fe$ (**8**) showing the labeling scheme employed. The atoms are drawn to include 30% of the electron density and hydrogen atoms have been given arbitrary isotropic thermal parameters.

Cyclic voltamograms of **4** and **8** show the reversible oxidation event to be shifted 0.30 V and 0.32 V, respectively, more positive than the ferrocene/ferrocenium redox couple. These data would suggest that the iron centers in **4** and **8** are much less capable of stabilizing a positive charge. It is logical to argue they would also form a much less stable α-ferrocenyl carbocation intermediate; hence, we tend to rule out such a species in the homopolymerization of complex **4**.

CONCLUDING REMARKS

In this study we have demonstrated that ferrocene complexes can be placed in a polymer solution and successfully aligned by the corona poling technique. In a qualitative terms the film showed significant SHG activity (*i.e.* a visible green beam emitted from the film) and as expected, lost this activity over time. In addition, the synthesis of a new copolymer was presented. This latter material has been submitted for poling and subsequent SHG activity evaluation.

EXPERIMENTAL

General. All manipulations of compounds and solvents were carried out by using standard Schlenk techniques. Solvents were degassed and purified by distillation under nitrogen from standard drying agents. Spectroscopic measurements utilized the following instrumentation: ^1H NMR, Varian XL 300; ^{13}C NMR, Varian XL 300 (at 75.4 MHz). NMR chemical shifts are reported in δ versus Me_4Si in ^1H NMR and assigning the $CDCl_3$ resonance at 77.00 ppm in ^{13}C spectra. The $(\eta^5\text{-}C_5H_4CHO)(\eta^5\text{-}C_5H_4SnBu_3)Fe$ was prepared by the literature method.[9] Polymer analyses were performed using a duPont 9900 thermal analysis data station. Elemental analyses were performed at Atlantic Microlab Inc, Norcross, Georia.

$(\eta^5\text{-}C_5H_4CH_2OH)(\eta^5\text{-}C_5H_4SnBu_3)Fe$ (**2**). ^1H NMR $(CDCl_3)$ δ 4.33 (t, J = 1.7 Hz, 2 H), 4.32 (d, J = 5.8 Hz, 2 H), 4.18 (t, J = 1.8 Hz, 2 H), 4.09 (d, J = 1.8 Hz, 2 H), 4.03 (t, J = 1.7 Hz, 2 H), 1.56 (m, 7 H, OH & CH_2), 1.35 (m, 6 H, CH_2), 1.01 (apparent triplet, 6 H, CH_2), 0.91 (t, J = 7.3 Hz, 9 H, CH_3); ^{13}C NMR $(CDCl_3)$ δ 88.0 (*ipso*-Cp),

74.7, 70.6 (Cp-SnBu$_3$), 69.5 (*ipso*-Cp), 68.4, 67.9 (Cp-CH$_2$OH), 60.9 (CH$_2$OH), 29.2 (CH$_2$), 27.4 (CH$_2$), 13.7 (CH$_3$), 10.2 (Sn-CH$_2$). Anal. Calcd for C$_{23}$H$_{38}$FeOSn: C, 54.69; H, 7.58%. Found: C, 54.78; H, 7.60%.

(η^5-C$_5$H$_4$CH$_2$OH)(η^5-C$_5$H$_4$CHO)Fe (3). ^{1}H NMR (CDCl$_3$) δ 9.98 (s, 1 H, CHO), 4.81 (t, *J* = 1.8 Hz, 2 H), 4.64 (t, *J* = 1.8 Hz, 2 H), 4.36 (t, *J* = 1.8 Hz, 2 H), 4.32 (d, *J* = 5.6 Hz, 2 H), 4.25 (t, *J* = 1.8 Hz, 2 H), 2.08 (br s, 1 H, OH); ^{13}C NMR (CDCl$_3$) δ 194.3 (C=O), 90.3 (*ipso*-CpCHO), 79.4 (*ipso*-CpCH$_2$OH), 73.7 (Cp), 70.1 (Cp), 69.6 (Cp), 69.0 (Cp), 59.9 (CH$_2$OH); IR (CH$_2$Cl$_2$) $\nu_{C=O}$ 1702 cm^{-1}. Anal. Calcd for C$_{12}$H$_{12}$FeO$_2$: C, 59.05; H, 4.96%. Found: C, 58.78; H, 4.93%.

(η^5-C$_5$H$_4$CH$_2$OH)(η^5-C$_5$H$_4$CH=C(CN)CO$_2$Et)Fe (4). ^{1}H NMR (CDCl$_3$) δ 8.17 (s, 1 H, =C<u>H</u>(CN)CO$_2$Et), 5.03 (t, *J* = 1.8 Hz, 2 H), 4.74 (t, *J* = 1.8 Hz, 2 H), 4.33 (t, *J* = 7.1 Hz, 2 H), 4.32 (m, 4 H), 4.27 (t, *J* = 1.8 Hz, 2 H), 1.70 (br s, 1 H, O<u>H</u>), 1.39 (t, *J* = 7.1 Hz, 3 H); ^{13}C NMR (CDCl$_3$) δ 163.2 (<u>C</u>O$_2$Et), 158.4 (C=<u>C</u>H), 116.8 (<u>C</u>N or =<u>C</u>(CN)CO$_2$Et carbon), 97.2 (*ipso*-Cp), 90.0 (*ipso*-Cp), 74.5 (Cp), 72.1 (Cp), 70.6 (Cp), 70.0 (Cp), 62.1 (O<u>C</u>H$_2$CH$_3$), 59.7 (<u>C</u>H$_2$OH); IR (film) $\nu_{C≡N}$ 2220, $\nu_{C=O}$ 1718, and $\nu_{C=C}$ 1588 cm^{-1}; UV-Vis (CH$_2$Cl$_2$) 322 (ε = 1.37 x 10^4) and 516 (ε = 2.23 x 10^3) nm. Anal. Calcd for C$_{17}$H$_{17}$FeNO$_3$: C, 60.18; H, 5.06%. Found: C, 60.24; H, 5.06%.

Polymer 7. IR (film) 2200, 1718, 1588 cm^{-1}. A new IR band appeared at 1400 cm^{-1}. Direct side-by-side comparison of films showed no change in color (the human eye as the spectrophotometry) before and after polymerization. Anal. Found: C, 60.80; H, 4.22; N, 4.09%.

ACKNOWLEDGEMENT

We wish to express our gratitude to donors of the Petroleum Research Fund, administered by the American Chemical Society and the Office of Naval Research for their financial support of this research. We also wish to thank the Utah State University Research Office and the NSF (grant CHE-9002379) for funding the purchase of the single-crystal X-ray diffractometer. We also express our appreciation to Drs. Robert Kubin and Michael Seltzer for carrying out the poling and qualitative SHG measurements.

REFERENCES

1. M. Rosenblum, "Chemistry of the Iron Group Metallocenes" Wiley, New York, (1965).
2. E. W. Neuse, J. R. Woodhouse, G. Montaudo, and C. Puglis, <u>Appl. Organomet. Chem.</u> 2:53 (1988) and references cited therein.
3. J. W. Harwood, "Industrial Applications of Organometallic Compounds," Reinhold, New York (1963). J. C. Johnson, jr. "Metallocene Technology," Noyes Data Corporation, Park Ridge, New Jersey (1973).
4. C. Iwakura, T. Kawai, M. Nojima, and H. Yoneyama, <u>J. Electrochem. Soc.</u> 134:791 (1987). For general treatments of "Organometallic polymers" see: M. Zeldin, K. J. Wynne, Allcock, H. R., Eds., "Inorganic and Organometallic Polymers: Macromolecule, Containing Silicon, Phosphorus, and other Inorganic Elements," ACS Symp. Ser., Washington D. C. (1987). C. U. Pittman, jr., M. D. Rausch, <u>Pure Appl. Chem.</u> 58:617 (1986). J. E. Sheats, C. E. Carraher, C. U. Pittman, jr., Eds., "Metal-Containing Polymer Systems," Plenum, New York (1985).

5. P. Singh, M. D. Rausch, and R. W. Lenz, <u>Polym. Bulletin</u> 22:247 (1989).

6. For a general treatment of NLO materials see: "Nonlinear Optical and Electroactive Polymers", eds. P. N. Prasad, D. R. Ulrich, Plenum Press, New York (1988).

7. M. L. H. Green, S. R. Marder, M. E. Thompson, J. A. Bandy, D. Bloor, P. V. Kolinsky, and R. J. Jones, <u>Nature</u> 330:360 (1987). J. W. Perry, A. E. Stiegman, S. R. Marder, D. R. Coulter, <u>in</u>: "Organic Materials for Nonlinear Optics," R. A. Hann and D. Bloor, eds.; Spec. Publ. No. 69, The Royal Society of Chemistry: London, England (1989). B. J. Coe, C. J. Jones, J. A. McCleverty, D. Bloor, P. V. Kolinsky, and R. J. Jones, <u>J. Chem. Soc., Chem. Commun.</u> 1485 (1989).

8. For a theoretical (SCF-LCAO MECI formalism) treatment of organometallic NLO materials see: D. R. Kanis, M. A. Ratner, and T. J. Marks, <u>J. Am. Chem. Soc.</u> 112:8203 (1990).

9. M. E. Wright, <u>Organometallics</u> 9:853 (1990).

10. I. K. Barben, <u>J. Chem. Soc.</u> 1827 (1961).

11. This work was completed at the Polymer Science Division of the Naval Weapons Center, China Lake, California. Division Head: Dr. Geoff Lindsay.

12. (a) E. W. Neuse and H. Rosenberg, "Metallocene Polymers," Marcel Dekker, New York (1970) and references cited therein. (b) For a more recent example see: L. Zhan-Ru, K. Gonsalves, R. W. Lenz, and M. D. Rausch, <u>J. Polym. Sci. A</u> 24:347 (1986) and references cited therein.

15. The α-ferrocenylisopropyl carbocation was isolated in ref. 12(b) and a α-(octamethylferrocenyl)methyl carbocation has also been isolated (C. Zou and M. S. Wrighton, <u>J. Am. Chem. Soc.</u> 112:7578 (1990). Furthermore, α-(nonamethyl-ruthenocenyl)methyl carbocation has been generated and utilized to prepare derivatives via nucleophilic reactions (U. Kölle and J. Grub, <u>J. Organomet. Chem.</u> 289:133 (1985). From these examples it is apparent that substitution at the α-methyl carbon or on the η^5-cyclopentadienyl rings are both very important in stabilizing the carbocation. To our knowledge no ferrocenyl carbocation derivatives bearing electron-withdrawing groups have been postulated or observed.

16. March, J. "Advanced Organic Chemistry" John Wiley & Sons, New York, 3rd edition (1985) pp 670-671 and references cited therein.

NONLINEAR OPTICAL THIN FILMS OF PLATINUM POLY-YNES

Pamela L. Porter, Shekhar Guha, Keith Kang, and Claude C. Frazier

Martin Marietta Laboratories
1450 S. Rolling Road
Baltimore, MD   21227

ABSTRACT

The imaginary part of the third-order hyperpolarizability (γ) was measured for a series of platinum poly-ynes in both solution and thin-film form, and for a group of related platinum-organic species.

INTRODUCTION

Transition metal poly-ynes have large third-order optical susceptibilities (χ^3) which consist of contributions from the real and imaginary parts of the third-order hyperpolarizabilities (γ' and γ'', respectively). In previous studies, various optical measurements were used to determine the nonlinear hyperpolarizabilities of the polymers.[1-6] Most recently, an extensive study of the structural dependence of the hyperpolarizabilities of the platinum and palladium poly-ynes was reported.[6] This study provides us with guidelines for designing polymers having high values of nonlinearity. From those polymers that exhibited the largest values of the hyperpolarizability in dilute-solution form, we selected several for study as free-standing films, in an attempt to obtain materials with the highest values of macroscopic nonlinearity. We present here a study of free-standing films made from these platinum poly-ynes. The aim of this work was to obtain materials that not only possess very high values of nonlinearity but also are well suited for the fabrication of optical devices. We have selected the approaches of alteration of polymer repeat structure, variation of chain length, and end-group modification in an effort to both enhance the nonlinearity and improve thin-film formation or allow the attachment of the polymers to surfaces.

Previously, the effects of modifying the polymer repeat unit and chain length on optical nonlinearity were explored.[6] Here, we report on the effect of polymer chain length on thin-film formation. Also studied was one of the simple platinum-organic complexes from which the polymers are derived. The nonlinearity of such complexes was previously observed to be sensitive to the nature of the end group.[6] We include here the preliminary results of additional end-group substitutions which will be used as a basis for determining the most promising candidates for attachment to the polymers.

SYNTHESIS

The simple organometallic complexes used for this study were prepared by the method of Takahashi et al.[7] Chlorine end groups were replaced by refluxing Bis[*trans*-chlorobis(tri-n-butylphosphine)platinum]-1,4-phenylenediethynylene with the appropriate compounds in acetone or toluene in the presence of piperidine. Purification procedures for the modified complexes were similar to that[7] for the parent compound. Polymers were prepared as previously described.[6]

EXPERIMENTAL

Polymer solutions were prepared with deoxygenated, spectral-grade tetrahydrofuran and placed in air-tight glass cuvettes with 2-mm pathlengths. Films were cast from concentrated solutions of the poly-ynes in either tetrahydrofuran or methylene chloride onto glass substrates. Solvent was usually removed under vacuum. The resultant films were removed from the glass and stored in film holders. Film thicknesses were measured with a Tencor Alpha-Step 200 surface profiler.

The two-photon absorption coefficients, β, for the solutions and films were determined by the intensity-dependent transmission technique that has been described earlier.[5] As before, the laser beam used was obtained from a frequency-doubled Nd:YAG laser (of 532 nm wavelength) that was mode locked to produce 30-picosecond pulses.

For the thin films, the susceptibility, χ^3, was obtained from the β values via the formula

$$\mathrm{Im}\!\left(\chi^3\right) = \frac{\beta n^2 c^2}{96\pi^2\omega}$$

where ω = frequency of light (radians)
   n = refractive index of the medium
   c = speed of light.

For the poly-yne solutions, the hyperpolarizability value per polymer repeat unit, γ'', was obtained from the formula

$$\gamma'' = \mathrm{Im}\!\left(\chi^3\right) N_p L$$

where N_p = number of repeating units per given volume

   L = local field factor = $\left[\left(\dfrac{n^2 + 2}{3}\right)\right]^4$.

The experimental error in the measurement of β was estimated to be at most 30%.

RESULTS AND DISCUSSION

Since β is concentration dependent, only samples with similar concentrations may be directly compared. A better correlation may be obtained by comparison of γ'' values; there, too, the concentrations should be fairly equivalent since we previously noted a concentration dependence of γ'' for some polymers[6] (that is, γ'' values increased for progressive dilutions of a given sample).

TABLE 1

Pt Poly-yne Solution Measurements

Ref. #	Sample	Moles/liter	β (cm/GW)	γ'' (10^{-36} esu)
2	(Pt poly-yne structure with OCH₃ substituents)	0.036 / 0.048 / 0.080	2.14 / 2.67 / 4.20	1,724 / 1,586 / 1,492
3	(Pt poly-yne structure with CH₃ substituents)	0.03 / 0.034	2.6 / 3.2	2,432 / 2,687
4	(Pt poly-yne structure with C₂H₅ substituents)	0.003 / 0.007	0.52 / 1.14	4,933 / 4,649
5	(Pt poly-yne structure)	0.007	1.09	4,466
6	(Pt poly-yne structure with pyridine N rings)	0.004	1.80	12,730
7	(Pt poly-yne structure) n=223 / n=97	0.010 / 0.010	1.4 / 1.6	4,558 / 4,025
8	(Cl-terminated Pt poly-yne structure)	0.072	5.5	2,167

Table 1 shows the solution measurements for the polymers which were later cast as thin films. A comparison of samples 4 and 6, which were measured as very dilute solutions, shows that nonlinearity increases by more than 2.5 times when pyridine is substituted for both benzene rings. This effect was seen in our earlier work;[6] however, the increase was not as dramatic when substitutions in analogous single-arene repeat unit polymers were compared.

Polymers 4, 5, and 7, in the 0.007-0.01M range, exhibit approximately equal values for nonlinear absorption.

Comparison of polymer 2 (0.036M solution) and polymer 3 (0.034M solution) shows that a poly-yne with two diethynylxylenes in the repeat unit has a 40% greater γ'' value than does the single diethynyldimethoxybenzene polymer.

We expect to see the same trends for the free-standing films. However, with thin films, optical clarity, color, uniformity and mechanical strength, as well as nonlinearity, must also be taken into account. We therefore used ($\chi 3/\alpha$) as the figure of merit to compare different films.

To date, our initial efforts to form good optical-quality films from most of the polymers with a single diethynylarene in the repeat unit have not been successful, with the exception of film 1 (Table 2). Additionally, polymers with chain lengths of less than 100 units were usually poorer film-formers. The best films were formed from the longer chain polymers which also demonstrated good solubility in tetrahydrofuran or dichloromethane. Addition of biphenyl to the repeat unit produced films with some of the best mechanical properties in this series. For example, although most of our thin polymer films were very fragile, polymer 5 could be gently stretched taut in a film holder. Film 5 also showed improved optical quality with good uniformity. Films from polymer 6, especially the thicker films in the 15-30-μm range, were also very flexible and could be easily handled. However, they possessed the poorest optical quality in the series. Many polymer 6 films had strong absorptions at 532-nm which limited their usefulness.

Films 3-5 exhibited the highest β values, as well as figures of merit, of the series. In solution, however, polymers 2 and 3 had the lowest β values of the series yet films of each yielded figures of merit comparable to those of polymers 4 and 5, whose solution β values doubled those of 2 and 3. Also noteworthy in this study was the low β values seen for film 6. The sizeable β value obtained in solution form may not have been realized for the film due to its poor optical quality. Clearly, though, other factors are significant.

As was previously postulated,[6] film β values were much higher than those for solutions. In each case, β increased by at least two orders of magnitude and by greater than 3 orders of magnitude for polymers 4-5.

The platinum poly-yne films presented here have figures of merit similar to that of film 1 (Table 2), a well-known film with high nonlinearity. While the nonlinearities of the platinum poly-yne films are usually higher than that of film 1, the absorption coefficients, α, are also larger.

TABLE 2

Pt Poly-yne Films

	Sample	thickness (μm)	β (cm/MW)	α (cm⁻¹)	χ^3 (esu)	χ^3/α (esu-cm)
1	Foster Miller Biaxially-oriented PBT (poly-p-phenylenebenzobisthiazole)	20	1.3	1530	7.0×10^{11}	4.6×10^{-14}
2	[Pt(PBu$_3$)$_2$–C≡C–C$_6$H$_2$(OCH$_3$)$_2$–C≡C–]$_n$	13.5	1.1	1333	5.9×10^{-11}	4.4×10^{-14}
3	[Pt(PBu$_3$)$_2$–C≡C–C$_6$H$_2$(CH$_3$)$_2$–C≡C–C≡C–C$_6$H$_2$(CH$_3$)$_2$–C≡C–]$_n$	8 / 15.2	1.4 / 1.8	1830 / 1808	7.5×10^{-11} / 9.6×10^{-11}	4.1×10^{-14} / 5.3×10^{-14}
4	[Pt(PBu$_3$)$_2$–C≡C–C$_6$H$_2$(C$_2$H$_5$)$_2$–C≡C–C≡C–C$_6$H$_2$(C$_2$H$_5$)$_2$–C≡C–]$_n$	8.2	1.5	2094	8.0×10^{-11}	3.8×10^{-14}
5	[Pt(PBu$_3$)$_2$–C≡C–C$_6$H$_4$–C≡C–Pt(PBu$_3$)$_2$–C≡C–C$_6$H$_4$–C$_6$H$_4$–C≡C–]$_n$	10.7	2.0	2876	1.1×10^{-10}	3.7×10^{-14}
6	[Pt(PBu$_3$)$_2$–C≡C–C$_5$H$_3$N–C≡C–C≡C–C$_5$H$_3$N–C≡C–]$_n$	8 / 2	0.6 / 0.7	2187 / 5108	3.2×10^{-11} / 3.7×10^{-11}	1.5×10^{-14} / 7.3×10^{-15}
7	[Pt(PBu$_3$)$_2$–C≡C–C$_6$H$_4$–C≡C–C≡C–C$_6$H$_4$–C≡C–]$_n$	1	0.8	1980	4.3×10^{-11}	2.2×10^{-14}

TABLE 2

End-Group Modifications

Ref. #	Platinum-organic Complex	Ratio of β values	Ratio of α values
I	Cl—Pt—C≡C—⟨O⟩—C≡C—Pt—Cl (PBu₃)	1.0	1.0
II	I—Pt—C≡C—⟨O⟩—C≡C—Pt—I (PBu₃)	0.86	8.6
III	NCS—Pt—C≡C—⟨O⟩—C≡C—Pt—NCS (PBu₃)	1.30	1.0
IV	⟨O⟩—C≡C—Pt—C≡C—⟨O⟩—C≡C—Pt—C≡C—⟨O⟩ (PBu₃)	1.33	1.0
V	Cl—Pt—C≡C—⟨O⟩—C≡C—C≡C—⟨O⟩—C≡C—Pt—Cl (PBu₃)	2.79	2.5

β and α values have been normalized for complex I to show relative increases

To further improve the film characteristics of these poly-ynes, we conducted a preliminary study of the effect of modifying the end groups on the platinum-organic complex analogous to the polymer repeat unit, dichlorobis[(tri-n-butylphosphine)platinum]-1,4-phenylenediethynylene. From Table 3, we see that the replacement of chlorine (complex I) with iodine (complex II) decreases nonlinearity and greatly increases linear absorption. Substitution of isothiocyanate (III) and phenylacetylene (IV) in the end groups produces no change in absorption at 532 nm but results in a 30% and 33% larger value for β, respectively. Still, the largest nonlinearity is observed for complex V, which contains two diethynylarenes. We are currently replacing the chlorines in this compound with additional end groups, some of which are known film-formers.

CONCLUSIONS

The work presented here showed that β increased significantly for the thin polymer films compared with the solution values.[6] A comparison of figures of merit indicated that the thin poly-yne films are competitive with Foster Miller's biaxially-oriented PBT film. Development of improved film quality should produce films with figures of merit surpassing those presented at this time.

The initial study of end-group modifications of a simple platinum-organic complex showed that appropriate substitutions increased nonlinearity without a concomitant rise in absorption. We will continue to explore further modifications of the simple organo-platinum complexes plus alterations of the platinum poly-yne end-groups to enhance film formation, film quality, and nonlinearity.

The usefulness of highly nonlinear organic thin films for applications in the field of integrated optics has been described in detail by Stegeman et al.[8]. For the organometallic polymer thin films and solutions described in this paper, the imaginary (absorptive) part of the nonlinearity dominates the real (refractive) part. As a result, these materials are not useful for applications which depend only on the phase change of light beams as light intensity changes; in these cases, the nonlinear absorption of light is an undesirable feature. However, the thin films may be of use in optical switching devices that are based on nonlinear absorption, such as the two-photon optical bistable switch proposed by Kothari and Kobayashi.[9] The favorable values of χ^3/α and the ultrafast response time of the nonlinearity of the poly-yne thin films described here may be exploited for the fabrication of such devices.

ACKNOWLEDGMENTS

We wish to thank NADC for the partial support of this work under contract N62269-90-R-0207.

REFERENCES

Frazier, C.C., Guha, S., Chen, W.P., Cockerham, M.P., Porter, P.L., Chauchard, E.A., and Lee, C.H., "Third-order optical non-linearity in metal-containing organic polymers," *Polymer* 28:553 (1987).

Frazier, C.C., Chauchard, E.A., Cockerham, M.P., and Porter, P.L., "Four-wave mixing in metal poly-ynes," *Proceedings of the Materials Research Society Symposium* 109:323 (1988).

Frazier, C.C., Guha, S., Porter, P.L., Cockerham, M.P., and Chauchard, E.A., "Nonlinear optical properties of transition metal poly-ynes," *Proceedings of the SPIE* 971:187 (1988).

Guha, S., Frazier, C.C., Chen, W.P., Porter, P.L., Kang, K., and Finberg, S., "Nonlinear devices using organo-metallic polymers, " *Proceedings of the SPIE* 1105:14 (1989).

Guha, S., Frazier, C.C., Porter, P.L., Kang, K., and Finberg, S., "Measurement of the third-order hyperpolarizability of platinum poly-ynes," *Optics Letters* 14:952 (1989).

Porter, P.L., Guha, S., Kang, K. and Frazier, C.C., "Structural dependence of platinum and palladium poly-ynes on third-order nonlinearity," *Polymer* 32:1756 (1991).

Takahashi, S., Ohyama, Y., Murata, E., Sonogashira, K., and Hagihara, N., Studies of poly-ynes polymers containing transition metals in the main chain," *Journal of Polymer Science: Polymer Chem. Ed.* 18:349 (1980).

Stegeman, G.I., Seaton, C.T., and Zanoni, R., "Organic films in non-linear integrated optics structures," *Thin Solid Films* 152:231 (1987).

Kothari, N.C., and Kobayashi, T., "Single beam two-photon optical bistability in a submicron size Fabry-Perot cavity," *IEEE Journal of Quantum Electronics* QE-20:418 (1984).

DESIGN OF NEW TYPE OF LIQUID CRYSTALLINE POLYMERS

THROUGH INTERMOLECULAR HYDROGEN BONDING

Takashi Kato,[*] Hajime Adachi,
Norifumi Hirota, and Akira Fujishima

Department of Synthetic Chemistry
Faculty of Engineering
The University of Tokyo
Bunkyo-ku, Tokyo 113, Japan

Jean M. J. Fréchet

Department of Chemistry
Baker Laboratory
Cornell University
Ithaca, New York, 14853-1301

INTRODUCTION

Relationship between structure and properties of liquid crystalline (LC) polymers has been studied extensively.[1-3] In the design of these thermotropic polymers, hydrogen bonding, which may lead to nonlinear molecular association, has been considered to be deleterious, while the importance of dipole-dipole interactions has been established. For example, poly(phenyleneterephthalamide) does not show melting and thermotropic LC behavior due to the hydrogen bond network between the amido linkage.[4,5] It exhibits lyotropic liquid crystalline behavior in sulfuric acid which breaks hydrogen bonding. Hydrogen bonding between amido linkages was successfully used for thermotropic liquid crystallinity of a series of polyesteramides containing aliphatic soft segments.[6]

Recently, we have found[7-10] that novel type of liquid crystalline structures is built through intermolecular hydrogen bonding between donor and acceptor moieties. Well-defined structures of H bonded complexes are formed from different and independent molecules. In this case, each molecule functions like an element of molecular Lego because this hydrogen bonding is directional and selective. These results have clearly shown that hydrogen bonding interaction

---

[*]Present address: Institute of Industrial Science,
The University of Tokyo, Minato-ku, Tokyo 106, Japan

Contemporary Topics in Polymer Science, Vol. 7., Edited by
J.C. Salamone and J. Riffle, Plenum Press, New York, 1992

is useful for liquid crystallinity if it is designed properly.

Schematic illustration of the structure of a typical side-chain LC polymer is shown in Scheme I. Mesogenic units are covalently attached to the polymer backbone via a flexible spacer.[2,3] Our aim in the present study is to construct new type of side-chain liquid crystalline polymers through intermolecular hydrogen bonding in place of covalent bonding. Hydrogen bonding may be used to connect the part where the broken line shows in Sceme I. We describe self-assembly of LC polymers through selective recognition between a polymer side chain and low-molecular-weight molecules.

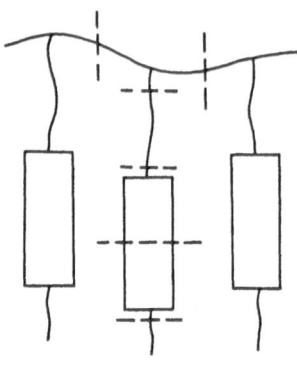

Scheme I

RESULTS AND DISCUSSION

A Novel Structure of Side-Chain Liquid Crystalline Polymer Built through Intermolecular Hydrogen Bonding

Polyacrylate **P6BA** containing a 4-oxybenzoic acid moiety has been desinged and used as H bonding donor polymer. **P6BA** shows a mesophase between 140 and 155 $^\circ$C.[7] Stilbazole derivatives, **8OSz** and **1PhOSz** have been selected as hydrogen bond acceptor which interacts with the polymer side chain through hydrogen bonding. **8OSz** has a longer aliphatic group and a shorter mesogen than those of **1OPhSz**. **8OSz** exhibits smectic phases between 77 and 89 $^\circ$C,[11] while **1OPhSz** shows only a nematic phase between 168 and 216 $^\circ$C.[7]

H-Bond Donor

$+CH_2CH+_n$ ... O−$(CH_2)_6$O−⟨⟩−C$\overset{O}{\underset{OH}{}}$ **P6BA**

H-Bond Acceptor

$CH_3+CH_2+_7$O−⟨⟩−CH=CH−⟨N⟩ **8OSz**

CH_3O−⟨⟩−C$\overset{O}{\underset{O}{}}$−⟨⟩−CH=CH−⟨N⟩ **1OPhSz**

Table I shows liquid crystalline behavior of the equimolar complexes between H-bond donor polymer **P6BA** and stilbazole derivative **8OSz** or **1OPhSz**. The polymeric complexes are denoted as **P6BA/8OSz** and **P6BA/1OPhSz**, respectively. Significant effects of the hydrogen bonding on the stability of the mesophases are observed for both complexes. For **P6BA/8OSz**, a smectic phase is observed between 125 and 183 $^\circ$C on heating, while each of single components of **P6BA** and **8OSz** exhibits a mesophase up to 155 and 89 $^\circ$C, respectively. Figure 1

Table I

Thermotropic behavior of H-bonded side-chain liquid crystalline polymeric complexes.

H-Bond Donor	H-Bond Acceptor	Transition Temperatures ($^{\circ}$C)				
P6BA	8OSz	k	125	s	183	i
P6BA	10PhSz	g	140	n	252	i

K:crystalline; g:glassy; s:smectic; n:nematic; i:isotropic.

illustrates DSC curves of the H-bonded polymeric complex between **P6BA** and **8OSz**. They clearly reveal reversible crystal-smectic and smectic-isotropic phase transitions. The endothermic peaks do not shift on repetition of heating and cooling. A focal-conic fan texture which is appearing from a isotropic phase at 177 $^{\circ}$C on cooling is shown in Fig. 2. **P6BA/10PhSz** exhibits significantly extended mesophase up to 252 $^{\circ}$C. These results clearly demonstrate that a new type of side-chain liquid crystalline polymer is built through the intermolecular hydrogen bonding and the H-bonded polymeric complex behaves as one side-chain LC polymeric complex, as shown in Fig. 3.

FT-IR measurements support the existence of the H-bonded polymeric complex. For **P6BA/8OSz**, the carbonyl band corresponding to the complexation between the benzoic acid and the pyridyl unit is observed at 1702 cm^{-1} while the band at 1685 cm^{-1} is seen for **P6BA** due to dimerization of the benzoic acid moiety. For the complex, the O-H band is observed at 2450 and 1900 cm^{-1}, which is indicative of a strong hydrogen bond.[12,13] For a phenol/pyridine complex which consists of a weaker hydrogen bond, the O-H band appears at 3010 cm^{-1}.[12]

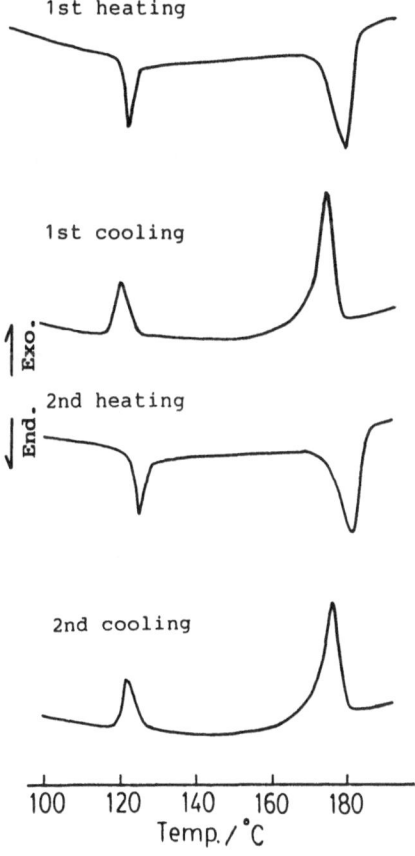

Fig. 1. DSC thermograms of hydrogen-bonded polymeric complex **P6BA/8OSz** on heating and cooling runs at the rate of 10 $^{\circ}$C/min.

Fig. 2. Photomicrograph of a focal-conic fan
texture of **P6BA/8OSz** appearing from an
isotropic phase at 177 °C on cooling viewed
between crossed polarizers.

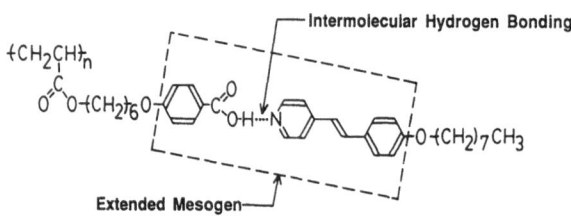

Fig. 3. The structure of the new mesogen formed
through intermolecular hydrogen bonding between
P6BA and **8OSz**.

The Effect of the Intermolecular Hydrogen Bonding on Liquid Crystalline Binary Mixtures between a Polyacrylate and a Low-Molecular-Weight Compound

The binary system involving the intermolecular hydrogen
bonding may be regarded as a new type of liquid crystalline
binary mixture as depicted in Scheme II.

Figure 4 shows a binary phase diagram of **P6BA** and **10PhSz**.[7]
The intermolecular hydrogen bonding interaction between the
polymer and the low-molecular-weight compound causes a
significantly strong enhancement of the mesophase. For
example, a binary mixture containing 25 mol% of **P6BA**
exhibits a nematic phase extended to 240 °C. Mesophase-
isotropization transition temperature is usually lower than
that of either component. Moreover, **P6BA** and **10PhSz** are
completely miscible over the whole range of composition. A
phase separation is usually observed for a binary mixture of

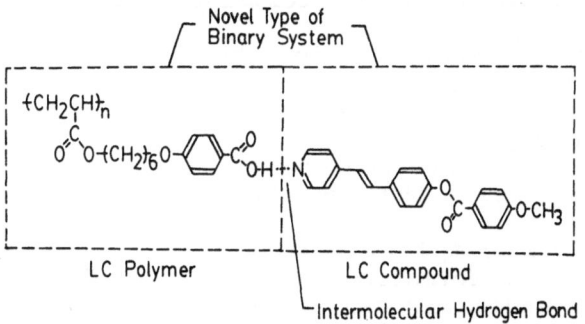

Scheme II

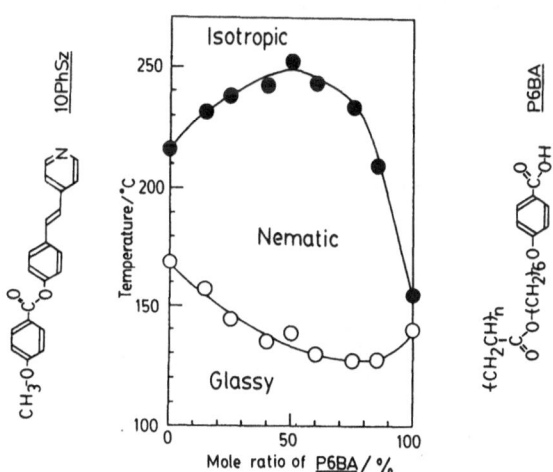

Fig. 4. Binary phase diagram of **P6BA** and **10PhSz**.

LC side-chain polymers and low-molecular-weight LC compounds.[14,15] Liquid crystalline polymers mixing with low-molecular-weight compounds have been used for functional materials.[16,17] The hydrogen bonding interaction in LC binary mixtures has great potential in the design of novel host-guest liquid crystalline materials.

Molecular Design through Intermoleculrar Hydrogen Bonding

The present results provide the concept for a new type of liquid crystalline polymers built through selective intermolecular hydrogen bonding between carboxylic acids and pyridyl units. This non-covalent bonding keeps the overall linear structure of the core unit of the complex, which results in the formation of a new and extended mesogen. In the structure, the hydrogen bonding functions like a connecting part of Lego, as illustrated in Scheme III.

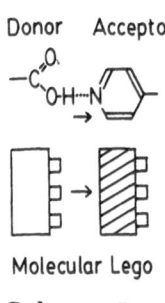

Molecular Lego

Scheme III

Moreover, molecules are self-assembled and the formation of a stable LC complex occurs through selective hydrogen bonding. Schematic illustration of the process between the H-bond donor polymer and the H-bond acceptor molecule is shown in Fig. 5. This is a new type of molecular recognition in molecular aggregates.

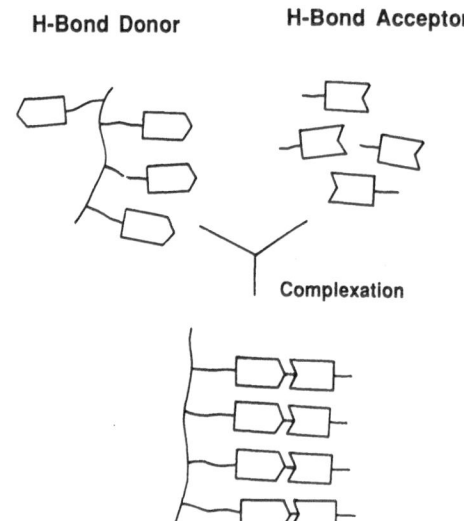

Fig. 5. Schematic illustration of self-assembly of a side-chain LC polymeric complex from H-bond donor and acceptor.

EXPERIMENTAL SECTION

Polymer **P6BA** was prepared by radical polymerization of 4-[(6-acryloxyhexyl)oxybenzoic acid in DMF with AIBN. trans-4-Octyloxy-4'-stilbazole 8OSz was obtained by condensation reaction from 4-octyloxybenzaldehyde and 4-picoline. trans-4-[(4-Methoxybenzoyl)oxy]-4'-stilbazole 10PhSz was prepared from 4-methoxybenzoyl chloride and trans-4-hydroxy-4'-stilbazole which was obtained according to the procedure of Chiang.[18]

The hydrogen-bonded complex was prepared by slow evaporation of pyridine solution dissolving an equimolar amount of H-bond donor and acceptor moieties. Liquid crystalline properties were examined by differential scanning calorimetry (Perkin Elmer DSC-2) and a plarizing microscope equipped with a Mettler FP82 hot stage.

REFERENCES

1. A. Cifferi, W. R. Krigbaum, and R. B. Meyer, Eds., "Polymer Liquid Crystals", Academic, New York (1982).
2. H. Finkelmann and G. Rehage, Adv. Polym. Sci., 60/61, 99 (1984).
3. V. P. Shibaev and N. A. Plate, Adv. Polym. Sci., 60/61, 173 (1984).
4. P. W. Morgan, Macromolecules, 6, 1381 (1977).
5. S. L. Kwolek, P. W. Morgan, J. R. Schaefgen, and L. W. Gulrich, Macromolecules, 6, 1390 (1977).
6. S. A. Aharoni, Macromolecules, 21, 1941 (1988).
7. T. Kato and J. M. J. Frêchet, Macromolecules, 22, 3818 (1989); 23, 360 (1990).
8. T. Kato and J. M. J. Frêchet, J. Am. Chem. Soc., 111, 8533 (1989).
9. T. Kato, A. Fujishima, and J. M. J. Frêchet, Chem. Lett., 1990, 919.
10. T. Kato, P. G. Wilson, A. Fujishima, and J. M. J. Frêchet, Chem. Lett. 1990, 2003.
11. D. W. Bruce, D. A. Dunmur, E. Lalinde, P. M. Maitlis, and P. Styring, Liq. Cryst., 3, 385 (1988).
12. S. E. Odinokov, A. A. Mashkovsky, V. P. Glazunov, A. V. Iogansen, and B. V. Rassadin, Spectrochim. Acta, 32A, 1355 (1976).
13. J. Y. Lee, P. C. Painter, and M. M. Coleman, Macromolecules, 21, 954 (1988).
14. H. Ringsdorf, H.-W. Schmidt, and A. Schneller, Makromol. Chem., Rapid Commun., 3, 745 (1982).
15. H. Benthack-Thoms and H. Finkelmann, Makromol. Chem., 186, 1895 (1985).
16. A. I. Hopwood and H. J. Coles, Polymer, 26, 1312 (1985).
17. G. R. Meredith, J. G. VanDusen, and D. J. Williams, Macromolecules, 15, 1385 (1982).
18. M.-C. Chiang and W. H. Hartung, J. Org. Chem., 10, 21 (1944).

PRODUCTION OF NON-NATURAL AND FUNCTIONAL POLY-β-HYDROXYALKANOATES

BY BACTERIA

Robert W. Lenz,* Herbert Ulmer,* Young-Baek Kim,* and
R. Clinton Fuller**

Depts. of Polymer Science & Engineering* and Biochemistry**
University of Massachusetts, Amherst, MA   01003

INTRODUCTION

Poly-β-hydroxyalkanoates, PHAs, are produced as storage granules or inclusion bodies by a wide variety of bacteria under conditions of metabolic stress in the presence of excess carbon. Normally the PHAs produced are a family of aliphatic polyesters with n-alkyl groups at the β-position, which is a chiral center that exists only in the [R] absolute configuration in these biopolymers, as follows:

$$\left[\overset{*}{O}CHCH_2\overset{\overset{O}{\|}}{C} \right] \quad ; \quad \text{where } R = (CH_2)_x CH_3, \; x = 0\text{-}8 \text{ and possibly higher}$$

with R below the bracket structure.

As indicated above, the substituent R can vary from CH_3 (x = 0) to at least C_9H_{19} (x = 8) depending on the bacterium and the growth conditions. In general, however, the types of bacteria which produce such reserve polymers can be divided into two groups: (1) those that produce PHAs in which x is 0 or 1 (that is, R is CH_3 for poly-β-hydroxybutyrate, PHB, and/or C_2H_5 for poly-β-hydroxyvalerate, PHV), and (2) those where x is greater than 3 and R is C_4H_9 or higher. In our studies we have found that there is little overlap in the PHA compositions between these two sets of polymer compositions. For example, both A. eutrophus, an aerobic bacterium and Rb. sphaeroides, an anaerobic, phototrophic bacterium under most natural conditions produce storage granules which contain the PHB homopolymer,[1,2] but P. oleovorans produces only copolymers in which R is C_4H_9 and higher.[3,4] R. rubrum, another anaerobic, phototrophic bacterium, under most growth conditions also produces copolymers, which generally contain both β-hydroxybutyrate, HB, and β-hydroxyvalerate, HV, units with a wide variety of carbon substrates, but this bacterium can be induced to produce

PHAs which include small amounts of units with R as C_3H_7 and even units with R as C_4H_9 in very small quantities, although it probably will only do so when it is fed specific carbon compounds having either 6- or 7-carbon atoms as the sole organic substrate.[2] In our limited studies, A. eutrophus did not produce PHAs containing higher alkyl substituents, R, even on such substrates, and it produced only HB/HV copolymers under all conditions we used.

STEREOREGULARITY AND CRYSTALLINITY

As can be seen in the structure above, and as mentioned, these bacterial PHAs contain a chiral center in every repeating unit, and all indications are that the configuration of that center is always the same regardless of either the specific bacterium involved or the substrate on which it grows or the structure of the polymer which it produces.[5] That is, as noted, only units with an absolute R configuration have been found in such polymers, so in the terminology of polymer science, these polymers are always 100% isotactic in their configurational structure. An interesting result of this behavior was observed in our study of the storage polymers produced by R. rubrum when grown on either the racemic mixture or pure [S]-β-hydroxybutyric acid.[2] In both cases the PHA produced was entirely of the [R] configuration, and the [R] and the [S] substrates appeared to be taken up from the medium by the bacterium at about the same rate.

Because these bacterial PHAs are perfectly isotactic in structure, the lower members of the series can achieve very high degrees of crystallinity, depending on their composition. This ability to achieve high degrees of crystallinity is true not only for the homopolymer containing only HB units, PHB, but also for the copolymers containing both HB and HV units, P(HB/HV); however, it is not the case for the family of copolymers in which R is C_6H_{13} and higher. The HB/HV copolymers, even though random in compositional sequence distribution, crystallize with an unusually high content of the crystalline fraction (an 80% degree of crystallinity or higher) for a copolymer, but that occurs because the HB and HV units are isodimorphous (that is, copolymers containing both units can crystallize to form a single crystalline phase).[6] Such is not the case for copolymers containing the higher alkyl groups, so these copolymers crystallize to only a relatively small degree, probably with less than 20% or so degree of crystallinity.[4] However, for both types of copolymers, those with either short or long alkyl groups, the rate of crystallization is unusually slow, and this behavior can be a serious problem in the melt processing of these polymeric materials unless nucleating agents are added.[7,8]

The P(HB/HV) copolymers have been commercialized by the Marlboro Plastics Division of ICI Ltd. under the tradename of "Biopol."[9] These copolyesters are true thermoplastics and can be melt processed without thermal degradation at temperatures below 180°C by a variety of methods, including melt spinning to form multifilament and monofilament fibers, melt extrusion to form films, and either injection molding, compression molding or blow molding to form shaped plastics. Such products are being evaluated for a wide variety of applications especially in applications such as biomaterials and disposable packaging materials, for which the inherent biodegradability and bioresorbability of these polymers is important and can be put to advantage in off-setting their presently high cost.

NON-NATURAL POLYMERS

While it has been emphasized in the discussion above that the PHAs produced by bacteria under natural conditions are always of the β-hydroxyalkanoate structure, nevertheless, by forcing the bacteria to grow on non-natural organic compounds as the sole carbon substrate, it is also possible to obtain PHAs which have never been produced in nature and which can contain functional groups on the R substituent and even γ- and δ-hydroxyalkanoate units as discussed below. Since these polymers are produced as storage materials, they are all inherently biodegradable, at least within the cell, although it is not known if such polymers can also be degraded by exocellular enzymes. In many cases the bacteria prefer to produce PHA copolymers, but the reason for this preference is unknown to us at present. Similarly, the bacteria produce polymers of different molecular weights from different substrates, also for reasons unknown to us.

Doi and coworkers found that A. eutrophus was particularly effective for producing copolymers containing the γ- or δ-hydroxyalkanoate units, and when this bacterium was fed 4-hydroxybutyric acid as the sole carbon source, it produced a copolymer containing up to 50 mole % of γ-HB with the β-HB units as shown below:[10,11]

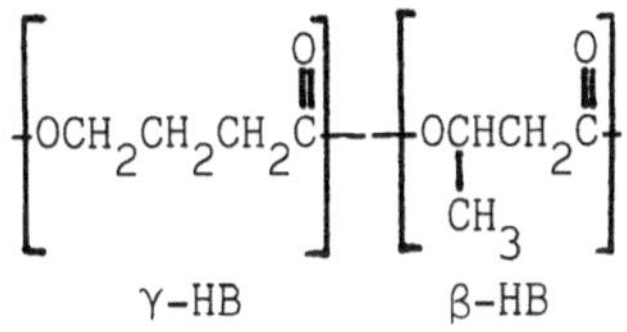

$$\left[OCH_2CH_2CH_2\overset{O}{\underset{\|}{C}}\right]\left[O\underset{\underset{CH_3}{|}}{C}HCH_2\overset{O}{\underset{\|}{C}}\right]$$

$$\quad\quad\gamma\text{-HB}\quad\quad\quad\quad\beta\text{-HB}$$

Similarly, when A. eutrophus was grown on 5-chlorovaleric acid it produced a terpolymer containing β-HB, β-HV and δ-HV units of the following structure:[10]

$$\left[\begin{array}{c} O \\ \| \\ OCHCH_2C \\ | \\ CH_3 \end{array}\right] \left[\begin{array}{c} O \\ \| \\ OCHCH_2C \\ | \\ CH_2CH_3 \end{array}\right] \left[\begin{array}{c} O \\ \| \\ OCH_2CH_2CH_2CH_2C \end{array}\right]$$

β-HB   β-HV   δ-HV

These non-natural types of copolymers are of interest for their different physical properties and for their higher rates of biodegradation as compared to natural P(HB/HV) copolymers. In this regard, we are studying two bacteria, which have also been found to be particularly capable of producing polyesters with unusual repeating units of the type that have never been found in PHAs obtained under natural conditions; these bacteria are R. rubrum and P. oleovorans. When grown on a single carbon source, these bacteria can produce polymers containing R substituents with groups such as either alkenyl, branched alkyl, bromo or phenyl groups, and others.[12,13]

R. rubrum is a versatile phototrophic bacterium which can grow on a wide variety of organic substrates, especially on alkanoic acids, to produce copolymers containing β-hydroxyalkanoate units with methyl, ethyl, propyl and even butyl R pendant groups, so we initially set out to determine whether this bacterium could produce a PHA with an unsaturated R group when grown on an alkenoic acid. For this purpose R. rubrum was grown on 4-pentenoic acid, 4-PEA, as the only carbon source, and for comparison, A. eutrophus was also grown on this single organic substrate. The polymers obtained from both bacteria were analyzed by methanolysis and gas chromatography for the compositions of the PHAs produced, with the results shown in Table 1. Both microorganisms were also grown on mixtures of 4-PEA and valeric acid, VA, and the compositions of PHAs so produced are also recorded in Table 1, as are the molecular weights of all of the PHAs produced.[12,13]

It is seen in the results in Table 1 that, of the two bacteria grown on 4-PEA, only R. rubrum was capable of producing a polymer with a vinyl substituent group, although even with 4-PEA as the sole carbon source, the saturated repeating units (the HB and HV units) were the principal components of the copolymer, and A. eutrophus produced polymers with only HB and HV units. According to our initial analyses, by both methanolysis-gas chromatography and NMR, the average composition of the storage material obtained with 4-PEA alone was approximately 30 mole percent HP (β-hydroxypentenoate) units, 75 mole percent HV units and 5 mole percent HB units. However, our more recent analysis by mass spectrometry of the oligomers formed from this storage PHA, by either controlled methanolysis or partial pyrolysis, indicated that the polymeric product was not a single random copolymer of that composition, but it was instead a mixture of two or

Table 1. Compositions and molecular weights of PHAs produced by
R. rubrum and A. eutrophus using 4-PEA and 4-PEA/VA mixtures
as carbon sources

Microorganism	Carbon[a] Source	Carbon Source conc.(mmol)	PHA Composition[b] mole %			$\overline{M}_w$[c] (g/mol)	$\overline{M}_n$[c] (g/mol)	$\overline{M}_w/\overline{M}_n$
			HB	HPE	HV			
R. rubrum	4-PEA	50	11	30	59	340,000	110,000	3.1
R. rubrum	4-PEA/VA	30/30	14	14	72	480,000	155,000	3.1
A. eutrophus	4-PEA	30	56	0	44	456,000	147,000	3.1
A. eutrophus	4-PEA/VA	10/10	45	0	55	315,000	158,000	2.0

[a] 4-PEA is 4-pentenoic acid; VA is valeric acid.

[b] HB is 3-hydroxybutyric acid; HPE is 3-hydroxy-4-penteneoic acid;
HV is 3-hydroxyvaleric acid.

[c] Determined from polystyrene standards by GPC.

three different copolymers, at least one of which did not contain the HP
units, only HB and HV units. Similar results were recently obtained when
we analyzed a copolymer produced by P. oleovorans grown on a mixture of 5-
phenylvaleric acid and n-nonanoic acid. Again, that storage material was
found to be a mixture of at least two different copolymers, which varied
in amount and possibly in composition depending on the time in the growth
cycle at which the polymer was harvested.[14]

In other studies we have been investigating the production of non-
natural PHAs by an aerobic bacterium, P. oleovorans, and are presently
testing the limits to which this bacterium will incorporate either
sterically-hindered or functionalized units into the storage copolyesters
it produces. In our study of steric effects this bacterium was fed a
series of methyl-substituted octanoic acids to determine how rapidly and
to what cell densities it could grow on these substrates, when they were
used as either the sole carbon source or in combination with n-octanoic
acids, and the yields and structures of the PHAs produced as a function of
the substrate.[15]

When grown on either n-octane or n-octanoic acid as the sole carbon
source, P. oleovorans always produces random copolymers containing prin-
cipally the 6-, 8- and 10-carbon atom containing repeating units; that is,
the units with R = C_3H_7, C_5H_{11} and C_7H_{15}, respectively, of the following
structure:[3,4]

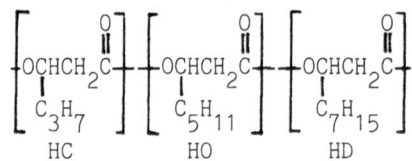

These units, which are designated as HC for β-hydroxycaproate, HO for β-hydroxyoctanoate and HD for β-hydroxydecanoate in the structure above, are present in the approximate molar ratio of 10/80/10:HC/HO/HD with n-octanoic acid as the substrate, depending somewhat on the growth conditions. The polymer yields obtained from such a substrate can be well above 50% of the cell dry weight, and the yield obtained with 7-methyloctanoic acid as the sole carbon source was even higher. Also, with the latter substrate a copolymer was produced which contained both 7-methyl-3-hydroxyoctanoate and 5-methyl-3-hydroxyhexanoate units in about an 8:1 molar ratio.[15]

In contrast to the results obtained with the 7-methyloctanoic acid as the sole carbon source, when P. oleovorans was fed 6-methyloctanoic acid as the only organic substrate it grew much more slowly to lower cell densities, and it did not produce a storage polymer. Furthermore, when 5-methyloctanoic acid was the only organic compound available no growth was observed. These results were quite surprising, so this bacterium was grown on mixtures of each of the last two methyl-substituted substrates with n-octanoic acid in order to evaluate the ability of P. oleovorans to cometabolize a good growth substrate, which produces PHAs, with a non-producing substrate. In both cases, quite remarkably, storage copolymers were produced which contained units from the non-producing substrate. As much as 50 mole percent of the methyl-branched units were included in the copolymers obtained by cofeeding with 6-methyloctanoate, and 30 mole percent of such units were included on cofeeding with 5-methyloctanoate. Also of considerable interest to organic chemists, in both cases the substrates were fed as racemic mixtures, and with 5-methyloctanoate the PHA produced was enriched with one of the optical isomers by a ratio of approximately 7:1, as indicated by analysis of the copolymer by NMR spectroscopy.[15]

We have obtained similar results for P. oleovorans when this bacterium was grown with mixtures of ω-bromoalkanoic acids and n-nonanoic acid, and by this type of cometabolism we have been able to induce P. oleovorans to produce copolymers with units containing either ω-cyano groups, ω-unsaturated groups or ω-ester groups.[16] However, possibly the most unusual PHA produced to date by P. oleovorans was the homopolymer of β-hydroxy-5-

phenylvalerate units (HPV) as shown below, which was obtained when P. oleo-vorans was grown with 5-phenylvaleric acid as the sole carbon source:[17]

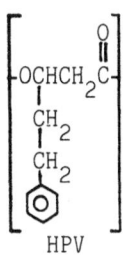

HPV

The evaluation of the ability of P. oleovorans to utilize ω-bromoacids included studies with 6-bromohexanoic acid, 8-bromooctanoic acid and 11-bromoundecanoic acid.[16] Growth was much faster with 11-bromoundecanoic acid than with 8-bromooctanoic acid, and there was no growth with 6-bromo-hexanoic acid. This result showed that cell growth occurred only when the distance between the polar substituent (Br) and the bioreactive group (COOH) was sufficiently long, but with none of these substrates alone were PHA granules observed in the cells obtained. For that reason, as in the studies described above, the three bromoalkanoic acids were combined with equimolar amounts of nonanoic acid and fed to P. oleovorans, which had been grown initially on nonanoic acid alone. Surprisingly, the growth rates were faster for 6-bromohexanoic acid than 11-bromoundecanoic acid in

Table 2. Fermentation results from mixtures of ω-bromoalkanoic acids and nonanoic acid

Growth Substrate, ω-bromoalkanoic acid (ratio to nonanoic acid)	Dry Cell yield (g/L)	PHA yield (g/L)	PHA, wt% in Dry Cell	Functional Group, mol% in Copolymer
11-Bromoundecanoic acid (1:1)	0.68	0.092	13.5	37.5
8-Bromooctanoic acid (1:1)	0.63	0.082	13.0	25
8-Bromooctanoic acid (1:2)	0.55	0.21	38.2	4.2
6-Bromohexanoic acid (1:1)	0.42	0.019	4.5	25

combination with nonanoic acid, and PHA granules were observed in the cells grown on all of the 1:1 molar mixtures. The cell yields, PHA yields and brominated β-hydroxyacid compositions are collected in Table 2. As seen from the data in this table, the amount of ω-bromo-β-hydroxyalkanoate units was as high as approximately 38 mole percent when a mixture of 1:1 moles of nonanoic acid to 11-bromoundecanoic acid was used as the carbon source. Somewhat smaller amounts of bromoacid units were present in the PHAs produced on the 8-bromooctanoic and 6-bromohexanoic acid mixtures, and decreasing the molar ratio of 8-bromooctanoic acid to nonanoic acid from 1:1 to 1:2 caused a sharp drop in bromoacid unit content from 25 to 4 mole percent.[16]

SUMMARY

In summary, these studies taken together have shown that bacteria can produce at least six different types of polymeric products, depending on the type of bacterium and the organic substrate, including: (1) homopolymers from a single good substrate, (2) random copolymers from a single good substrate, (3) random copolymers from two good substrates, (4) two different polymers from a single good substrate, (5) two different polymers from two good substrates, and (6) random copolymers from a good and a non-producing substrate.

Acknowledgement

The authors are grateful to the Office of Naval Research, Molecular Biology Program, for the support of this work under Grant No. N00014-86K-0369.

References

1. H. Brandl, R. A. Gross, R. W. Lenz, and R. C. Fuller, in: "Advances in Biochemical Engineering/Biotechnology. Vol. 41," A. Fiechter, ed., Springer-Verlag, Berlin-Heidelberg (1990); pp. 77-93.

2. H. Brandl, E. Knee, Jr., R. C. Fuller, R. A. Gross, and R. W. Lenz, Int. J. Biol. Macromol. 11:49 (1989).

3. H. Brandl, R. A. Gross, R. W. Lenz, and R. C. Fuller, Appl. Environ. Microbiol. 54:1977 (1988).

4. R. A. Gross, C. DeMello, R. W. Lenz, H. Brandl, and R. C. Fuller, Macromolecules 22:1106 (1989).

5. E. A. Dawes and P. J. Senior, Adv. Microb. Physiol. 10:135 (1973).

6. T. L. Bluhm, G. K. Hamer, R. H. Marchessault, C. A. Fyfe, and R. P. Veregin, Macromolecules 19:2871 (1986).

7. S. Bloembergen, D. A. Holden, G. K. Hamer, T. L. Bluhm, and R. H. Marchessault, Macromolecules 19:2865 (1986).

8. R. H. Marchessault, T. L. Bluhm, Y. Deslandes, G. K. Hamer, W. J. Orts, P. R. Sundararajan, M. G. Taylor, S. Bloembergen, and D. A. Holden, Makromol. Chem., Macromol. Symp. 19:235 (1988).

9. E. A. Dawes, Bioscience Reports 8:537 (1988).

10. M. Kunioka, A. Tamaki, and Y. Doi, Macromolecules 22:694 (1989); Y. Doi, M. Kunioka, Y. Nakamura, and K. Soga, Macromolecules 21:2722 (1988).

11. Y. Doi, A. Tamaki, M. Kunioka, K. Soga, Macromol. Chem., Rapid Commun. 8:631 (1987).

12. R. W. Lenz, B. W. Kim, H. W. Ulmer, K. Fritzsche, E. Knee, Jr., and R. C. Fuller in: "New Biosynthetic Biodegradable Polymers of Industrial Interest from Microorganisms," Proceedings of the 1990 NATO Workshop in Sitges, Spain, E. Dawes, ed., Kluwer Academic Publishers (in press).

13. R. W. Lenz, B. W. Kim, H. Ulmer, K. Fritzsche, and R. C. Fuller, Polymer Preprints 31(2):408 (1990); American Chemical Society, Washington, DC.

14. K. Fritzsche and B. W. Kim, University of Massachusetts, unpublished results.

15. K. Fritzsche, R. W. Lenz, and R. C. Fuller, Int. J. Biol. Macromol. 12:92 (1990).

16. B. W. Kim, Ph.D. Thesis, University of Massachusetts, 1991.

17. K. Fritzsche, R. W. Lenz, and R. C. Fuller, Makromol. Chem. 191:1957 (1990).

GENETIC ENGINEERING OF MOLECULAR AND SUPRAMOLECULAR STRUCTURE IN POLYMERS

David A. Tirrell,[a,c] Thomas L. Mason[b,c] and Maurille J. Fournier[b,c]

Departments of Polymer Science and Engineering[a] and Biochemistry[b]
and Program in Molecular and Cellular Biology[c]
University of Massachusetts, Amherst, MA 01003 USA

The development of new advanced materials with useful optical, electronic and mass transport properties will require increasingly precise control of solid state organization at the supramolecular level. Polymeric materials offer special advantages in this regard as a result of their characteristic extended molecular connectivity, which limits the number of possible configurations of the system. On the other hand, polymeric materials suffer from the statistical nature of the processes used to prepare them, and from the associated heterogeneity in their molecular structures. New methods of synthesis that afford homogeneous populations of chains will offer important advantages in the control of supramolecular organization. At present, biological polymerization processes provide the only clear route to polymeric materials characterized by uniformity of chain length, sequence and stereochemistry. With the advent of two related technologies (gene synthesis and recombinant DNA), it has become possible to imagine the exploitation of biological polymerizations – and protein biosynthesis in particular – to produce new materials of predictable and desirable properties. We describe herein our own first steps toward this objective.

Concepts. The conceptual approach to this problem is outlined in Figure 1, and begins with the design of the artificial protein of interest. The ideas that drive this design will in general be drawn in part from polymer materials science and in part from structural biology, and will include considerations of both molecular and supramolecular structure. As an example, consider the hypothetical ultrathin membrane shown in Figure 2. The design of such a membrane might include consideration of its thickness, surface functionality (to control wettability) and porosity, each a critical factor in the control of permeability and selectivity. The structure shown in Figure 2 exploits regular chain folding to define the thickness of the membrane, functionalized folding elements to decorate the membrane surfaces, and controlled fluctuations in molecular cross section to dictate porosity. At this level, the guiding concepts are those of polymer materials science. At the molecular level, though, e.g., in the design of extended and folded conformational units, one looks to biology (and specifically to the protein folding problem) for clues to the selection of appropriate amino acid sequences.

Design of the amino acid sequence is followed by the encoding of that sequence into a corresponding duplex of DNA. The degeneracy of the genetic code (i.e., the fact that organisms use more than a single codon to specify most amino acids) raises interesting questions at this stage. The choice of codons for each amino acid reflects at least three considerations: i). the pattern of codon use in the host organism, ii). the avoidance of strict periodicity in the coding sequence, and iii). the details of the strategies selected for cloning and expression. The avoidance of strict periodicity is motivated by concerns about the limited genetic stability of repetitive DNAs in many host organisms.[1]

Contemporary Topics in Polymer Science, Vol. 7., Edited by
J.C. Salamone and J. Riffle, Plenum Press, New York, 1992

The coding sequence is constructed by solid phase organic synthesis,[2] and ordinarily involves the preparation and polymerization of oligomeric DNAs 50-100 nucleotides long.[3] The length of the oligomeric DNA is dictated in large part by the efficiency of present methodology for automated synthesis of nucleic acids; oligomers of degree of polymerization ≤ 100 can be prepared routinely in acceptable purity.

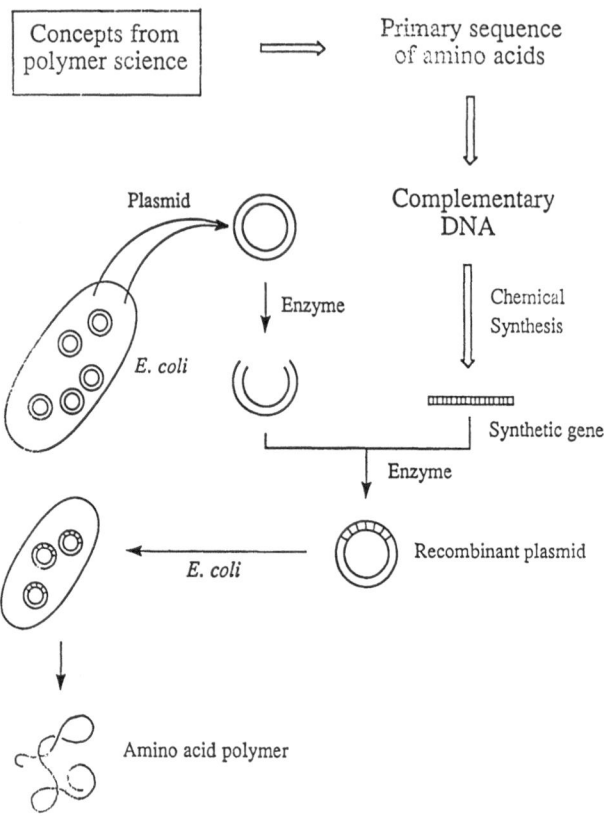

Figure 1. Outline of the application of genetic engineering to the synthesis of new polymeric materials.

The coding sequence must then be converted to the target artificial protein via the steps shown in Figure 3. The DNA must be replicated stably at each generation of growth of the host organism, and then transcribed into a messenger RNA template for protein synthesis. Translation of the message occurs on the ribosome, a catalytic unit constructed from a complex array of RNAs and proteins. The coding sequence must therefore be equipped with binding sites for the various enzymes and associated factors required for replication, transcription and translation. This is readily accomplished through the use of recombinant DNA methods (Figure 1): the coding sequence is inserted into a circular, non-chromosomal DNA (a "plasmid") that has been engineered to carry the requisite binding

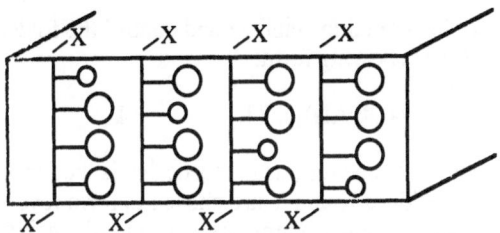

Figure 2. Hypothetical ultrathin membrane constructed from stacked β-sheets of a periodic polypeptide. Large and small circles represent steric requirements of amino acid side chains. Variation in side chain size creates local vacancies – or pores – in the structure; functional groups X decorate the membrane surface and control wettability.

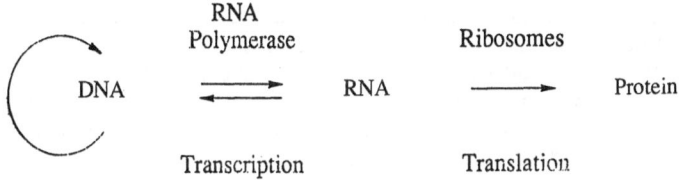

Figure 3. Elementary steps in protein biosynthesis.

sites. A large – and rapidly expanding – set of plasmid vectors is available,[4] and will accommodate a wide variety of cloning and expression strategies. The recombinant plasmid is then placed in the host organism – most often a bacterium – and the host is grown under conditions where protein accumulation is anticipated. Synthesis of multigram quantities of material is now relatively routine using laboratory-scale fermentation apparatus.

Control of Molecular and Supramolecular Structure. The concepts described above have been applied in our own laboratories to the control of supramolecular order in polymeric solids and solutions. A central theme of this work has been the use of sequence-dependent secondary structures (α-helices, β sheets and reverse turns) to dictate interchain organization. In addition, we have been concerned with the prospects for incorporating unnatural amino acids into artificial proteins, with the objective of expanding the range of properties that might be achieved in such materials. We discuss each of these objectives briefly below.

Design of Folded-Chain Lamellae. Folded-chain lamellar crystals are a dominant morphological feature of natural[5] and synthetic[6] polymers of repetitive chemical structure. In synthetic polymers, the folded-chain structure is dictated by the kinetics, rather than the thermodynamics, of the crystallization process, and is metastable. In our early experiments with artificial proteins, we have tried to stabilize and control the folded-chain arrangement by preparing sequences of the following form:

$$\{(GlyAla)_x XY\}_n \qquad \mathbf{1}$$

The intended role of the glycylalanine repeats in this design is the formation of β-sheet crystalline stems,[7,8] while residues X and Y are chosen to form reverse turns.[9,10] In each case, the selection of local sequence elements was based on information about the conformational properties of known polypeptides: poly(glycylalanine) has been reported to form β-sheet crystals,[7,8] and Sibanda and Thornton have identified five X,Y pairs that appear to favor β–hairpin formation in globular proteins.[9,10] As of this writing, we have expressed six different sequence variants of **1**, and several chain-length variants as well. Detailed structural analysis by x-ray scattering and spectroscopic methods is underway and will be reported elsewhere.

Synthesis of Monodisperse Derivatives of Poly(L-glutamic acid). Poly(L-glutamic acid) (PLGA) and poly(γ-benzyl-L-glutamate) (PBLG) are known to adopt α-helical conformations in solution (PLGA and PBLG) and in the solid state (PBLG).[11] Recently Morris and coworkers have published striking scanning tunnelling micrographs of PBLG deposited on highly oriented pyrolytic graphite,[12] illustrating directly the helical (though not necessarily α-helical) trajectory of the chain. The possibility of preparing crystalline,[13] liquid crystalline[14] and oriented arrays[15] of PBLG and related polymers, coupled with the large dipole moment of the helix[16] has generated substantial interest in these polymers. Using a strategy of the kind outlined above, we have recently prepared apparently monodisperse derivatives of PLGA.[17] Conversion to PBLG and related polymers is underway, and will add a new element of control to the process of preparing advanced materials based on highly polar macromolecules.

Incorporation of Unnatural Amino Acids. The methods described in this chapter are most readily applied to the construction of polymers of the 20 "natural" amino acids. Within this set of monomers, one has access to diverse chemical functionality: alkyl and aryl groups, hydroxyls, acids, amines and others. Nevertheless, the impact of this methodology on polymer materials science will be dependent in part on the extent to which additional monomers can be used. A portion of our research program has therefore been devoted to an exploration of the incorporation of unnatural amino acids into artificial proteins.

As a starting point for this exploration, we have examined the replacement of methionine (Met, **2**) by selenomethionine (SeMet, **3**) in proteins of repeating unit sequence **4**. The substitution of **3** for **2** in native bacterial proteins was demonstrated many years

$$-(GlyAla)_3GlyX-$$

$$X = Met, SeMet$$

2   **3**   **4**

ago by Cowie and Cohen,[18] and preliminary results are consistent with high levels of replacement in artificial proteins built up from repeating unit **4**.[19]

Conclusions. Strategies for the design and synthesis of artificial proteins for materials applications are emerging rapidly. The precise control of molecular structure provided by these strategies offers entirely new prospects for the engineering of advanced materials with useful optical, electronic and mass transport properties. Preliminary successes have been achieved in the control of chain folding, in the synthesis of monodisperse helical polymers and in the incorporation of unnatural amino acid monomers.

Acknowledgments

Important contributions to the work described herein have been made by Howard Creel, Yoshikuni Deguchi, Michael Dougherty, Srinivas Kothakota, Mark Krejchi, Kevin McGrath, Ajay Parkhe, Guanghui Zhang, Edward Atkins, Ronald Beavis, Joseph Cappello and Brian Chait. Our program is supported by grants from the National Science Foundation (DMR8914359), from the Center for University of Massachusetts–Industry Research on Polymers, and from the NSF Materials Research Laboratory of the University of Massachusetts.

References

1. R. Kucherlapati, G.R. Smith, eds., *Genetic Recombination*, American Society for Microbiology (1988), Ch. 3.
2. L.J. McBride and M.H. Caruthers, *Tetrahedron Lett.* 24:425 (1983).
3. H.S. Creel, M.J. Fournier, T.L. Mason and D.A. Tirrell, *Macromolecules* 24:1213 (1991).
4. For typical procedures, consult J. Sambrook, E.F. Frisch, T. Maniatis, *Molecular Cloning, A Laboratory Manual, 2nd ed.*, Cold Spring Harbor Press, Cold Spring Harbor (1989).
5. A.J. Geddes, K.D. Parker, E.D.T. Atkins and E. Beighton, *J. Mol. Biol.* 32:343 (1968).
6. D.C. Bassett, *Principles of Polymer Morphology*, Cambridge University Press, Cambridge (1981).
7. R.D.B. Fraser, T.P. MacRae, F.H.C. Stewart and E. Suzuki, *J. Mol. Biol.* 11:706 (1965).
8. J.M. Anderson, H.H. Chen, W.B. Rippon and A.G. Walton, *J. Mol. Biol.* 67:459 (1972).
9. B.L. Sibanda and J.L. Thornton, *Nature* 316:170 (1985).
10. B.L. Sibanda, T.L. Blundell and J.M. Thornton, *J. Mol. Biol.* 206:759 (1989).
11. A.G. Walton and J. Blackwell, *Biopolymers*, Academic Press, New York (1973).
12. T.J. McMaster, H.J. Carr, M.J. Miles, P. Cairns and V.J. Morris, *Macromolecules* 24:1428 (1991).
13. J. Watanabe and I. Vematsu, *Polymer* 25:1711 (1984).
14. P. Russo and W.G. Miller, *Macromolecules* 16:1690 (1988).
15. R. Jones and R.H. Tredgold, *J. Phys. D: Appl. Phys.* 21:449 (1988).
16. A. Wada, *Adv. Biophys.* (1976), 1.
17. G. Zhang, M.J. Fournier, T.L. Mason and D.A. Tirrell, unpublished results.
18. D.B. Cowie and G.N. Cohen, *Biochim. Biophys. Acta* 26:252 (1957), p. 252.
19. M.J. Dougherty, S. Kothakota, M.J. Fournier, T.L. Mason and D.A. Tirrell, manuscript in preparation.

FUNCTIONALIZED POLYOLEFINS via COPOLYMERIZATION

OF BORANE MONOMERS IN ZIEGLER-NATTA PROCESS

T. C. Chung

Department of Materials Science and Engineering
The Pennsylvania State University
University Park, PA 16802

INTRODUCTION

Polyolefins, especially polyethylene and polypropylene, are used in a wide range of applications, since they incorporate an excellent combination of mechanical, chemical and electronic properties and processibility. Nevertheless, deficiencies, such as the lack of reactive groups in the polymer structure, have limited some of their end uses, particularly those in which adhesion, dyeability, paintability, printability or compatibility with other functional polymers is paramount. Accordingly, the chemical modification of polyolefins has been an area of increasing interest as a route to higher value products and various methods of functionalization (1-3) have been employed to alter their chemical and physical properties.

It is well-known that the Ziegler-Natta process is the most important method for preparing polyolefins (4), but the direct polymerization of functional monomers by this method is normally very difficult, because of catalyst poisoning and other reactions(5). The Lewis acid components (Ti, V, Zr and Al) of this catalyst will tend to complex with nonbonded electron pairs on N, O, and X of functional monomers, in preference to complexation with the π-electrons of double bonds. The net result is the deactivation of the active polymerization sites by formation of stable complexes between catalysts and functional groups, thus inhibiting polymerization.

Theoretically, there are two approaches in addressing this problem. One is the inorganic approach by decreasing the acidity of Z-N catalysts to minimize the acid-base interaction between catalyst and functional group. This approach has been suffering from some fundamental problems, mainly the simultaneous decrease in both of functional group sensitivity and catalyst reactivity. The other approach is the organic one by using the protecting groups. There are some reports of using bucky groups to sterically protect heteroatom during the copolymerization process. However, this process is unattractive because of the resulting polymers are as inert as pure hydrocarbon polymers. The other protection chemistry involved the pretreatment of functional group with an organo-aluminum to prevent catalyst poisoning. Ester copolymers (6) are especially useful because they are relatively stable to such polymerizations and easily converted to an acid functionality after polymerization. However, there are several drawbacks, mainly the use of large excess of lewis acid protecting agent, the low levels of ester functional group incorporated to polymer, the deprotecting reaction and the high viscosity of polymer solution in the polymerization process, due to inonic association of polar groups in hydrocarbon solvent. Moreover, the ester polymer is not a versatile intermediate. It is difficult to convert the ester functional polymer to other functional (e.g., -OH, -NH2) polymers by simple mild chemical conditions. It is clear that there is a fundamental need to develop a new chemistry, which can address the challenge of preparing functional polyolefins with wide varieties and a wide concentration range of functional groups by the process of Ziegler-Natta polymerization.

BORAN MONOMER APPROACH

Our initial idea to use borane monomers, ω-alkenylboranes (7-8), in Ziegler-Natta polymerization was based on three considerations, (a) the stability of borane to transition metal catalyst. Because trialkylborane is a Lewis acid, it offers an very good chance for them to coexist with the catalyst. In addition, the boron atom is relatively small, steric protection can be effectively applied if needed. (b) the solubility of borane monomers and polymers in the hydrocarbon solvents (hexane and toluene) used in Ziegler-Natta polymerization. A soluble growing polymer chain is essential to obtain high molecular weight polymer. (c) the versatility of borane groups, which can be transformed to a remarkably fruitful variety of functionalities, as shown by H. C. Brown (9).

The feasibility of polymerizing alkenylboranes by Ziegler-Natta catalysts can be deduced from the stability of such catalysts in organoborane solutions. Several reactions were used to examine this stability. One of them was homopolymerization of 1-octene with and without the presence of equimolar amount of Et_3B. The typical reaction condition used was $TiCl_3/10Et_2AlCl/100Et_3B/100$ 1-octene. GPC results Shown in Figure 1 conclusively prove the absence of any retarding effect on the molecular weight due to the presence of trialkylborane.

The other direct evidence comes from the polymerization of a borane containing α-olefin (B-alkenyl-9-BBN) as shown in Equation 1.

Equation 1

Several Ziegler-Natta catalysts, including both homogeneous and heterogeneous catalysts, have been used in this polymerization reaction. In most cases, high molecular weight polymer was obtained with high product yield in relatively short reaction time. The ^{11}B NMR spectrum of the polyborane always displays a single chemical shift at 88.2 ppm (relative to $BF_3 \cdot OEt_2$), the same as that of the monomer. The coexistence of the same trialkylborane in both monomer and polymer is crucial, and strongly supports two central points: the stability and the polymerizability of an alkenylborane with Ziegler-Natta catalysts. Moreover, the polyborane was as easily converted to many functional polymers as the corresponding small borane molecule. One example of polyhexene-8-ol was obtained by the oxidation of poly(B-7-hexenyl-9-BBN) using sodium hydroxide and hydrogen peroxide at 50°C for 2 h. It is interesting to note that this type of polymers is new and has a comb-like polymer structure, with the functional groups located at the end of side chains. In the past, many attempts to prepare the similar polymer structure by free radical polymerization

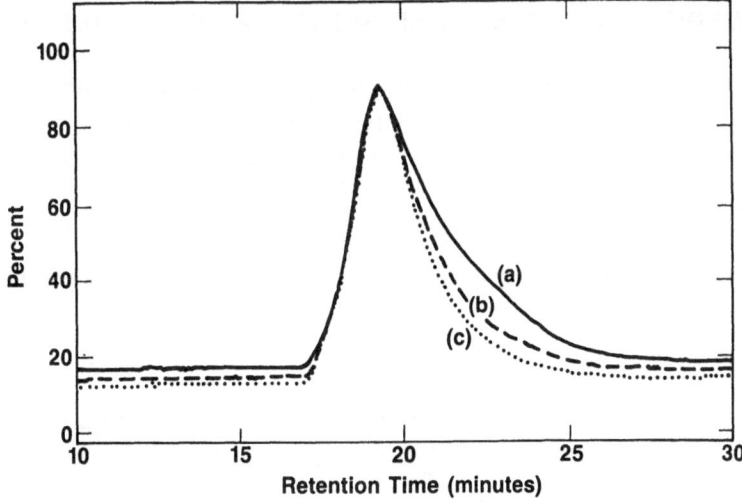

Figure 1. GPC curves of poly(1-octene), curves(a) and (c) for the samples prepared without Et₃B, at 83% and 23% conversions, and curve (b) for the sample prepared in the presence of Et₃B.

were failed because of the degradative chain transfer reaction (10) between allylic protons and active sites. The primary functional group at the end of side chain has many advantages, such as thermal stability, chemical reactivity and side chain group flexibility. As shown in Figure 2, the TGA results indicates the thermal decomposition temperature for polyhexene-8-ol is about 280°C in air. In the other hand, polyvinyl alcohol (11) is dehydrated below 170°C. In addition, the type of functional polymers has isotactic microstructure. The ordered molecular structures are no doubt due to the stereoregular propagation during Z-N polymerization.

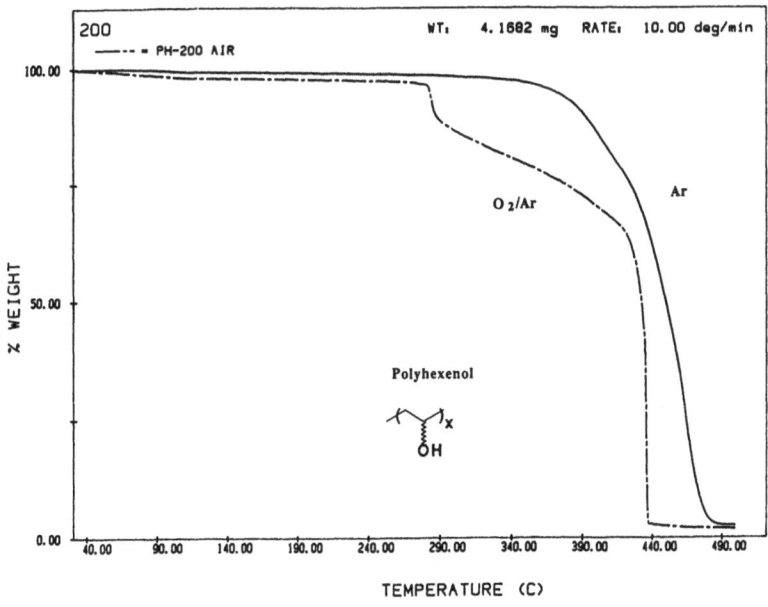

Figure 2. TGA curves of polyhexene-8-ol under (a) Ar, (b) O₂/Ar.

COPOLYMERIZATION OF BORANE MONOMER AND α-OLEFIN

Consequently, we also investigated the copolymerization of borane monomers and α-olefins. The experimental results indicated that a broad composition of copolymers could be achieved with an appropriate choice of monomer pairs. One example is the copolymerization of B-5-hexenyl-9-BBN and 1-octene (12). As shown in Figure 3, a broad range of compositions were obtained. The percent incorporation of the borane monomer is almost linearly proportional to the corresponding monomer feed (Table I), and the polymer composition is relatively constant during the polymerization process.

Table I. A Summary of Poly(octene-co-hexenol)

Sample	mol % 1-octene in feed	in copolymer *	GPC data Mw x 10-3	Mw/Mn	Tg°C
poly(1-octene)	100	100	1678	5.1	-67
POH-31	75	85	1476	6.1	-51
POH-11	50	60	983	7.8	-26
POH-13	25	35	401	6.0	2
poly(1-hexenol)	0	0	145	2.6	15

* Copolymerizations were allowed to proceed to 40-60 % conversion. For experimental convenience, all borane groups in copolymers were oxidized to hydroxyl groups. The numbers of POH-13, POH-11 and POH-31 indicates the molar feed ratios of two monomers, 1-octene and hexenyl-9-BBN, respectively. The steady increase in the Tg is characteristic of the increase in the hexenol content with increasing borane monomer in the feed. The presence of only one glass transition is taken as evidence for the absence of macroscopic phase separation and therefore implies that the copolymer samples are fairly homogeneous. The same conclusion was reached on the basis of GPC results which were obtained by using both UV and RI detectors. The hydroxyl groups in copolymers were esterified by benzolic chloride to obtain THF soluble and UV-

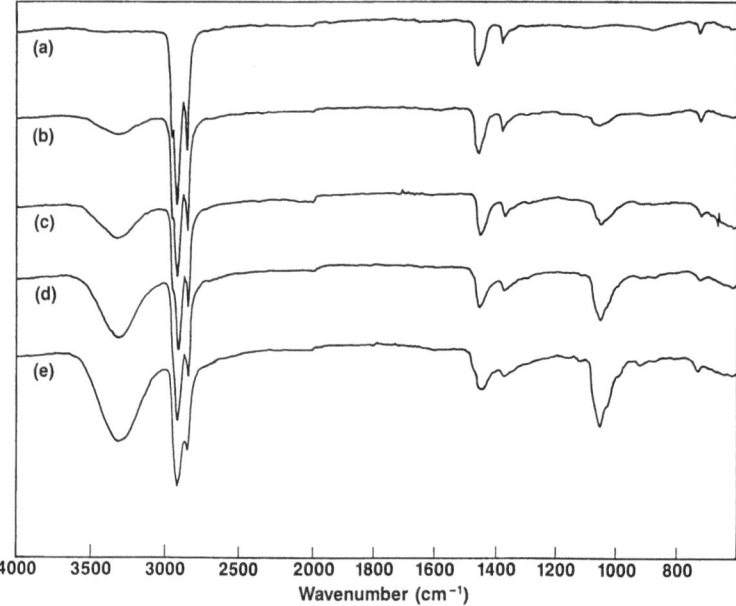

Figure 3. IR spectra of the homopolymers and copolymers; (a) poly(1-octene), (b) POH-31, (c) POH-11, (d) POH-13, (e) poly(1-hexenol).

sensitive polymers. The two UV and RI profiles almost exactly overlap. This is a strong indication of the homogeneity of the copolymer samples. The hydrophilicity of polymer is very dependent on the hydroxyl group content. Polyhexen-6-ol is soluble in methanol and its copolymers are only soluble in mixed solvents, such as THF/methanol and chloroform/methanol. The amounts of methanol in mixed solvent is proportional to that of hexen-6-ol in the copolymers.

FUNCTIONALIZED POLYPROPYLENE WITH BLOCKY STRUCTURE

Among the polyolefins, polypropylene is one of the most desirable polymers. The improvement of it's physical properties is both scientifically and technologically interesting. Unfortunately, polypropylene is also the most difficult one to be functionalized by the existing processes. Only a Ziegler-Natta polymerization can be used for the preparation of polypropylene. As being discussed, this type of catalysts is incapable of incorporating functional group-containing monomers . On the other hand, post-polymerization processes suffer other problems; e.g. degradation (13) of polypropylene backbone. The development of borane monomer approach certainly offers a new way to prepare this highly desirable polymer.

The polymerization reaction was carried out in the inert gas atmosphere at ambient temperature, using the catalysts for the preparation of isotactic polypropylene. Typically, two monomers of propylene and hexenyl-9-BBN were dissolved in toluene solution, then the polymerization was started by adding $TiCl_3AA$ and $Al(Et)_3$ which were pre-mixed and aged for 1/2 hour in of toluene. Figure 4 shows the ^{11}B NMR spectra (14) of the copolymers which were sampled during the copolymerization reaction. The intensity of trialkylborane peak (δ=87 ppm) increases continuously with the reaction time. The boron concentration in polymer can be calculated by comparing the peak intensity of trialkylborane to the internal reference of triethylborate (δ=19 ppm). Due to the significant difference in polymerization reactivity between two comonomers, the initial incorporation of borane monomer in copolymer was very small, only 1.6 mole % in the first 0.1 hour, and then increased to 3.5 and 6.3 mole % after 1 and 2 hours respectively. It is obvious that propylene was preferentially polymerized in the initial copolymerization reaction. In fact, the copolymer formed insoluble particle in the early stage of polymerization process. In fact, the resulting copolymer has the blocky microstructure, which is consisted of isotactic polypropylene and functional polymer. This polymer is completely insoluble in common organic solvents at room temperature, including THF which is a good solvent for polyhexenyl-9-BBN homopolymer. Subsequently, borane groups in propylene copolymer were interconverted to various functional groups under mild conditions. Figure 5 shows the oxidation product. Despite the heterogeneous nature of the reaction, insoluble borane containing polypropylene was completely converted to the corresponding hydroxy group containing polypropylene. The resulting polymer was boron free and completely soluble in xylene at high temperature (> 120 ^{0}C). Complete modification of an insoluble polymer by such mild reaction conditions is very unusual. This suggests a high surface area of borane groups available to the reagents. While the thermoplastic segments were crystallized, the borane groups might be expelled out to the surface of crystalline phase.While the thermoplastic segments were crystallized, the borane groups might be expelled out to the surface of crystalline phase as shown in Figure 6. This morphological picture is also consistent with the results in thermal studies. The DSC studies in Figure 7 show approximately the same melting point (~160 ^{0}C) for both isotactic

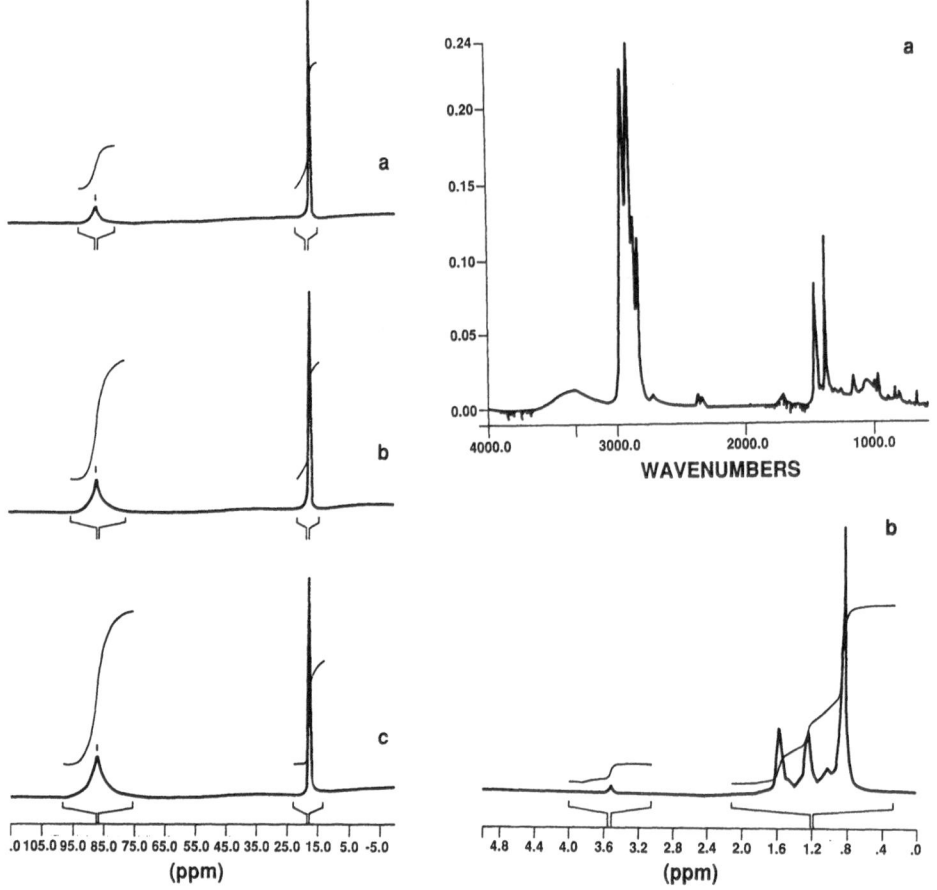

Figure 4. ^1B NMR spectra of poly (propylene-co-hexenyl-9-BBN) with various borane mole concentrations: (a) 1.5%, (b) 3.5%, (c) 6.3%.

Figure 5. (a) IR and (b) 1H NMR spectra of poly(propylene-co-hexenol)

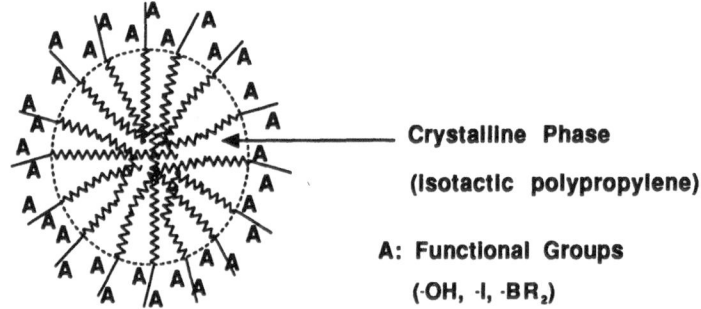

Crystalline Phase

(Isotactic polypropylene)

A: Functional Groups

(-OH, -I, -BR$_2$)

Figure 6. The sketch of poly(propylene-co-hexenol) microstructure.

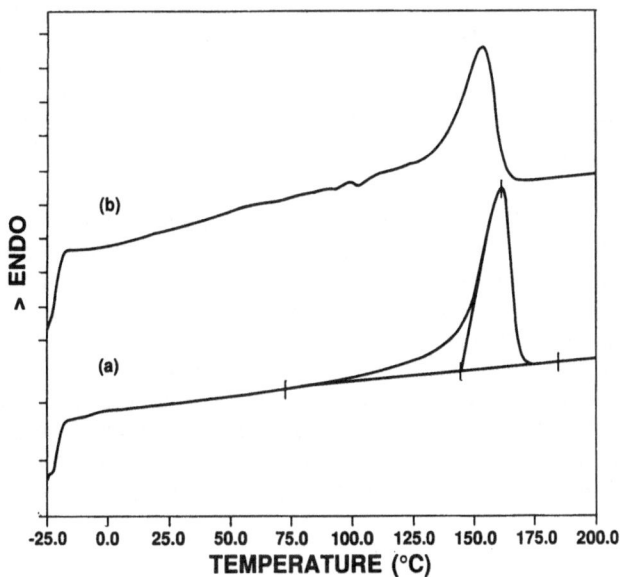

Figure 7. Comparison of DSC curves between (a) isotactic polypropylene and (b) poly(propylene-co-hexenol) with 4.2 mol% alcohol groups.

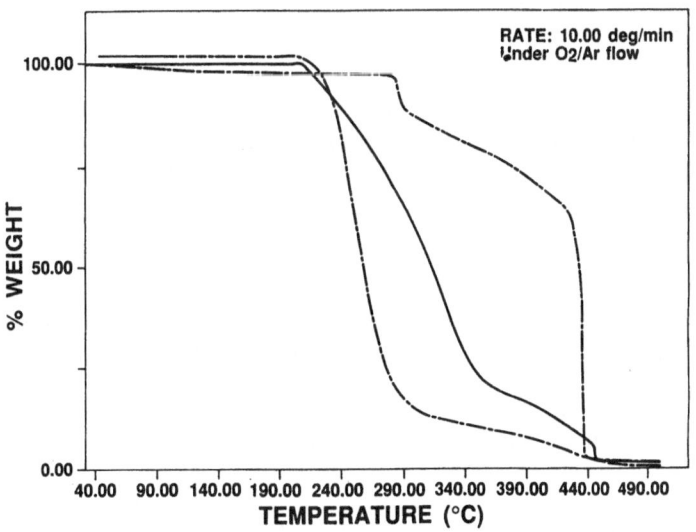

Figure 8. TGA results of (a) isotactic polypropylene and (b) poly(propylene-co-hexenol) with 4.2 mol% alcohol groups and (c) polyhexenol.

polypropylene and a 4.3 % hydroxy group containing polypropylene. The actual depression in melting point caused by functional groups is less than 5 ^{0}C. Preservation of crystallinity in the propylene copolymer apparently due to its blocky microstructure which offers enough consecutive sequences of propylene units in the polymer backbone to form a crystalline phase. In addition, the thermal stability of functionalized polypropylene (shown in Figure 8) is also very similar to that of pure isotactic polypropylene, both decomposition temperature in air are about 205°C. A sightly better resistence in decomposition process in hydroxy polypropylene may be contributed from the relatively high thermal stability of polyhexenol.

Functional polypropylene prepared by this route has many interesting physical properties. The polymer not only contains functional groups but also preserves some original physical properties of pure isotactic polypropylene. The combination of blocky molecular microstructure and crystallization of polypropylene results an unique morphological arrangement, in which the functional groups are located on the surface of the thermoplastic crystalline phases. This polymer has many potential uses. It has been demonstrated as a very effective interfacial modifier to improve the adhesion between pure polypropylene and substrates, such as aluminum and glass. The application of this functionalized polypropylene as an immobile phase (15) in supported catalysts has shown many unique advantages and will be reported in the future papers.

SUMMARY

The preparation of functional polyolefins by direct Ziegler-Natta polymerization has long been a goal of synthetic polymer chemists. A new method has been developed based on borane monomers and polymers. The value of this method rests with the "cleanliness" and versatility of its chemistry. Not only can the borane derivatives be quantitatively carried through the Ziegler-Natta process, but they can also be quantitatively converted to a host of other functionalities.

ACKNOWLEDGEMENT

We thanks the donors of the Petroleum Research Fund, administered by the American Chemical Society, and The Pennsylvania Research Corporation for financial support.

REFERENCES

1. C. Pinazzi, P. Guillaume and D. J. Reyx, Eur. Polym., 13, 711 (1977).
2. T. C. Chung, M. Raate, E. Berluche and D. N. Schulz, Macromolecules, 21, 1903 (1988).
3. T. C. Chung, J. Polym. Sci., Polym. Chem. Ed., 27. 3251 (1989).
4. J. Boor, Jr., "Ziegler-Natta Catalysts and Polymerizations", Academic Press, New York, (1979).
5. M. D. Purgett: Ph.D. Thesis, University of Massachusetts (1984).
6. K. J. Clark and W. G. City, U. S. Patent 3,492,277 (1970).
7. T. C. Chung, Macromolecules, 21, 865 (1988).
8. T. C. Chung, U. S. Patent 4,734,472 and 4,751,276 (1988).
9. H. C. Brown, "Organic Synthesis via Boranes", Wiley-Interscience, New York (1975).
10. R. H. Yocum and F. B. Nyquist, "Functional Monomers", Marcel Dekker, New York (1973).
11. J. G. Pritchard,"Polyvinyl Alcohol, Basic Properties and Uses", Gordon and Breach, New York (1971).

12. S. Ramakrishnan, E. Berluche and T. C. Chung, Macromolecules, 23, 378 (1990).

13. G. Ruggeri, M. Aglietto, A. Petragnani, F. Ciardelli, Eur. Polymer J. 1983,19, 863.

14. Each NMR sample was prepared by dissolving 70 mg of polymer in 0.7 g of d-toluene solvent, containing an internal standard of triethylborate (7×10^{-5} M), at 100 ^{0}C, then kept at 70 ^{0}C during the measurement.

15. T. C. Chung, U. S. Patent Application (1991).

DERIVATIVES OF POLY(ALKYL/ARYLPHOSPHAZENES)

Patty Wisian-Neilson

Department of Chemistry
Southern Methodist University
Dallas, Texas 75275-0314

ABSTRACT

Polyphosphazenes with simple alkyl and aryl substituents directly attached to the backbone by P-C linkages can be prepared by the condensation polymerization of N-silylphosphoranimine precursors. These simple polymers can then be converted to a variety of functionalized polyphosphazenes by derivatization reactions. In this paper, the synthesis and characterization of some derivatives of poly(methylphenyl-phosphazene), $[Me(Ph)P=N]_n$, and the copolymer, $[Me(Ph)P=N]_x[Me_2P=N]_y$, are discussed. These polymers include grafted copolymers, water soluble carboxylated polymers, and polymers with silyl, vinyl, alcohol, ester, ferrocene, phosphine, thiophene, and/or fluoroalkyl groups.

INTRODUCTION

Polyphosphazenes are a diverse class of polymers with an inorganic backbone of phosphorus and nitrogen atoms. Because these polymers have two substituents at each phosphorus, it is possible to widely vary the properties of this polymer system by simple changes in these substituents. In fact, several hundred such polymers are known and their potential applications include materials that are flame retardant, resistant to UV radiation and chemicals, semiconductors, conductors, insulators, water soluble, biologically inert or active, etc. (1,2)

The poly(alkyl/arylphosphazenes) are a special type of phosphazene polymer in which all substituents are attached to the backbone phosphorus by direct P-C linkages. Unlike the majority of polyphosphazenes, which are usually prepared by ring opening of fully (3 - 5) or partially (6 - 8) halogenated cyclic phosphazenes followed by nucleophilic substitution of the halogens, the poly(alkyl/arylphosphazene) homopolymers and simple copolymers are made by the condensation polymerization (9, 10) of Si-N-P compounds known as N-silylphosphoranimines (eq 1 and 2).

$$Me_3Si-N=\underset{\underset{Me}{|}}{\overset{\overset{Ph}{|}}{P}}-OCH_2CF_3 \xrightarrow{\Delta} \underset{\underset{Me}{|}}{\overset{\overset{Ph}{|}}{[N=P]_n}} + Me_3SiOCH_2CF_3 \qquad (1)$$

1

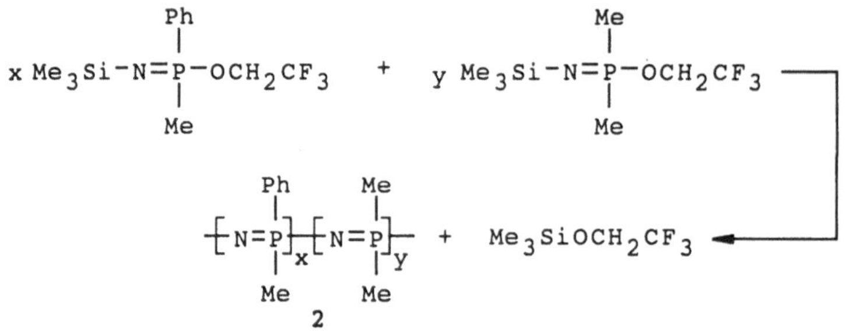

$$x \; Me_3Si-N=\overset{\underset{\displaystyle Me}{|}}{\overset{\displaystyle Ph}{|}}P-OCH_2CF_3 \;\; + \;\; y \; Me_3Si-N=\overset{\underset{\displaystyle Me}{|}}{\overset{\displaystyle Me}{|}}P-OCH_2CF_3 \longrightarrow$$

$$\left[N=\overset{\underset{\displaystyle Me}{|}}{\overset{\displaystyle Ph}{|}}P \right]_x \left[N=\overset{\underset{\displaystyle Me}{|}}{\overset{\displaystyle Me}{|}}P \right]_y \;\; + \;\; Me_3SiOCH_2CF_3 \longleftarrow$$

$$\textbf{2}$$

$$(2)$$

The Si-N-P precursors used in this process are prepared as shown in SCHEME I. (10) This approach allows for the incorporation of simple alkyl and aryl substituents at the small molecule stage and avoids the chain degradation reactions that occur when preformed halogenated phosphazenes are treated with organometallic reagents (e.g., RLi, RMgX, R_3Al). (11)

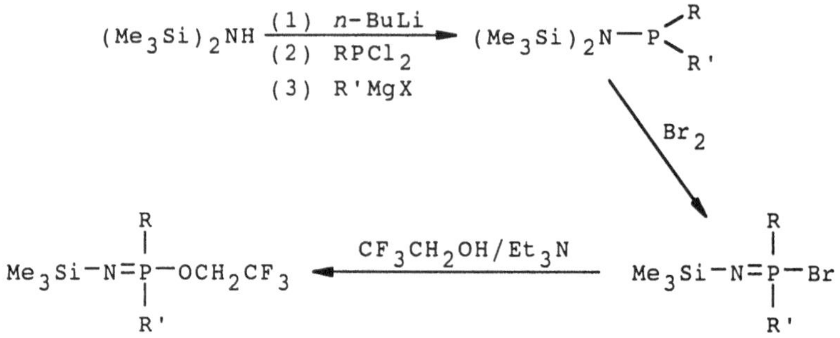

$$(Me_3Si)_2NH \;\; \xrightarrow[\substack{(2) \; RPCl_2 \\ (3) \; R'MgX}]{(1) \; n-BuLi} \;\; (Me_3Si)_2N-P\overset{\displaystyle R}{\underset{\displaystyle R'}{\Big\backslash}}$$

$$\Big\downarrow Br_2$$

$$Me_3Si-N=\overset{\underset{\displaystyle R'}{|}}{\overset{\displaystyle R}{|}}P-OCH_2CF_3 \;\; \xleftarrow{\;\; CF_3CH_2OH/Et_3N \;\;} \;\; Me_3Si-N=\overset{\underset{\displaystyle R'}{|}}{\overset{\displaystyle R}{|}}P-Br$$

SCHEME I

DERIVATIVES OF POLY(METHYLPHENYLPHOSPHAZENE)

While it is possible to introduce selected functional groups at the carbon attached to phosphorus (e.g., substituted vinyl groups) in the precursor synthesis, many groups do not withstand the rigorous thermolysis conditions or are not compatible with the reactive Si-N linkage. (12 - 14) Thus, we are investigating methods of introducing a variety of functional groups into the preformed poly(alky/arylphospha-enes). Our primary focus has been the deprotonation-substitution reactions of the methyl substituents on poly(methylphenylphosphazene), $[Me(Ph)PN]_n$, 1, and, more recently, on copolymers, $[Me(Ph)PN]_x[Me_2PN]_y$, 2. Treatment of a THF solution of the simple homopolymer, 1, with n-BuLi at -78°C produces a polymer anion intermediate, 3, which reacts with a variety of electrophiles (SCHEME II)

The reactions of the polymer anion with chlorosilanes established that, under these mild reaction conditions (-78 °C), treatment with n-BuLi does not cause chain degradation. This was confirmed both by intrinsic viscosity measurements and by GPC data on a series of silylated polymers, 4 [where E = $SiMe_3$, $SiMe_2H$, $SiMe_2(CH=CH_2)$, and $SiMe_2(CH_2)_3CN$], the latter of which showed slight increases in molecular weight with polydispersity values that were virtually identical to that of the parent polymers. (15) The polymers produced in this reaction contain reactive functional groups (i.e., vinyl, Si-H, and CN) that are sites for further derivatization reactions, including crosslinking and grafting. The polymer anion 3 also reacts with other element-halogen species, again giving polymers 4 (E = PPh_2, Br, $CH_2CH=CH_2$, Me) that contain sites for additional reaction. (16) For example, the Ph_2P group can serve as a coordination site for transition metals.

In essence, the polymer anion behaves as a "typical" organolithium reagent and can be used as such for the attachment of a variety of simple organic functional groups. In this manner, a wide variety of alcohols, 5, can be prepared by treating the anion with aldehydes or ketones. The degree of substitution in most of these reactions was about 35% (i.e., $x:y = 2:1$), but is as high as 45% for the ferrocene and thiophene derivatives. The T_g values of the alcohol derivatives showed the expected correlation with the size and number of groups R and R' and

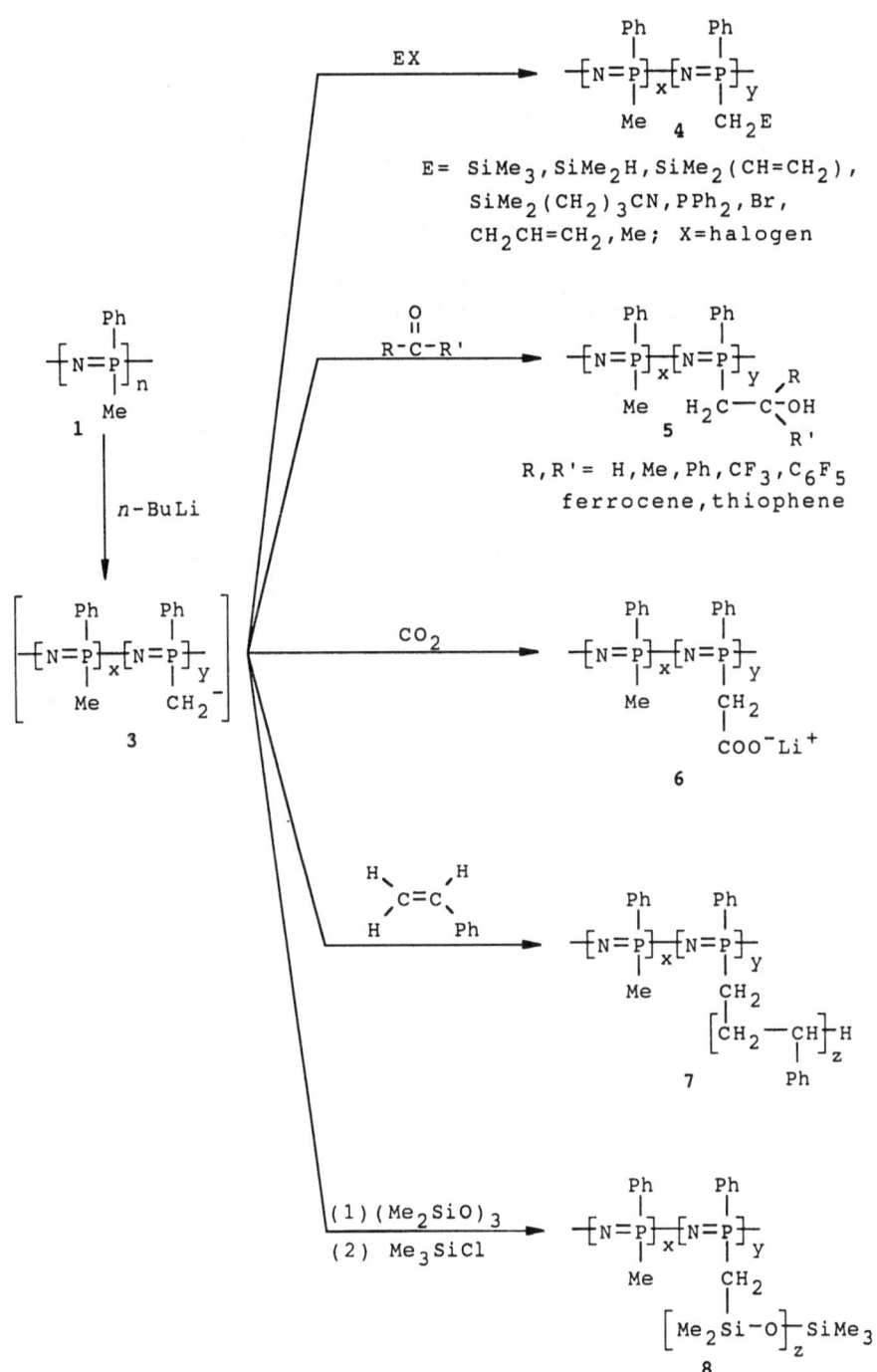

E= $SiMe_3$, $SiMe_2H$, $SiMe_2(CH=CH_2)$, $SiMe_2(CH_2)_3CN$, PPh_2, Br, $CH_2CH=CH_2$, Me; X=halogen

R,R'= H, Me, Ph, CF_3, C_6F_5 ferrocene, thiophene

SCHEME II

335

ranged from 49 to 102 °C (parent T_g = 37 °C). (17) The fluorinated alcohols had T_g values very near those of the non-fluorinated analogs. (18)

Carboxylic acids, salts, and esters can also be prepared by treatment of anion 3 with CO_2. (19) The 50% substituted salt 6 (where x = y) is the first water soluble derivative of a poly(alkyl/arylphosphazene). The salts are readily converted to the corresponding carboxylic acids upon protonation or to esters when treated with an activated species such as $4-NO_2C_6H_5CH_2Br$. Fluorescence studies of these simple acids indicate that they form a moderately hydrophobic environment in aqueous media. (20)

The polymer anion 3 has also been used to initiate anionic polymerization of styrene (21) and hexamethylcyclotrisiloxane (22) to produce both an organic (7) and inorganic (8) graft copolymers of polyphosphazenes (SCHEME II). Grafting has been demonstrated by marked changes in GPC data in terms of higher molecular weights and broadened polydispersity values. More importantly, absolute molecular weight determinations by membrane osmometry confirm the large increases in molecular weight upon grafting.

As part of our investigation of the chemistry of the new polymers, we have attempted to prepare both acid chloride derivatives and ester derivatives. The former is complicated by chain degradation which presumably occurs when HCl (or even R_3NHCl) is produced as a byproduct. The chemistry of the alcohol functionality has, however, been more straightforward. As shown in eq 3, the alcohol produced from the reaction of the anion 3 with acetone reacts smoothly with a variety of acid chlorides to give a series of esters 9. While these reactions were

$$R = (CH_2)_nCH_3; \; n = 2 - 16$$

(3)

initially conducted to test the feasibility of this reaction for attaching more elaborate groups R, including moieties that might impart liquid crystallinity to the polymers, an interesting trend was observed in the thermal data of these ester derivatized polymers. As the length of the chain increased from n = 2 - 6 the T_g decreased from 37 to 7 °C, but when the length was increased from n = 10 - 16, the T_g increases from 14 to ca. 45 °C. (23)

DERIVATIVES OF A PHOSPHAZENE COPOLYMER

In order to further control the properties of the polyphosphazenes, we are now investigating the materials obtained by derivatizing the copolymer $[Me(Ph)PN]_x[Me_2PN]_y$, 2, where x = y (T_g = ca. 0°C). The decreased solubility of this polymer in THF relative to the homopolymer 1 necessitates longer reaction times for deprotonation. Both the silylated (eq 4) and carboxylated (eq 5) derivatives have very different solubilities relative to the homopolymer derivatives. For example, the silyl derivatives are partially soluble in hexane (24) and both the 50% substituted acid and the lithium salt of the 25% substituted acid are water soluble. (25) Thus, the derivatization of the copolymers allows for further variation and control of the ultimate properties of the poly(alkyl/arylphosphazene) derivatives.

$$\left[N{=}\underset{\underset{Me}{|}}{\overset{\overset{Ph}{|}}{P}}\right]_x\left[N{=}\underset{\underset{Me}{|}}{\overset{\overset{Me}{|}}{P}}\right]_y \quad \xrightarrow[\text{(2)Me}_3\text{SiCl}]{\text{(1)}n\text{-BuLi}} \quad \left[N{=}\underset{\underset{Me}{|}}{\overset{\overset{Ph}{|}}{P}}\right]_x\left[N{=}\underset{\underset{Me}{|}}{\overset{\overset{Me}{|}}{P}}\right]_y\left[N{=}\underset{\underset{CH_2SiMe_3}{|}}{\overset{\overset{Me(Ph)}{|}}{P}}\right]_z \qquad (4)$$

$$\left[N{=}\underset{\underset{Me}{|}}{\overset{\overset{Ph}{|}}{P}}\right]_x\left[N{=}\underset{\underset{Me}{|}}{\overset{\overset{Me}{|}}{P}}\right]_y \quad \xrightarrow[\substack{\text{(2)CO}_2\\ \text{(3)H}^+}]{\text{(1)}n\text{-BuLi}} \quad \left[N{=}\underset{\underset{Me}{|}}{\overset{\overset{Ph}{|}}{P}}\right]_x\left[N{=}\underset{\underset{Me}{|}}{\overset{\overset{Me}{|}}{P}}\right]_y\left[N{=}\underset{\underset{CH_2COOH}{|}}{\overset{\overset{Me(Ph)}{|}}{P}}\right]_z \qquad (5)$$

Finally, ongoing work includes the derivatization of the phenyl group in $[Me(Ph)PN]_n$. We have found that simple acylation reactions are complicated by chain degradation, presumably due to the production of HCl in many procedures. However, the phenyl group on the polymer appears to be nitrated. (eq 6) Reduction of the nitro group has also not been

$$\left[N{=}\underset{\underset{Me}{|}}{\overset{\overset{C_6H_5}{|}}{P}}\right]_n \quad \xrightarrow{\text{HNO}_3/\text{H}_2\text{SO}_4} \quad \left[N{=}\underset{\underset{Me}{|}}{\overset{\overset{C_6H_4NO_2}{|}}{P}}\right]_n \qquad (6)$$

straightforward since many reactions involve high temperatures and acid which cause chain scission and/or metal catalysts which tend to coordinate to the backbone nitrogen. (26)

SUMMARY

The derivatization reactions of the poly(alkyl/arylphosphazenes) afford many new polyphosphazenes with a variety of functional groups and significantly extend the versatility of this class of polymers. Furthermore, the chemistry described above demonstrates that it will be possible to incorporate a number of side-groups into the polymer to enhance the thermal, mechanical, and electrical properties. A large number of graft copolymers of polyphosphazenes should also be accessible via initiation of anionic polymerizations by the polymer anions.

ACKNOWLEDGMENT

We thank the U. S. Army Research Office, The Robert A.Welch Foundation, and the Texas Advanced Technology Program for generous financial support of this project.

REFERENCES

1.   *ACS Symp. Series* **1988**, *360*, Chapters 19 - 25.
2.   See for example: (a) Singler, R. E.; Hagnauer, G. L.; Sicka, R. W.
     ACS Symp. Series **1984**, *260*, 143. (c) Blonsky, P. M.; Shriver, D. F.;
     Austin, P., Allcock, H. R. *J. Am. Chem. Soc.***1984**, *106*, 6854. (d)
     Allcock, H. R. *ACS Symp. Series* **1983**, *232*, 439.
3.   Allcock, H. R. *Angew Chem., Int. Ed. Engl.* **1977**, *16*, 147.
4.   Allcock, H. R. *Chem. Eng. News* **1985**, *63(11)*, 22.
5.   Allcock, H. R.; Kugel, R. L. *Inorg. Chem.***1966**, *5*, 1016 and 1716.
6.   Allcock, H. R.; Ritchie, R. J.; Harris, P. J. *Macromolecules* **1980**,
     13, 1332.
7.   Allcock, H. R.; Lavin, K. D.; *Macromolecules* **1980**, *13*, 1332.
8.   Allcock, H. R.; Brennan, D. J.; Graskamp, J. M. *Macromolecules* **1988**,
     21, 1.
9.   Neilson, R. H.; Wisian-Neilson, P. *Chem. Rev.* **1988**, *88*, 541.
10. (a) Neilson, R. H.; Hani, R.; Wisian-Neilson, P.; Meister, J. J.;
     Roy, A. K.; Hagnauer, G. L. *Macromolecules* **1987**, *20*, 910. (b)
     Wisian-Neilson, P.; Neilson, R. H. *Inorg. Synth.* **1989**, *25*, 69.
11. See for example: (a) Allcock, H. R.; Evans, T. L.; Patterson, D. B.
     Macromolecules **1980**, *13*, 201. (b) Allcock, H. R. *J. Am. Chem. Soc.*
     1983, *105*, 2814.
12. Wettermark, U. G.; Wisian-Neilson, P.; Scheide, G. M.; Neilson, R. H.
     Organometallics **1987**, *6*, 959.
13. Scheide, G. M.; Neilson, R. H. *Organometallics* **1989**, *8*, 1987.
14. Scheide, G. M.; Neilson, R. H. *Phosphorus, Sulfur, and Silicon* **1989**,
     39, 189.
15. Wisian-Neilson, P.; Ford, R. R.; Roy, A. K.; Neilson, R. H.
     Macromolecules **1986**, *19*, 2089.
16. Wisian-Neilson, P.; Ford, R. R.; Islam, M. S.; Ganapathiappan, S.;
     Haley, E. A., unpublished results.
17. (a) Wisian-Neilson, P.; Ford, R. R. *Organometallics* **1987**, *6*, 2258.
     (b) Wisian-Neilson, P.; Ford, R. R. *Macromolecules* **1989**, *22*, 72.
18.   Wisian-Neilson, P.; Wang. T., unpublished results.
19. Wisian-Neilson, P.; Islam, M. S.; Ganapathiappan, S.; Scott, D.L.;
     Raghuveer, K. S.; Ford, R. R. *Macromolecules* **1989**, *22*, 4382.
20. Hoyle, C. E.; Wisian-Neilson, P.; Chatterton, P. M.; Trapp, M. A.
     Macromolecules, in press.
21. Wisian-Neilson, P.; Schaefer, M. A. *Macromolecules* **1989**, *22*, 2003.
22. Wisian-Neilson, P.; Islam, M. S. *Macromolecules* **1989**, *22*, 2026.
23. Islam, M. Q.; Huang, L.-M.; Wisian-Neilson, P., unpublished results.
24. Wisian-Neilson, P.; Huang, L.-M.; Crane, R. A., unpublished results.
25. Wisian-Neilson, P.; Huang, L.-M.; Islam, M. S., unpublished results.
26. Wisian-Neilson, P.; Wood, C.; Islam, M. S.; Iriarte, J. M.,
     unpublished results.

CROSSLINKING OF PMMA ELASTOMER BLOCK COPOLYMERS
THROUGH PMMA STEREOCOMPLEXES

T. E. Hogen-Esch, J. R. Mason and B. J. Ladd

Loker Hydrocarbon Research Institute
Department of Chemistry
University of Southern California
Los Angeles, California, 90089-1661

G. Helary and G. Belorgey

Laboratoire de Recherches sur les Macromolecules
Centre Scientifique et Polytechnique
Avenue Jean Baptiste Clement
Villetaneuse, 93430, France

It is known for some time that in suitable solvents, mixtures of syndiotactic and isotactic PMMA form stereocomplexes.[1] The parameters that govern stereocomplex formation have been extensively studied by Challa,[2] Spevacek[3] and others and were recently reviewed by Spevacek.[4] Stereocomplex formation is believed to involve the formation of a type of double helix in which the larger (60/4) pitched syndiotactic PMMA helix is coiled around the smaller (30/4) pitched isotactic helix.[5] Formation of the complex is exothermic (~ −10.2 kJ/base mole) and is believed to be due to a sterically optimal fit between the two helices. Stereocomplex formation has been shown to be reversible both in solution and in solids. An approximate decomposition ("melting") temperature of 200°C has been demonstrated in solids.[5]

The use of stereocomplexes as thermally reversible crosslinkers[6] in block copolymers containing syndiotactic PMMA blocks is unprecedented and may result in interesting new properties.[7,8,9] This process is illustrated in Figure 1 for AB and ABA block copolymers in which B is an elastomeric block. In the following, we will very briefly describe the general methods of synthesis of these block copolymers and some of the characteristic of their mixtures with isotactic (I)-PMMA.

Synthesis.   Polybutadiene (PBD) and polydimethylsiloxane (PDMS) were selected as the elastomers. The synthesis of the corresponding AB and ABA block copolymers is shown in Schemes 1 and 2 respectively. The synthesis of the AB and ABA–PBD–block copolymers does not require the use of the acetal initiator **1** and could easily be carried out using simple one- and two-ended lithium initiators. However, the merit of the present synthetic method is that AB, ABA and (AB)$_3$ copolymers may be prepared from the same precursor, **12**. Similar considerations apply for the PDMS–PMMA copolymers. The use of 1,1-diphenylethylene as a capping agent is needed to prevent nucleophilic attack of the reactive allylanion of living PBD on the ester group of MMA.[10] The use of the masked OH initiator **1**

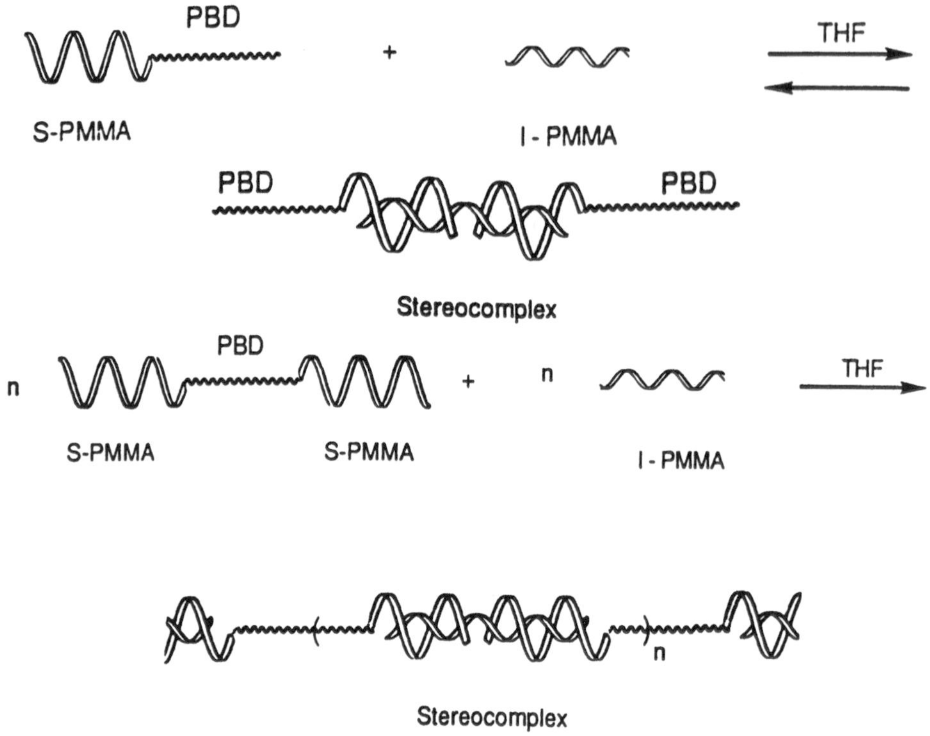

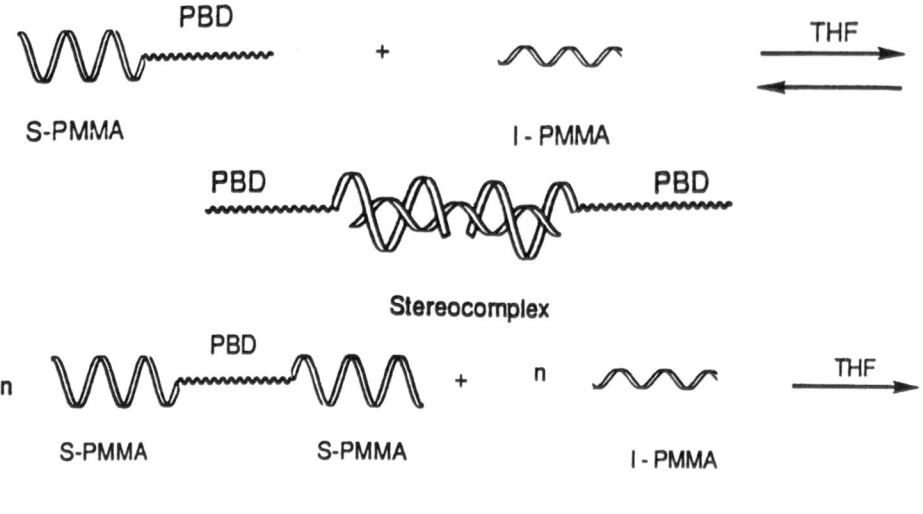

S-PMMA   I - PMMA

Stereocomplex

n S-PMMA   S-PMMA   n   I - PMMA

Stereocomplex

Figure 1. Schematic Representation of PMMA Stereocomplex Formation for AB and ABA Block Copolymers of PBD-S-PMMA with I-PMMA.

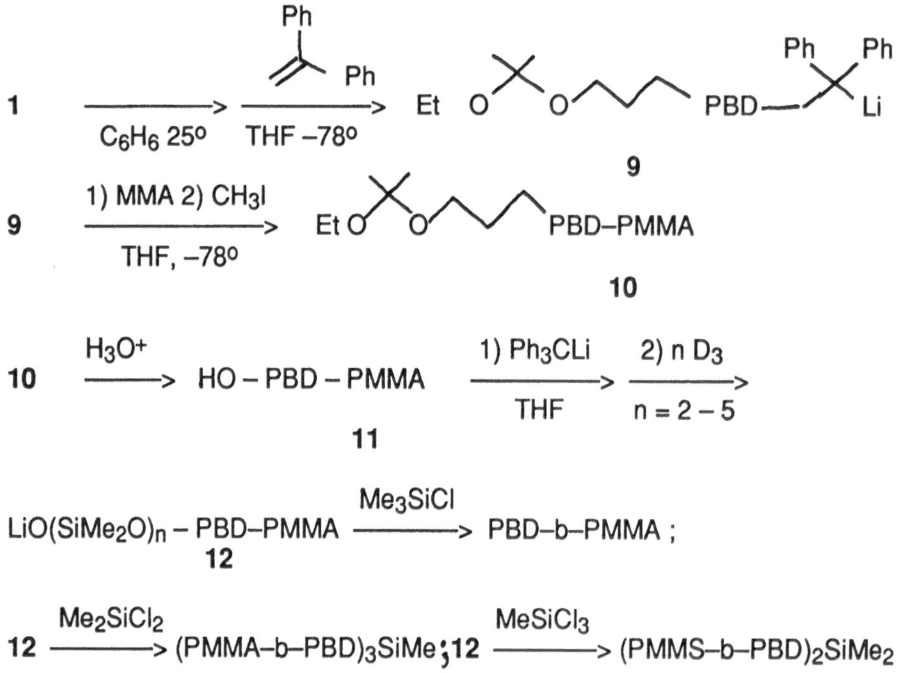

$$1 \xrightarrow[\text{C}_6\text{H}_6\ 25°]{} \xrightarrow[\text{THF} -78°]{} \quad \text{Et}\ \text{O} \diagdown \text{O} \diagup \text{PBD} \diagup \underset{\text{Li}}{\overset{\text{Ph}\ \text{Ph}}{\diagdown}}$$

9

$$9 \xrightarrow[\text{THF, } -78°]{\text{1) MMA 2) CH}_3\text{I}} \quad \text{Et}\ \text{O} \diagdown \text{O} \diagup \text{PBD–PMMA}$$

10

$$10 \xrightarrow{\text{H}_3\text{O}^+} \text{HO – PBD – PMMA} \xrightarrow[\text{THF}]{\text{1) Ph}_3\text{CLi}} \xrightarrow[\text{n = 2 – 5}]{\text{2) n D}_3}$$

11

$$\text{LiO(SiMe}_2\text{O)}_n\text{ – PBD–PMMA} \xrightarrow{\text{Me}_3\text{SiCl}} \text{PBD–b–PMMA ;}$$
12

$$12 \xrightarrow{\text{Me}_2\text{SiCl}_2} \text{(PMMA–b–PBD)}_3\text{SiMe ;} \quad 12 \xrightarrow{\text{MeSiCl}_3} \text{(PMMS–b–PBD)}_2\text{SiMe}_2$$

Scheme 1

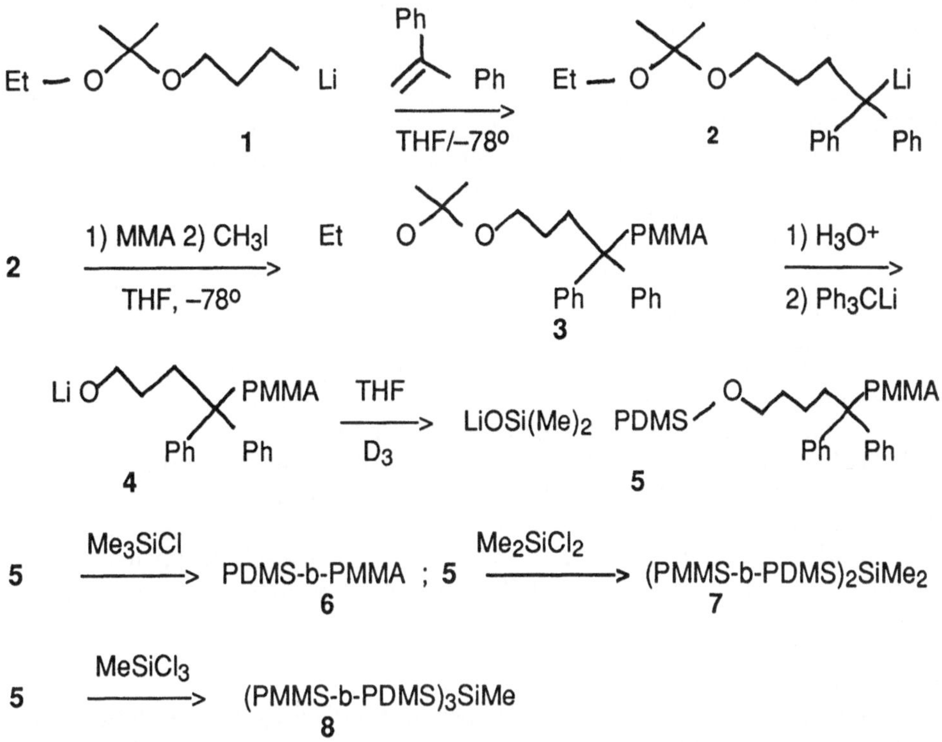

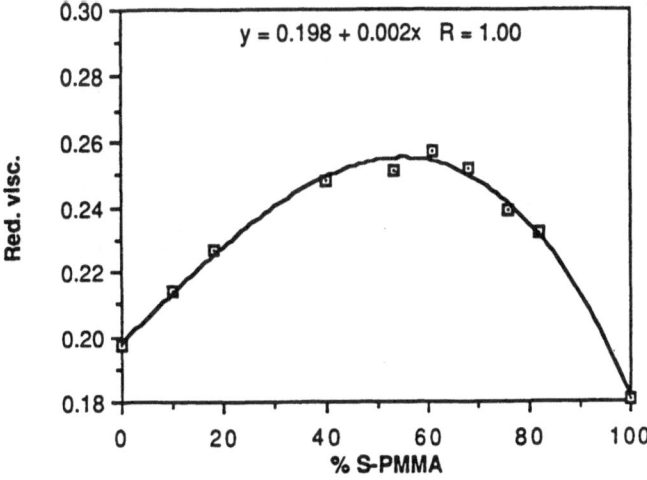

Scheme 2

Figure 2. Reduced Viscosity of I-PMMA (MW = 6000) and PBD-S-PMMA (MW's 67,000-8000) in THF at 30° as a function of S-PMMA content.

allows the generation of lithium alkoxides convertible to lithium silanolates that can be conveniently terminated with Me_3SiCl coupled with Me_2SiCl_2 or $MeSiCl_3$ to give the AB, ABA and $(AB)_3$ block copolymers respectively having identical block composition. Lithium alkoxides derived from **11** or alkoxide **4** were found to be inadequate for the coupling reactions.[8,9]

The coupling reactions, particularly those involving the preparation of the $(AB)_3$ copolymers were complicated by the occurrence of side reactions consistent with nucleophilic attack of the $-Si(Me)_2OLi$ group on the PMMA ester groups.

The coupling reactions, particularly those involving the preparation of the $(AB)_3$ copolymers were complicated by the occurrence of side reactions consistent with nucleophilic attack of the $-Si(Me)_2OLi$ group on the PMMA ester groups.

The synthesis of monodisperse PDMS–PMMA block copolymers was found to be quite challenging. Attempted direct reaction of living PMMA with D_3 was unsuccessful as was the attempted functionalization of living PMMA, Li with ethylene oxide followed by addition of D_3.[11]

Some of the AB and ABA polymers of PBD and PDMS are shown in Tables 1 and 2 respectively. Isolation of the precursor PBD or PMMA by protonation of a portion of the living PBD or PMMA allowed direct determination of the precursor block length in all of the copolymers. The mol. weight of the PMMA blocks for the PBD copolymers was obtained by ^{1}H NMR integration (Table 1). Inspection of Tables 1 and 2 shows that the AB and ABA block copolymers are quite monodisperse (Mw/Mn < 1.20) and that the polydispersity of the star-block copolymers is also quite reasonable (Mw/Mn < 1.30) if the molecular weights are not too high.[8]

<u>Solution Properties</u> of PBD–PMMA, AB and ABA block copolymers in the presence of isotactic–PMMA are illustrated in Figures 2 and 3.

The reduced viscosity (= η sp/c) of a mixture a PBD (67,000)-S-PMMA (8,000) copolymer with an isotactic PMMA (60,000) at a constant overall polymer concentration in THF is shown in Figure 2. The occurrence of the stereocomplex is indicated by a viscosity maximum at the expected S/I-PMMA ratio of about two. Apparently bridging is occurring of two or more PBD-S-PMMA polymers by the longer I-PMMA leading to higher viscosity. However, at lower concentrations of total polymer (c = 0.2 gr/dl) a reduced viscosity increase is not observed. Apparently, at this concentration, the critical DP for stereocomplex formation under our conditions is more than 80. Further confirmation for stereocomplex formation is provided by the analogous system involving the ABA copolymer (Figure 3). At or near the stoichiometry for stereocomplex formation, the bridging of the ABA copolymer by the I-PMMA now results in chain extension, thus causing a very large increase in reduced viscosity. This steep increase in apparent molecular weight is indeed expected on the basis of step-polymerization. Reducing the DP of the S-PMMA block of the ABA copolymer from 200 to about 35, did not result in viscosity increases under similar conditions indicating that at c = .50 gr/dl, the DP necessary for stereocomplex formation is greater than 35. This is consistent with the findings of Challa and coworkers who determined a critical DP of about 60 for stereocomplex formation.[2]

<u>Stereocomplex Formation in Bulk.</u> DSC measurements were carried out by mixing appropriate chloroform solutions of the block copolymers and I-PMMA followed by drying in a vacuum oven at 50°C for 24 hours prior to annealing at various temperatures. DSC thermograms of mixtures of I-PMMA (124,000) with S-PMMA (6800) and with AB-PDMS-S-PMMA (4000-6800) are shown in Figure 4. A single decomposition temperature of about 170° is observed in both cases and the ΔH values for the homo- and copolymer are also seen to be quite similar. This indicates that, at least in this case, the PDMS block does not appear to hinder stereocomplex

Table 1. Microstructure, block molecular weights and polydispersities of PBD-S-PMMA. AB, ABA and (AB)$_3$ copolymers.

Sample	$\overline{Mn}$[a,b]	1,4 Content[c] (%)	Syndiotactic[c] Content (%)	$\overline{Mw}/\overline{Mn}$[a]
P$_1$–PBD	9,300	67	0	1.06
P$_1$C$_1$–AB	11,400	67	80	1.11
P$_1$C$_1$–ABA	22,800	67	80	1.16
P$_1$C$_1$–(AB)$_3$	34,200	67	80	1.24
P$_2$–PBD	22,800	69	8	1.04
P$_2$C$_3$–AB	26,400	69	84	1.12
P$_2$C$_3$–ABA	48,300	69	84	1.15
P$_2$C$_3$–(AB)$_3$	72,500	69	84	1.22
P$_3$–PBD	48,000	74		1.05
P$_3$C$_1$–AB	51,500	74	85	1.04
P$_3$C$_1$–ABA	103,100	74	85	1.10
P$_3$C$_3$–ABA	107,000	74	81	1.30

a. Determined by SEC using PBD standards. b. Copolymer Mn determined using Mn of PBD precursor and ^{1}H NMR integration. c. By ^{1}H NMR.

Table 2. Block molecular weights and polydispersities of PDMS–S–PMMA, AB, ABA and (AB)$_3$ Copolymers.

Sample	$\overline{Mn}$[a,b]	$\overline{Mp}$[b]	$\overline{Mw}$[b]	$\overline{Mw}/\overline{Mn}$[b]	Properties[c] Copolymer
PMMA–11	3,500	—	3,700	1.07	—
AB 11a	10,300	10,500	11,000	1.08	cr
ABA 11a	19,300	20,300	22,000	1.14	tr
(AB)^3 11a	25,200	29,900	32,000	1.27	tr
PMMA–19	6,000	—	6,600	1.10	—
AB–19	18,400	22,100	21,700	1.18	cr
ABA–19	38,200	47,500	45,100	1.18	tr
(AB)$_3$–19	53,500	60,100	70,600	1.32	tr
PMMA–21	19,700	—	23,000	1.17	
AB–21	29,300	32,900	32,800	1.12	bp
ABA–21	44,500	55,100	49,400	1.11	tp
PMMA–20d	3,900	—	3,400	1.07	—
AB–20d	54,100	59,300	57,800	1.07	sw
ABA–20d	90,700	106,700	107,000	1.18	tr
(AB)$_3$–20d	130,200	168,800	169,200	1.30	tr

a. PMMA precursors determined independently by SEC using PMMA standards.
b. Copolymers mol. weight determined by SEC using polystyrene standards.
c. cr = crumbly rubber. tr = tough rubber, bp = p brittle plastic, tp = tough plastic.

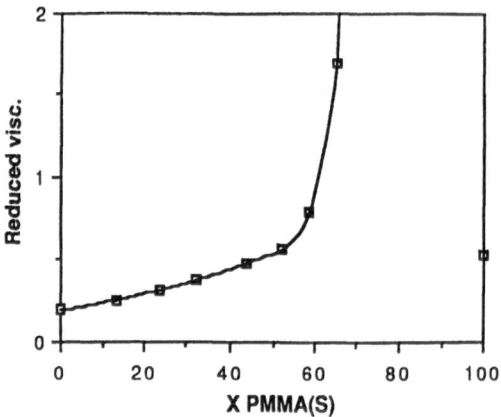

Figure 3. Reduced Viscosity of a S-PMMA-PBD-S-PMMA mixed with I-PMMA in THF at 30ºC MW I-PMMA = 6,000 ; Copolymer: S-PMMA blocks : 43,000 MW PBD block is 121,000.

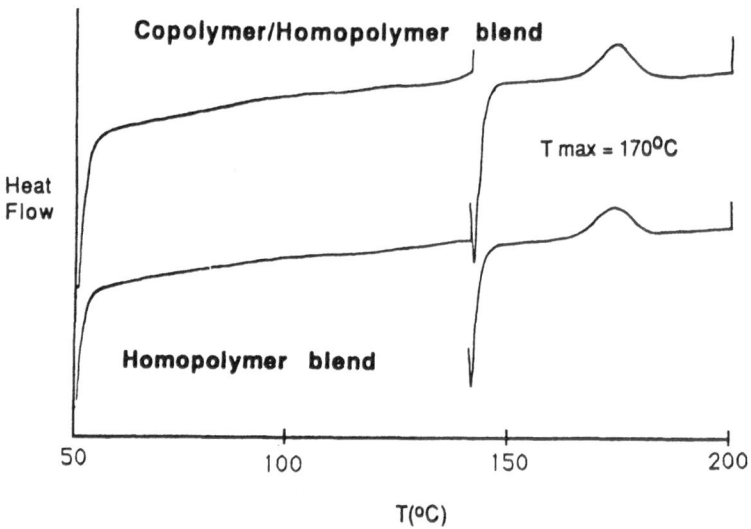

Figure 4. Copolymer Composition: MWS-PMMA is 6800 ; MW Copolymer 10,800 ; MW I-PMMA is 124,000. Homopolymer: MW S-PMMA > 6800, MW I-PMMA is 124,000.

Table 3. Effects of Composition and Annealing Temperature on the ΔH of Stereo-complex Decomposition in Mixtures of PBD–S–PMMA and I–PMMA.

I/S PMMA Composition	Annealing temperature T_a				
	100°C	110°C	120°C	130°C	140°C
25/75	6.7	9.6	10.3	8.7	1.0
35/65	18.3	16.2	16.0	11.2	1.4
42/58	23.4	24.6	21.6	18.4	6.0
50/50	22.2	20.5	19.7	15.6	6.3
58/42	19.2	17.7	14.4	12.6	5.5
79/21	10.8	9.5	8.2	7.0	3.8

a. Annealing time is 64 hours.

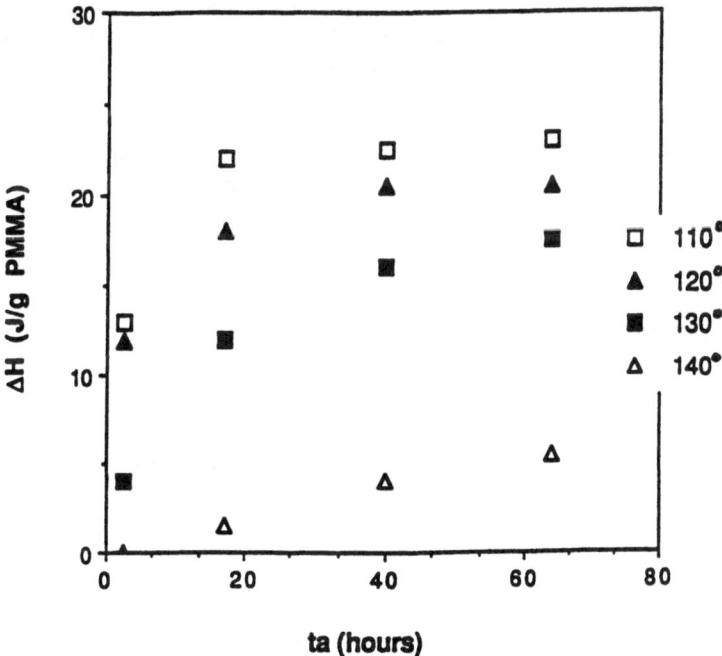

Figure 5. Heat of stereocomplex decomposition as a function of annealing time (t_a) for several annealing temperatures (T_a).

formation of the attached S-PMMA block. This finding also suggests the occurrence of PMMA domain formation necessary for the proper functioning of a thermoplastic elastomer.

Very similar results are obtained for the PBD–PMMA block copolymers indicating that the elastomer block does not hinder complex formation. Detailed DSC studies were carried out for PBD-S-PMMA/-I-PMMA blends. The I-PMMA sample had a Mn value of 44,000 and an isotactic content of 94 %. The PBD-S-PMMA (AB) sample had a molecular weight of 4500 for each of the blocks and the 1,4 and syndiotactic contents of the two blocks were 85 and 80 % respectively. The effect of annealing time (t_a) on the ΔH (Joules/gr PMMA) for various annealing temperatures (T_a) at 110, 120, 130 and 140° is shown in Figure 5. The results show that ΔH increases with t_a and decreases with T_a at constant t_a. Table 3 summarizes the I/S-PMMA compositions and annealing temperatures together with the corresponding ΔH values of stereocomplex decomposition for a t_a of 64 hrs. There is a decrease of ΔH with increasing T_a except for the 25/75 blend in the 100 – 120° range. This decrease is a general trend regardless of the S/I ratio. However, the value of the stereocomplex decomposition temperature (T_d) is observed to increase with increasing T_a. The variation of ΔH with PMMA I–S composition for various values of t_a shows a maximum at a I : S ratio of about 45 : 55.

These results point up the occurrence of PMMA stereocomplex formation in the PBD -S-PMMA I-PMMA copolymer systems. Although the extent of stereocomplex formation increases at lower Ta values, the value of Td decreases. This appears to indicate that at higher temperatures, a smaller fraction of higher-melting stereocomplexes persists. Whether this higher T_d value is related to the existence of more perfectly formed stereocomplexes or to more perfect crystallites remains to be seen. Additional studies on this and similar systems are in progress.

Acknowledgements: Support for the research of TEH was provided by DOD/Army Research Office and by NSF-CBTE/MCCP. TEH would like to thank the Universite Paris Nord for a visiting Professorship during July 1989.

References

1. W.H. Watanabe, C.F. Ryan, P.C. Fleischer and B.S. Garret, *J. Phys. Chem.,* **65,** 896 (1961).
   A.M. Liquori, G. Anzuino, V.M. Coirs, M. d'Alagni, P. DeSantis and M. Savino, *Nature,* **206**, 358 (1965).
2. Ref. A–2; E.L. Feitsma, A. de Boer and G. Challa, *Polymer, 16,* 515 (1975).
3. J. Spevacek and B. Schneider, *Die Makromol.Chem.,* **176,** 729 (1975); Ibid. **175,** 2939 (1974).
4. J. Spevacek and B. Schneider, *Advances in Colloid Interface Science,* **27,** 81 (1987) and references therein.
5. F. Bosscher, G. ten Brinke and G. Challa, *Macromolecules,* **15**, 1442 (1982).
6. T.E. Hogen-Esch and B.J. Ladd, *Proceed. 33rd IUPAC Symp. Macromolecules,* 1990; T.E. Hogen-Esch and J.P. Mason, *Proceed. 33rd IUPAC Symp. Macromolecules,* 1990.
7. J.P. Mason and T.E. Hogen Esch, *Polym. Preprts.,* **31** (1), 510 (1990); T.E. Hogen-Esch, B.J. Ladd, J.P. Mason, J. Helary and G. Belorgey, *Polym. Preprts.,* **31**(2) 405 (1990).
8. The Ph.D. Thesis, J.P. Mason, University of Southern California, December 1989.
9. The Ph.D. Thesis, B. J. Ladd, University of Southern California, May 1990.

10. D. Freyss, M. Leng and P. Rempp, *Bull. Soc. Chem. France,* 211 (1964).
11. G. Helary and T.E. Hogen-Esch, Unpublished Results.
12. E. Schomaker, H. Hoppen and G. Challa, *Macromolecules,* **21**, 2203 (1988).

SYNTHESIS AND CHARACTERIZATION OF MACROCYCLIC POLY (2-VINYL PYRIDINE) AND THE EFFECT OF A RIGID COUPLING AGENT ON ITS PROPERTIES

Janakiraman Sundararajan* and T.E. Hogen-Esch

Loker Hydrocarbon Institute and Department of Chemistry
University of Southern California, Los Angeles, CA 90089-1661

INTRODUCTION

Studies on macrocyclic polymers have been of fundamental interest to the polymer community for some time now. Since, unlike linear polymers, macrocyclic polymers lack chain-ends, a comparison of the two over a broad spectrum of properties is expected to lead to an enhanced understanding of polymer properties in general.

Our group has been involved in the synthesis of macrocyclic poly (2-vinyl pyridine) for the past few years.[1,2] We recently reported that the glass transition temperature (Tg) of a series of macrocyclic poly (2-vinyl pyridines) in contrast to the usual decrease, __increases__ with decreasing molecular weight as seen in figure 1. A similar but less pronounced trend has been reported by Semylen and coworkers[3] recently for poly dimethyl-siloxane. The increase of Tg with decreasing molecular weight has been theoretically predicted by DiMarzio and Guttmann[4] for polystyrene. In this publication we wish to report the effect of structure of the coupling agent on the Tg (glass transistion temperature) and on the hydrodynamic volume as observed by SEC (size exclusion chromatography). For example, by changing the coupling agent from the 1,4-α,α'-Dibromo xylene [1,4-DBX] to 1,2-α,α'-Dibromo xylene [1,2-DBX] intriguing changes occur in the properties of the macrocyclic polymer.

EXPERIMENTAL

General Consideration

As high dilution techniques were used, involving very low carbanion concentrations (10^{-4}-10^{-6} M), reactions were carried

---

*Current address: Baker Laboratory, Cornell University, Ithaca, NY, 14853-1301.

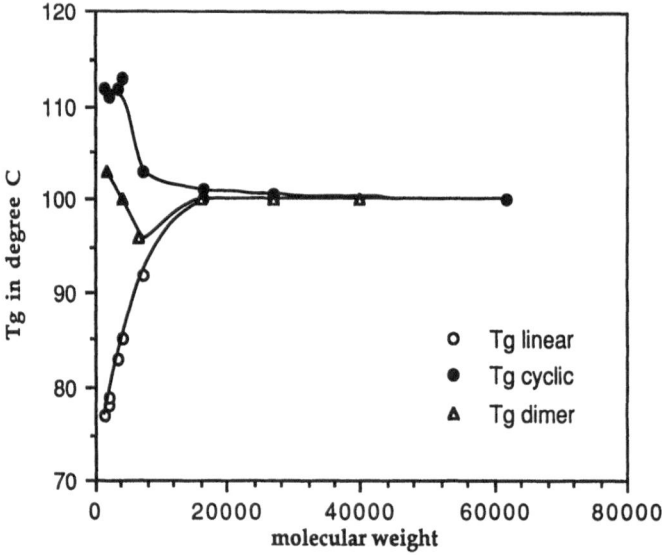

Figure 1. Tg versus molecular weight for linear, dimer and macrocyclic
poly (2-vinyl pyridine) coupling agent: 1,4-DBX.

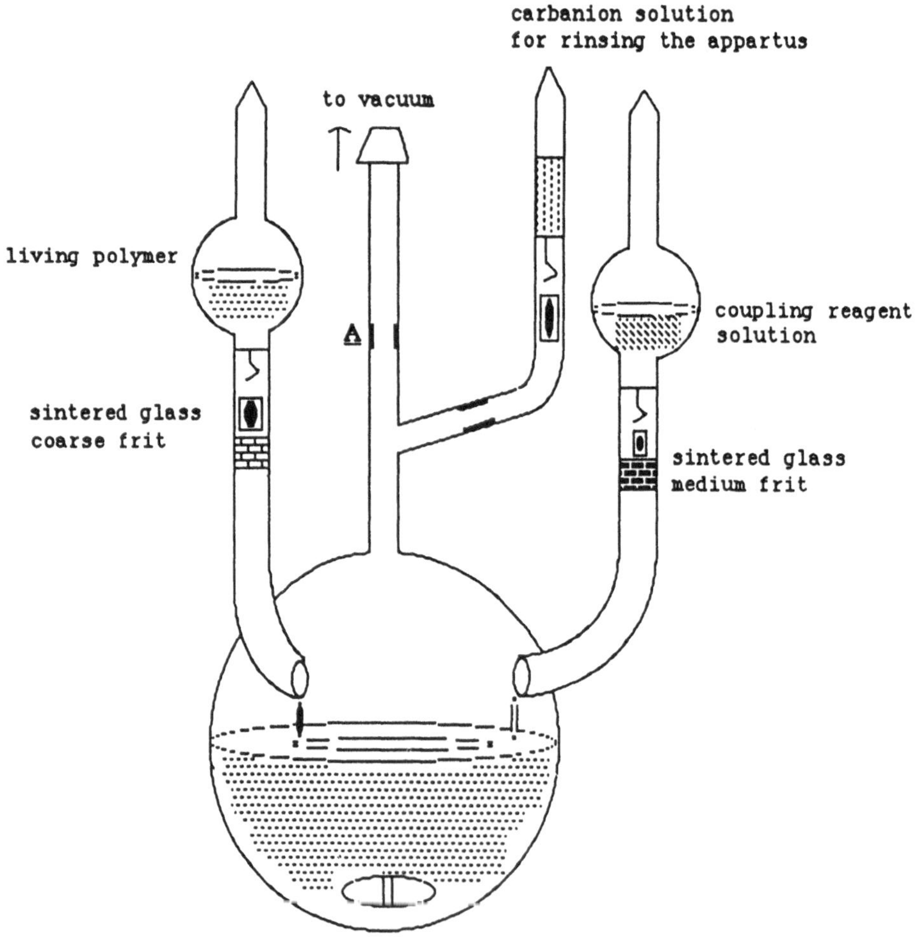

Figure 2. Cyclization Reaction

out on a high vacuum line using break seal techniques. The reaction vessels were rinsed with solutions of lithio-2-ethyl pyridine, and all solvents used were distilled from carbanion solutions to remove traces of reactive impurities.

Polymerizations

The two-ended living poly (2-vinyl pyridine) was synthesized by slow distillation of 2-vinyl pyridine onto a THF solution of 1,4-dilithio-1,1,4,4-tetra phenyl butane (DD^2-Li$_2^+$) kept at $-78°C$. After each polymerization a small (~10%) sample was reacted with methanol to obtain the molecular weight of the linear precursor.

Coupling Reactions

Dimerization. The one-ended living polymer was synthesized as mentioned above with t-BuLi as the initiator. A small (~10%) part was removed protonated to obtain molecular weight of the linear precursor. To the remaining living polymer a dilute solution of the coupling reagent (1,4-α,α'-Dibromo xylene) in THF was added. The concentration of the one-ended living polymer was always kept in excess compared to that of the coupling agent in order to ensure that the two ends of the coupling agent reacted quantitatively.

Cyclization. The cyclization reactions were carried out in a glass reaction vessel, as seen in figure 2, under high vacuum. Low concentrations of the living polymer (~10^{-5} to 10^{-6} M) and coupling agent were maintained by continuous addition of carbanion and the coupling agent solutions (~10^3 M) through sintered glass frits. The rates of addition of the two reactants were controlled by cooling and heating various sections of the glass vessels.

Precipitation and Fractionation of the Cyclic Polymer

The polymer was precipitated from THF by adding it dropwise to three times excess hexane. The fractional precipitation of the cyclic polymer from the crude mixture (~3g) was performed by dissolving it in 10ml of chloroform and adding it to hexane (~25ml) dropwise until a turbid solution was observed. Further addition of hexane gave a white precipitate which was centrifuged and the precipitate was checked by SEC. The supernatant solution was removed and hexane was added, precipitation was performed, and this was continued to give 10 to 12 fractions. The various fractions isolated were checked by the SEC. The fraction with the narrowest molecular weight distribution and free of linear contaminants was used for further characterization. Preparative SEC was also attempted on a Waters preparative GPC set up with THF containing 1% triethyl amine as the solvent, on a 500 A° Styragel preparative SEC column using an automated sample collection system.

Size Exclusion Chromatography (SEC)

SEC analysis was performed on a Waters SEC system containing a Waters 510 pump, Shodex KF-800, 802.5, 804 (of effective range of mol. wt. of 500 to 10^6) columns in series, with THF containing 1% triethyl amine as the solvent. Narrow molecular

Table 1. Apparent Molecular Weights, Poly Dispersities and Glass Transition Temperatures of Linear and Cyclic P2VP coupled with 1,4-DBX

MWb Linear	D Linear	Tg Linear (°C)	MWCyclicb,c Apparent	D Cyclic	Tg Cyclic (°C)	<G>a
1550	1.17	77	1320	1.07	112	0.85
2040	1.14	78	1800	1.13	111	0.88
3330	1.13	83	2900	1.11	112	0.87
4200	1.08	85	3570	1.08	113	0.85
7230	1.14	92	5490	1.14	103	0.76
7448	1.10	92	5727	1.10	103.5	0.77
16540	1.10	100	12800	1.07	101	0.77
26800	1.07	100	20200	1.04	100.5	0.75
62000	1.12	100	46500	1.13	100	0.75

aRatio cyclic/linear apparent MW.
bApparent molecular weights determined by SEC.
cPrepared by using 1,4-α,α'-Dibromo xylene.

Table 2. Apparent Molecular Weights, Poly Dispersities and Glass Transition
Temperatures of Linear and Cyclic P2VP coupled with 1,2-DBX

MW[b] Linear	D Linear <G>[a]	Tg Linear (°C)	MWcyclic[b,c]	D Cyclic	Tg Cyclic (°C)	<G>[a]
1730	1.12	78	1590	1.10	112	0.92
2940	1.07	82	2400	1.13	112	0.82
4250	1.06	85	3100	1.10	116	0.73
7448	1.10	92	5349	1.09	103	0.72
11900	1.10	94	8690	1.10	102	0.73
17740	1.05	100	12960	1.08	101	0.73
32300	1.11	101	23500	1.09	101	0.73
58340	1.12	101	42580	1.12	100	0.73

[a]Ratio MWcyc/ MWlin.
[b]Apparent MW determined by SEC.
[c]Prepared by using 1,2-α,α'-Dibromo xylene.

Table 3. Apparent Weights, Mono Dispersities and Glass Transition
Temperatures of Linear and Dimers of P2VP

Mw[a] Linear °C	D Linear	Tg Linear °C	MWDimer[a,b]	D Dimer	Tg Dimer °C
800	1.05	77	1575	1.06	103
2100	1.07	85	4120	1.08	100
3410	1.07	92	6700	1.07	96
8020	1.09	100	15970	1.10	100
13540	1.06	99	26900	1.08	100
19900	1.06	100	39900	1.07	100

[a]Apparent molecular weights determined by SEC.
[b]Prepared by using 1,4-α,α'-Dibromo xylene.

weight poly (2-vinyl pyridine) standards (Polysciences Inc., Pressure Chemical Company) were used for calibration. Applied Biosystems 757 Absorbance detector (λ max 268nm) was used as the detector. The flow rates ranged from 0.8 - 1.10 ml/min. The detector was interfaced with a Zenith PC-100 computer equipped with a Dascon-1 data collection system. A BASIC computer program was used for data accumula-tion, processing, calcula-tion of ·molecular weights, distribu-tion and graphic display of

Thermal Analysis

Differential Scanning Caloriemetry (DSC) was carried out on a Perkin-Elmer DSC-4 Thermal analysis system. The Tg's were measured on samples of ~30mg in sealed aluminum pans. The data was collected, stored and processed on TADS software. The experiments were performed under nitrogen atmosphere. All samples were scanned at a heating and cooling rate of 20°C/min. The samples were thermally preconditioned by heating from 50°C to 150°C, and cooling back to 55°C and heating them to 135°C for Tg measurements. The system was calibrated with Indium standards prior to usage. The reproducibility of the measure-ments were roughly +/- 1°C.

RESULTS AND DISCUSSION

The synthetic procedures discussed above were used to prepare macrocyclic poly (2-vinyl pyridine) in the molecular weight range of 1500 to 100,000 with 1,4-DBX and from 1500 to 58,000 for 1, 2-DBX. The yields of the cyclic polymer ranged from 40 to 75% of the remaining product being the higher molec-ular weight "polycondensates". Linear coupled "dimers" of poly (2-vinyl pyridine) were prepared by reaaction of one-ended living polymer with 1,4-DBX in the molecular weight range of 1600 to 40,000. The results are summarized in tables 1, 2 and 3. The SEC chromatograms of a pair of linear and cyclic crude P2VP are seen in figure 3. The cyclic polymer obtained by fractionational precipitation and preparative SEC of the same cyclic polymer is shown in figure 4. It is clearly evident that cyclic polymer obtained by the fractional precipitation technique is quite monodisperse (D<1.06) and comparable poly-dipersity to the linear precursor. The SEC of a linear precur-sor and its dimer are shown in figure 5. It is clear that the dimerization is quantitative and does not lead to the presence of the linear polymer in the dimer. The presence of the precur-sor would be expected to reduce the Tg.

The Tg versus molecular weight curve for the linear, coupled dimer and the macrocyclic poly (2-vinyl pyridine) with 1,4-DBX as the coupling agent is seen in table 4 and figure 1. For a molecular weight of about 1600 the placement of a xylyl unit in the middle of the linear gives an increase in Tg of about 26°C. The cyclization of the ring raises the Tg an additional 11°C. The calculations of DiMarzio and Guttman's treat the macrocycle as a single unit but without the effect of the coupling agent. This approach leads to a calculated increase in Tg of about 90°C of the macrocyclic polystyrene compared to the linear precursor. This discrepancy may be due to the reason that the inter-molecular factors were not con-sidered. Our data show that the Tg seems to reach a maximum at

Table 4. Glass Transition Temperature of P2VP as a
Function of the Molecular Architecture and the
Coupling Agent.

Architecture	MW Apparent by SEC	Tg in °C
Linear	1550	77
Linear dimer coupling agent 1,4-α,α'-Dibromo xylene	1600	103
Macrocyclic polymer coupling agent 1,4-α,α'-Dibromo xylene	1550	114
Macrocyclic dimer coupling agent 1,4-α,α'-Dibromo xylene	2450	119

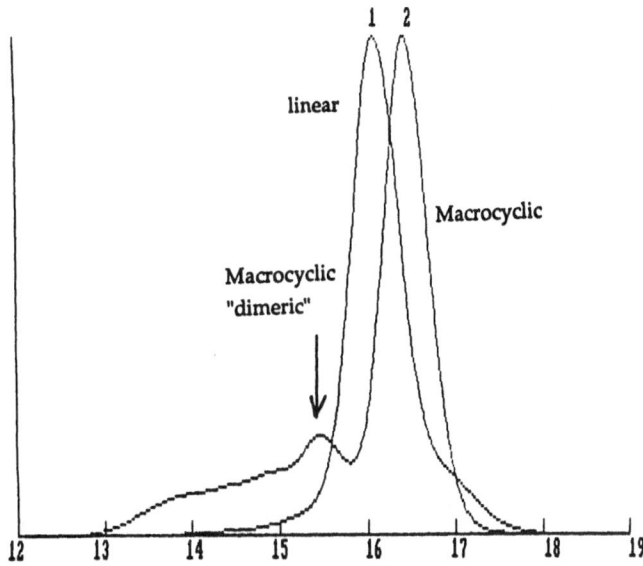

Figure 3. SEC of a pair of Macrocyclic P2VP and its linear precursor P2VP of DPn 70.

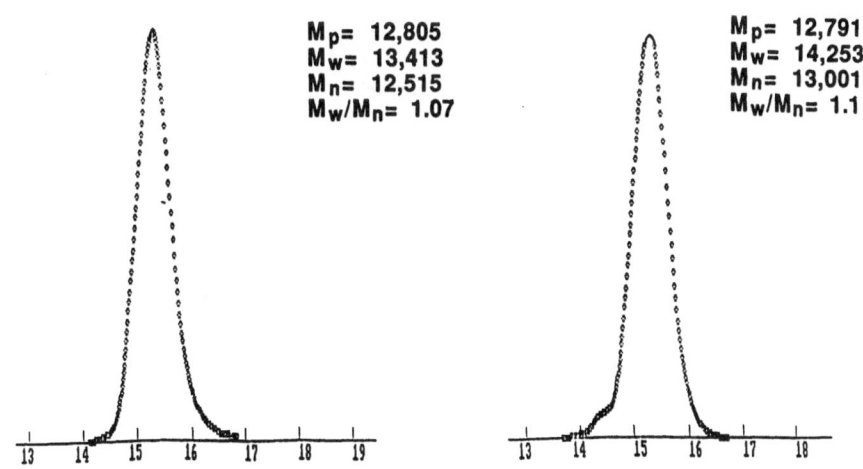

M_p= 12,805
M_w= 13,413
M_n= 12,515
M_w/M_n= 1.07

M_p= 12,791
M_w= 14,253
M_n= 13,001
M_w/M_n= 1.1

Figure 4. Fractional precipitation versus preparative SEC.

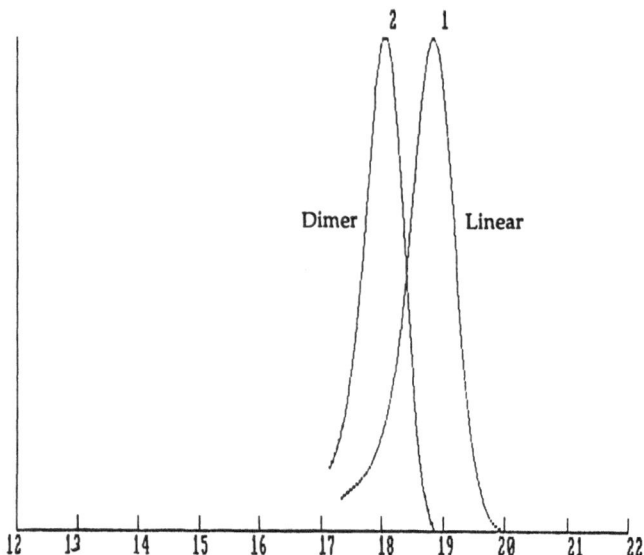

Figure 5. SEC trace of "dimer" P2VP and its linear precursor of DPn 8. Coupling agent P-DBX.

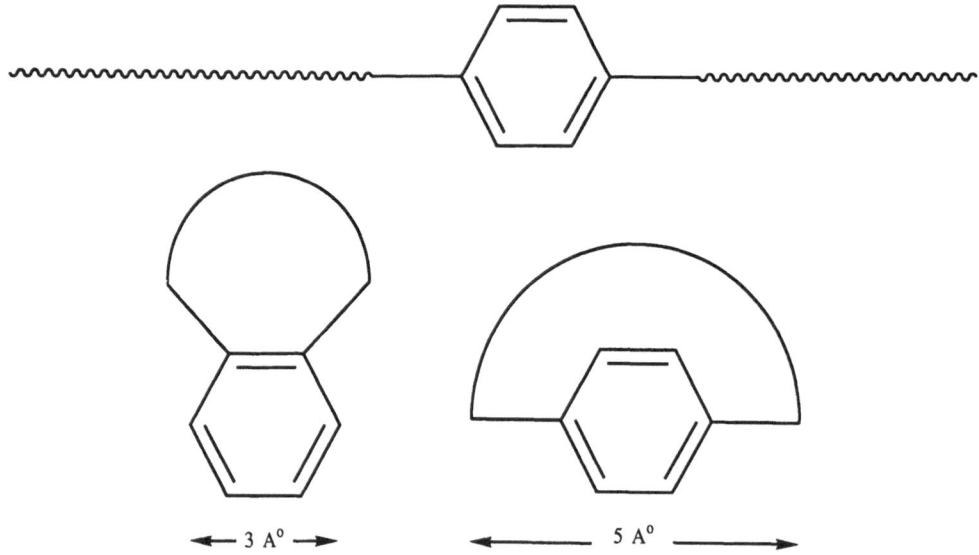

Figure 6. Comparison of 1,4-DBX, 1,2-DBX and the "dimer" of P2VP.

the results. The program used for this work is quite similar to the one described by Yau and coworkers.[5]
$DP_n=40$ and then appears to drop off at even lower molecular weights, one would expect to find oligomers of lower Tg that are expected to be waxy or liquid. From figure 1 it is clear that the effect of the coupling agent drops rapidly with increasing molecular weight for both the macrocycle and the coupled dimer. In the case of the coupled dimer, there is a minimum in Tg at DP of about 80 and the Tg then approaches the value for the linear polymer at about DP=200.

Interesting differences in hydrodynamic radius (as determined by SEC) were observed for macrocycles prepared from the same dianion precursor using 1,4-DBX and 1,2-DBX. This spacer lengths differ by only about 2 A° (5 - 3A°) but this difference seems to significantly affect hydrodynamic size of the smaller macrocycles (DP<80). For example, at a molecular weight of 1550, the ratio of apparent molecular weights of cyclic to linear polymer is 0.85 for 1,4-DBX while it is 0.92 for a comparable molecular weight of macrocycle coupled with 1,2-DBX ($MW_n=1730$). This could perhaps be due to the fact that the presence of the longer 1,4- coupling agent could result in an elongated shape of the macrocycle, compared to the more compact spherical shape of the 1,2- coupled macrocycle. As the average pore shape of the SEC column packing is cylindrical,[6] an elongated macrocycle would penetrate the pores more extensively than a more spherical macrocycle thus leading to increased elution volume.

CONCLUSIONS

Macrocyclic poly (2-vinyl pyridine) in the 1500-62,000 molecular weight range and of the highest structural integrity has been synthesized in vacuo by high dilution technique. The Tg values show a dramatic increase when the molecular weight of the macrocycles is lowered. The effect of the rigid spacer group from the coupling agent has been studied. The change in the coupling agent from 1,4-DBX to 1,2-DBX does not seem to play a big role in the Tg measurements while it seems to have a pronounced effect on hydrodynamic volume as judged by SEC.

ACKNOWLEDGMENTS

Support for this research was provided by NSF/DMR and Polymers Program.

REFERENCES

1. W. Toreki and T. E. Hogen-Esch, Polymer Preprints, 28(2):343 (1987).
2. T. E. Hogen-Esch and W. Toreki, Polymer Preprints, 30(1):129 (1988).
3. J. A. Semylen, S. J. Clarson, and K. Dodgson, Polymer Preprints, 31(1):563 (1990).
4. E. DiMarzio and C. Guttmann, Macromolecules, 20:1403 (1987).
5. W. W. Yau, H. J. Stoklosa, and D. D. Bly, J.Appl.Poly.Sci., 21:1911 (1977).
6. W. W. Yau, D. D. Bly in: "Size Exclusion Chromatograpy (GPC)" ACS Symposium series #138, T. Provder, ed., ACS, Washington, DC (1980).

TOPOCHEMICALLY CONTROLLED POLYMERIZATION OF A SMECTIC LIQUID

CRYSTALLINE DIACRYLATE MONOMER

Xuejun Qian and Morton Litt

Department of Macromolecular Science
Case Western Reserve University
Cleveland, OH 44106

INTRODUCTION

In the past two decades, the field of polymerization in the mesomorphic phase has drawn the attention of an increasing number of both chemists and physicists due to the possibility of using a liquid crystalline matrix to control both the polymerization kinetics and the structure of resulting polymers (topochemical control). Macromolecules possessing unique chemical and physical characteristics can be obtained this way. The combination of local molecular orientation and molecular mobility exhibited by mesophases is considered to be of prime importance in achieving both kinds of control. Most of the systems studied so far concern only thermally initiated polymerization of mono-functional monomers in the nematic phase; both in bulk and in nematic media (or solvents). Generally no real enhancement of polymerization rate and no substantial structural difference of the resulting polymers were observed when compared with polymerization in isotropic state.[1,2,3] This is probably due to the fact that there is no marked structural difference between the isotropic liquid and nematic liquid crystal schematically illustrated in Figure 1. Both of them lack long range and even medium range positional correlation between the molecules, which in our view is necessary to attain both kinetic and structural control during polymerization.

In the smectic mesophase, the monomer molecules are organized in two or three dimensions, as schematically shown in Figure 1c. Monomer molecules lie within the plane of a stratum, with their long axes mutually parallel. Reactive groups in the molecules can be polymerized to near completion with little or no molecular diffusion because all the reactive groups are concentrated in a small volume fraction of the system. Therefore, fast polymerization rate and high degree of order of the resulting polymer could

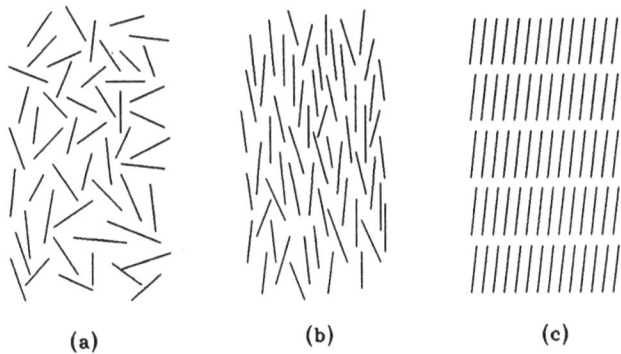

(a)   (b)   (c)

Figure 1.   Schematic representation of molecular arrangement in (a) isotropic
state; (b) nematic state; and (c) smectic state.

be achieved even though the system as a whole has high viscosity or becomes glassy, if polymerization is carried out in the smectic state. Literature on this subject was scarce because monomers exhibiting smectic mesophases were limited due to the destabilizing effect of the acrylate functional group on mesomorphic phase formation. However, polymerizations in the smectic mesophases have been reported by a number of authors;[4,5,6,7] Most of them were concerned with the influence of the positional and orientational order (on a molecular and supermolecular scale) on the reaction kinetics of the systems. Only a limited amount of detailed work was carried out for the purpose of analyzing the actual structure of the original monomer phase and of the polymer phase.

The objective of this work was to investigate polymerization in highly ordered smectic phases from the standpoint of structural characterization, with the aid of wide angle x-ray diffraction and cross-polarized optical microscopy. Mesomorphic diacrylate monomers were selected for this study because they offer the distinct advantage of locking in the liquid crystalline structure of the resulting polymer. Molecular relaxation of as-formed LC network is avoided because the molecules are cross-linked, and therefore the LC order is stabilized. The LC structures will be stable in a wide temperature range and can be conveniently studied at room temperature without worrying about the structural changes usually accompanying phase transitions or glass transition for linear side chain polyacrylates. Biphenyl was chosen as the mesogen due to its tendency to form highly ordered smectic mesophase.[8,9,10]

The monomer selected for this investigation was 4,4'-bis[5-(acryloyloxy)pentyloxy]biphenyl (**1**) (abbreviated as B5A, where B=biphenyl; 5=number of CH_2's in the flexible methylenic spacer; and A=acrylate functional group).

1

The syntheses of monomer B5A and its structural characterization will be described elsewhere.[11] Monomer B5A exhibits two enantiotropic smectic mesophases (S_E66S_B75I) with a very wide temperature range for the smectic E phase (from below -20 °C to 66 °C). In this investigation, we concentrate on the photopolymerization of B5A at room temperature (i.e., in its SmE phase) and structural characterization of the resulting polymeric network. The methods used for characterizing the structure were wide-angle x-ray diffraction (WAXD), cross-polarized optical microscopy(CP-OM) and FT-IR. WAXD was of prime importance in providing information about the type of liquid crystalline order in the polymer, the arrangement of the side chains and the orientation of the side chains on the macroscopic scale.

EXPERIMENTAL

Photopolymerization

Smectic E phase is a highly ordered smectic mesophase with very high viscosity. In order to obtain homogeneous polymer film, the monomer was heated to the isotropic state to flow and consolidate into coherent film, which was then cooled to room temperature (SmE state) for polymerization. Photopolymerization was used for that reason to avoid premature thermal curing. Monomer B5A was mixed with 2 mole percent photo-initiator, 2,2-dimethoxy-2-phenylacetophenone (Irgacure 651, Ciba-Geigy) in chloroform, and the solution was evaporated to dryness under vacuum. The blend was then put between two Pyrex glass sheets. The thickness was controlled by a PET film spacer. Sample was prepared by heating to the isotropic state in the sample cell under vacuum to remove bubbles and then cooling to room temperature. The sample was UV cured at room temperature with a mercury lamp (GE SUNLIGHT™ lamp, 275 W) for 2 hours from a distance of about 5 inches.

Characterization

Differential scanning calorimetry was carried out with a Perkin-Elmer 7500 DSC-7 Scanning Calorimeter at a heating and cooling rate of 10 °C/min. Optical observations were made using a Carl Zeiss polarizing microscope equipped with a Mettler FP 82 hot stage, and camera. Heating rates of 5 and

10 °C/min were used. IR spectra were recorded with a Bomem Michelson 110 FT-IR spectrophotometer at a resolution of 4 cm⁻¹. The sample was dispersed in a KBr pellet; 50 scans were used for each spectrum. Flat plate X-ray patterns were recorded under vacuum on Kodak No-Screen film using Ni-filtered Cu $K\alpha$ radiation (λ = 1.5418 Å) and a Searle toroidal focusing camera with pinhole collimation. WAXD spectra were also recorded as diffractometer scans, using a Phillips APD 3520 automated diffractometer using Ni-filtered Cu $K\alpha$ radiation in the reflection mode.

RESULTS AND DISCUSSIONS

Monomer B5A exhibits two enantiotropic smectic mesomorphic phases (S_E66S_B75I). The S_E phase does not crystallize even when cooled down to -20 °C. In this study, we will concentrate on the photopolymerization of B5A at room temperature (S_E phase) and the detailed structural characterization of the resulting polymeric network (subsequently called PB5A-RT). The synthesis and structural characterization for monomer B5A will be presented in a separate paper[11] and only summarized results are presented here for comparison with its polymer. Monomer B5A in the smectic E phase has an orthorhombic structure with unit cell dimensions, a = 6.609 Å, b = 7.704 Å and c = 51.92 Å. It adopts a tilted molecular bilayer structure within the unit cell in the c direction. Within each smectic layer the molecules are tilted, making an angle of about 30° with the smectic layer normal (or c direction).

Figure 2a is the microphotograph of PB5A (~ 50 μm thick) photopolymerized at room temperature for 2 hours, showing liquid crystalline textures very similar to that shown by monomer B5A (Figure 2b). The liquid crystalline structure was retained upon polymerization of monomer in SmE state. Upon heating, the liquid crystalline texture was maintained up to 220 °C where thermal degradation started. PB5A-RT has a Tg of about 120 °C (measured by DMTA). Therefore the LC organization was successfully locked into the PB5A-RT network upon polymerization of B5A in the S_E phase. The polymerization reached almost 100% conversion. This is confirmed by FTIR measurements which show almost complete disappearance of the vinyl C=C stretching band at 1636 cm⁻¹ and =C-H stretching band at 2905 cm⁻¹. To determine the mesomorphic nature of the polymeric network formed, WAXD experiments (both flat plate and diffractometer scan) were performed on thin plate samples (~ 0.3 mm thick) cured between two glass plates. The plates were treated with trimethylchlorosilane to facilitate the removal of the polymers.

The diffractometer WAXD scan of PB5A-RT is shown in Figure 3 (curve a). The spectrum has great resemblance to that of monomer B5A before curing (curve b in Figure 3). The well defined three dimensional order within the structure is quite obvious, manifested by the presence of several sharp small

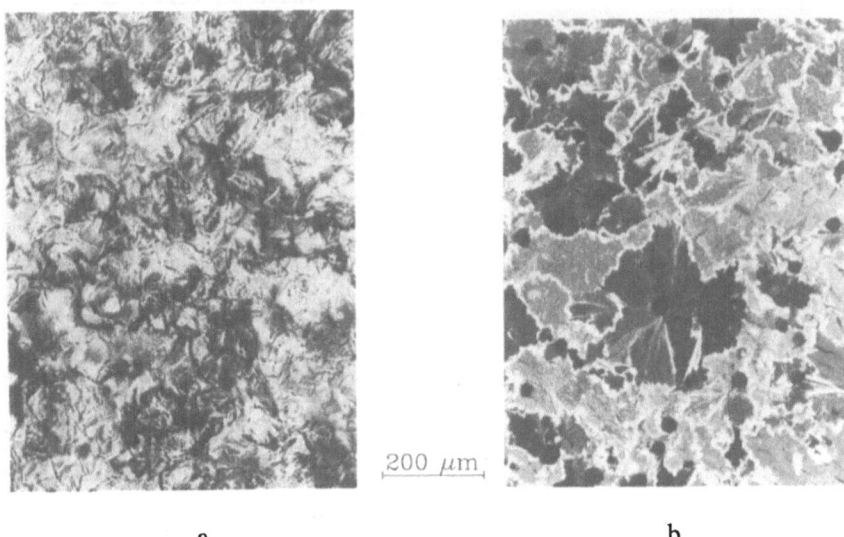

200 μm

| a | b |

Figure 2. Cross-polarized OM microphotographs: (a) polymer PB5A cured at room temperature; and (b) monomer B5A at room temperature.

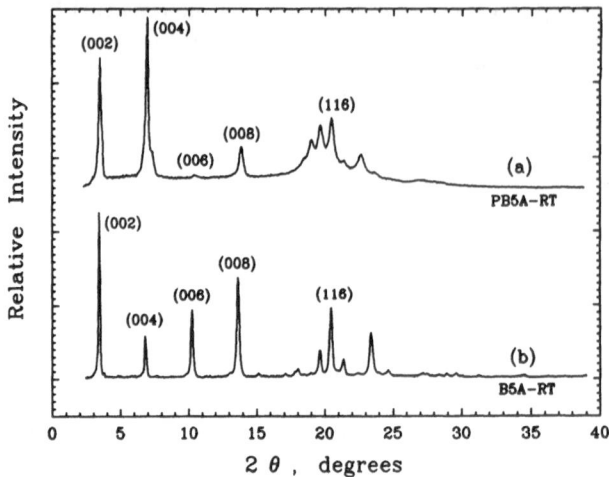

Figure 3. WAXD diffractometer scans of (a) PB5A cured at room temperature; and (b) B5A at room temperature.

angle reflections (2θ < 16°, due to smectic layer periodicity) and several sharp large angle reflections (2θ > 16°, due to regular intermolecular packing within smectic layer). Like B5A, the WAXD data for PB5A-RT can be indexed using an orthorhombic unit cell with unit cell parameters of a = 6.547 Å, b = 7.800 Å, and c = 50.95 Å. Table 1 lists the indexing for PB5A-RT WAXD data; excellent fits between observed and calculated d-spacings are obtained. Like B5A-RT, space group **Pna2₁** was selected for PB5A-RT

Table 1. Indexing of WAXD Data for PB5A Cured at Room Temperature

No.	d_{obs} , Å	Intensity	d_{calc} , Å	h k l	Error , %
1	25.54	M	25.48	0 0 2	-0.24
2	12.74	S	12.74	0 0 4	0.00
3	8.489	W	8.492	0 0 6	0.04
4	6.371	M	6.369	0 0 8	-0.03
5	4.796	W	4.809	1 1 3	0.28
6	4.666	M	4.666	1 1 4	0.00
7	4.506	S	4.499	1 1 5	-0.15
8	4.332	S	4.318	1 1 6	-0.32
9	4.141	W	4.130	1 1 7	-0.28
10	3.917	S	3.900	0 2 0	-0.44
			3.940	1 1 8	0.58
11	3.754	W	3.754	1 1 9	0.00
			3.729	0 2 4	-0.67
12	3.309	W	3.326	0 2 8	0.51
13	3.281	W	3.273	2 0 0	-0.23

Unit cell parameters (Orthorhombic):
a = 6.547 Å   b = 7.800 Å   c = 50.95 Å
Density$_{calc}$ = 1.200 gm/cm³   Density$_{obs}$ = 1.217 gm/cm³

according to the extinction rules. The calculated density for PB5A-RT, based on above unit cell parameters, is 1.200 gm/cm^3 which is close to the measured density of 1.217 gm/cm^3, supporting the proposed unit cell. Smectic E phase is assigned to PB5A-RT due to its orthorhombic nature. For PB5A the backbone or main chain (originally the vinyl group in the monomer molecule) is in the smectic layer surface, and the side chain (the portion of monomer molecule excluding the vinyl groups) is oriented in the smectic layer thickness direction. The small angle reflections correspond to the smectic layer thickness; reflections with only even order reflections (up to the 8th order) were observed and a value of 50.95 Å was obtained for the layer thickness. The large angle reflections correspond to the periodicities of molecular (side chain) packing in the lateral direction within each smectic layer. Since the layer thickness of 50.95 Å is much larger than the calculated side chain length of 28.6 Å (based on fully extended molecular conformation) but about 6 Å short of twice of the side chain length, a tilted side chain bilayer structure is proposed for PB5A, as for B5A. The calculated side chain tilt angle with respect to smectic layer normal is about 27°, which is supported by the facts that reflections (115) and (116) have the highest intensity in (11l) series, and, (115) and (116) planes make an average angle of approximate 28° to the smectic layer normal. Figure 4 shows the schematic diagram of the molecular (side chain) arrangement within the smectic layer in PB5A-RT.

Since a reflection mode was used for WAXD diffractometer scans on thin plate samples ($\sim$ 0.3 μm thick), the reflections correspond to the periodicities in the sample thickness direction. The extent of smectic layer orientation can be evaluated by comparing the relative intensities of reflections due to smectic layer thickness (small angle reflections) with those due to lateral packing within smectic layers (large angle reflections). As one can see in Figure 3 (both for monomer and polymer) the smectic layer reflection intensities are noticeably much stronger than those due to the lateral packing. However, the intensities due to smectic layer reflections for powder samples are weak compared to those due to the lateral packing. Therefore it is quite reasonable to say that there is strong orientation of smectic layer normal in the sample thickness direction. To further substantiate this observation, flat plate wide angle x-ray patterns were recorded using two different sample orientations with respect to the incident x-ray direction, schematically shown in Figure 5. Figures 6a and 6b show the WAXD patterns for parallel and perpendicular (x-ray with respect to plate surface) orientations respectively. Due to the preferred lamellar orientation, the pattern (for parallel orientation in Figure 6a) resembles that of a conventional polymer fiber diagram. The reflections, due to the smectic layer thickness up to the 8th order (odd order reflections are either absent or too weak to see) are concentrated around the meridian with strong spotty reflections sitting exactly on the meridian. The WAXD diffractometer scan in reflection mode shown in Figure 3 is actually a meridional scan of the flat plate pattern in Figure 6a. The three strong reflections

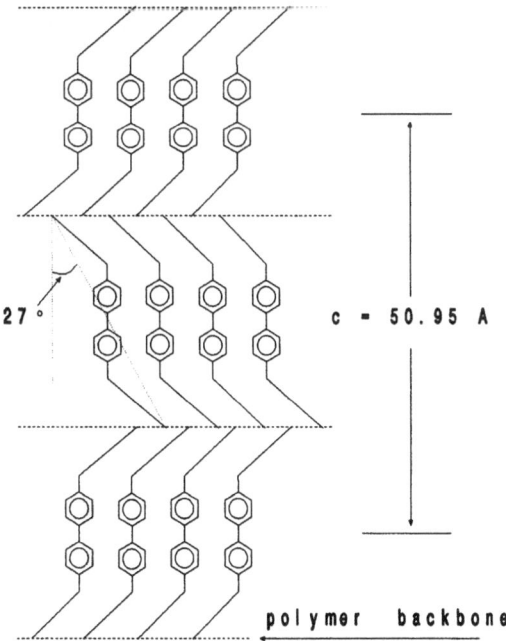

Figure 4. Schematic diagram of molecular (side chain) packing in PB5A-SmE.

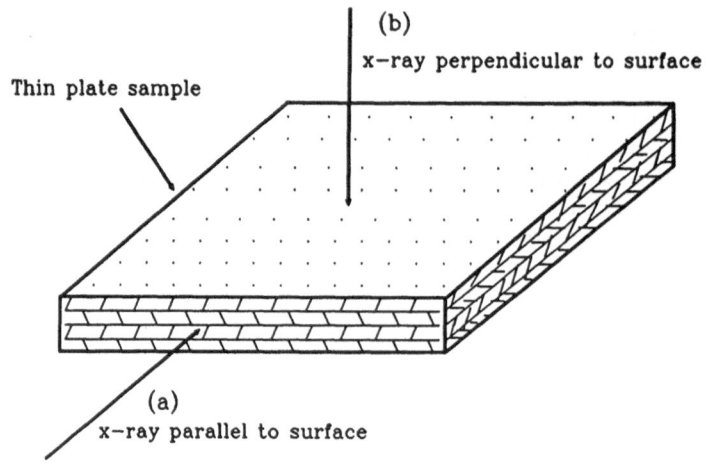

(b)
x-ray perpendicular to surface

Thin plate sample

(a)
x-ray parallel to surface

Figure 5. Schematic diagram of WAXD experimental setup : (a) x-ray ∥ sample surface; and (b) x-ray ⊥ sample surface.

(114), (115), and (116), being very close to each other, are concentrated near the equator on both sides making an average angle of 27° to the equator. This is in accordance with calculated side chain tilt angle of 27° with respect to the smectic layer normal. The strong (020) reflection is concentrated on the equator as it should be when the smectic layer normal is oriented in the sample thickness direction. Therefore there is strong lamellar orientation in the sample thickness direction. Figure 6b shows the flat plate WAXD pattern of PB5A-RT when the x-ray is perpendicular to the sample surface. One can clearly see that here the small angle reflections due to smectic layer periodicity are much weaker than the large angle reflections due to lateral packing, supporting the conclusion that there is preferred smectic layer orientation normal to the sample thickness direction. Concentric rings were obtained for the small angle reflections as well as for the large angle reflections, indicating that there is orientational disorder among individual microdomains or crystallites across the sample thickness direction (cylindrically distributed).

If one examines Figure 6a closely, one will notice that the rings are not really homogeneous; rather they are composed of numerous spots, indicating that the liquid crystalline domains are of considerable size. The dependence of the reflection width on the order of the reflection was analyzed according to the theory of paracrystalline distortions of the second kind.[12,13] Information about the crystallite or domain size and distortion parameter can be obtained from such analysis. The breadth due to instrumental

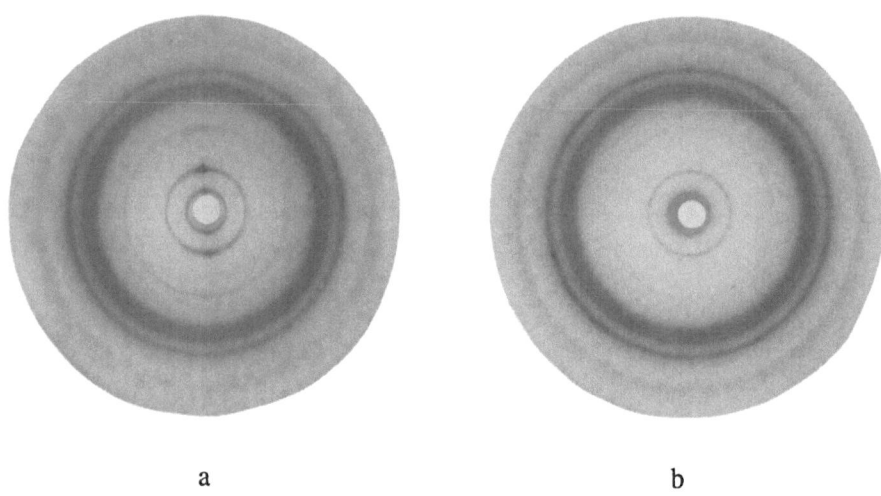

a   b

Figure 6. Flat plate WAXD patterns of PB5A-SmE : (a) x-ray ∥ sample surface; and (b) x-ray ⊥ sample surface.

broadening is usually negligible and is not corrected in this analysis. The reflection breadth δs ($s = 2\sin\theta/\lambda$) can be related to the relative fluctuation $g_{\parallel}$ of the lattice dimension (smectic layer dimension) d_{001} (i.e. $\delta d_{001}/d_{001}$) by the expression

$$\delta s = (\delta s)_0 + (\delta s)_{\parallel} = \frac{1}{\overline{L}_{001}} + \frac{(\pi g_{\parallel})^2 m^2}{d_{001}} \tag{1}$$

where $(\delta s)_{\parallel}$ is the integral breadth contributed by the lattice distortion of the second kind, $(\delta s)_0$ is the breadth corrected for the distortion of the second kind, L_{001} is the size of smectic aggregates (crystallite or domain size in the direction of the smectic layer normal), and m is the order of the reflection. The reflection breadths (δs)'s are plotted against m^2 in Figure 7, which yields a straight line; the reflections of the 2nd, 4th and 8th orders are used for the plot. From the intercept and the slope of the line a value of 482 Å was obtained for L_{001} and a value of 1.14 % for $g_{\parallel}$. Therefore rather long range order exists in the smectic layer normal direction; up to 19 molecular layers are within one smectic aggregate. The relative fluctuation of the lattice dimension, 1.14%, is very small in the smectic layer normal direction compared with 2% usually observed in the semi-crystalline polymers[14].

The crystallite or domain size in the lateral direction was determined using the familiar Scherrer equation[15]. The mean dimension of the crystallites L_{hkl} perpendicular to the (hkl) planes can be obtained from β_0 the peak breadth at half-maximum intensity of the reflection profile in radians. The (116) reflection was selected for the calculation due to its well defined peak profile and relatively strong intensity. A value of 262 Å was obtained for L_{116}, corresponding to 66 sets of molecular planes. Therefore, there is very long range order both in the smectic layer normal direction and in the lateral direction. The highly ordered S_E structure of B5A was preserved upon polymerization.

The unit cell parameters, calculated and observed densities, liquid crystalline domain size both in smectic layer normal and lateral directions, and relative fluctuation of lattice dimension in the smectic layer normal direction, for both B5A-SmE and PB5A-SmE, are listed in Table II. Upon polymerization of B5A in smectic E phase, the liquid crystallinity was retained in PB5A-RT and the smectic phase type did not change. The liquid crystalline order of PB5A did decrease upon polymerization (smectic domain sizes decreased in both the smectic layer normal direction and in the smectic layer lateral direction, and the relative fluctuation of lattice dimension increased from 0.92% to 1.14% in the smectic layer normal direction). Interestingly, the percent decrease in domain size in both directions are almost the same: 33% in the smectic layer normal direction (from 720 Å to 482 Å) and 32% in the lateral direction (from 386 Å to 262 Å). Although the correlation length decreased considerably upon polymerization, PB5A-RT still has a higher degree of long range order than most conventional semi-crystalline polymers and side-chain liquid crystalline polymers so far studied. This is probably due to the quasi-topochemical reaction during monomer-to-polymer conversion with only slight changes in unit cell parameters, -0.94% in a, 1.23% in b, -1.90% in c and -1.61% in total unit cell volume change. One point worth mentioning is that although

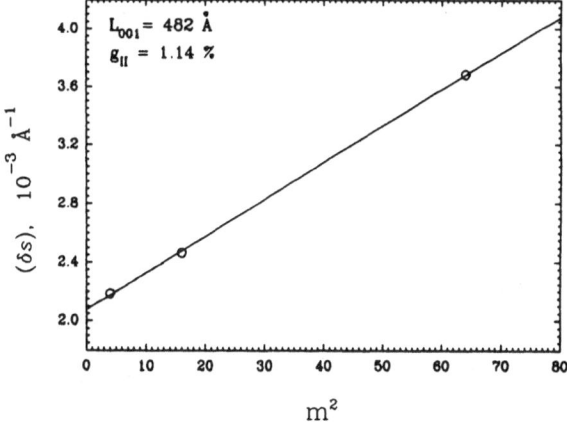

Figure 7. Reflection breadth as a function of reflection order squared.

Table 2. Structural Comparison between B5A-SmE and PB5A-SmE

Structural Characterization		B5A - RT	PB5A - RT
Unit cell type		Orthorhombic	Orthorhombic
Phase type		SmE	SmE
Unit	a , Å	6.609	6.547
Cell	b , Å	7.704	7.800
	c , Å	51.92	50.95
Parameters	V , Å³	2644	2602
Number of mol's/unit cell		4	4
Density gm/cm³	Calculated	1.183	1.200
	Observed	1.189	1.217
Smectic Aggregate Size, Å	Layer normal, L_{001}	720	482
	Layer lateral, L_{116}	386	262
Relative lattice fluctuation g_{II} , %		0.92	1.14
Unit Cell Dimensional Change Upon Polymerization (%)	a	-0.94	
	b	+1.23	
	a-b net area	+0.30	
	c	-1.90	
	V	-1.61	

the **a** dimension shrunk by 0.94% and the **b** dimension expanded by 1.23%, the overall **a - b** net cross-sectional area expanded only by 0.30%, i.e., virtually unchanged. Therefore the minimal molecular movement involved in the polymerization process accounts for the retention of the highly ordered liquid crystalline organization of PB5A-RT. A full characterization of the liquid crystalline structure of both B5A and PB5A-RT, to account for the quasi-topochemical reaction of the acrylate group is in progress.

Like most semi-crystalline polymers, several crystalline reflections reside on a relatively broad hump or peak in the spectrum in Figure 3. The broad peak is probably due to some poorly defined regions present in the structure such as boundaries of ordered regions. But by no means can those poorly defined regions be classified into the same category as the amorphous regions usually encountered in semi-crystalline polymers. The calculated correlation or persistence length from the measured peak breadth at half-maximum intensity using the Scherrer equation is 32 ± 0.6 Å, corresponding to 7 ~ 8 molecular planes. Although the calculated correlation length for the less defined region is about one eighth that present in the ordered region (262 Å), it is still longer than that found in nematic and low order smectic liquid crystalline materials such as the smectic A phase. Therefore PB5A-RT can be considered as being composed of two liquid crystalline structures, one of them with a highly ordered smectic structure and the other with intermediate order.

CONCLUSIONS

Monomer B5A can be photopolymerized at room temperature in its S_E phase to near completion with little volume change. The smectic liquid crystalline structure is retained and locked into the resulting polymeric network upon polymerization. This can be attributed to the topochemical control exerted by the LC monomer matrix on the as-formed polymer; a quasi-topochemical polymerization reaction is involved (B5A-RT and PB5A-RT has almost the same unit cell parameters). Fixation of the LC structure is realized through cross-linking, which severely restricts molecular motion and relaxation. The LC structure is maintained up to very high temperatures. This could extend the application temperature of the linear side chain liquid crystal polymers which usually have a clearing temperature of < 150 °C. It provides means of obtaining a "profile" of the molecular organization of the liquid crystal, a technique complementary to rapid quenching of mesophases below their glass transition temperature with the added advantage of permanency at room temperature and above.

PB5A-RT, like B5A-RT, is in the SmE phase, which has well defined 3-dimensional order. The WAXD data can be indexed with an orthorhombic unit cell whose parameters are very close to that of B5A-SmE. A tilted smectic bilayer arrangement was found from the WAXD study. The broadening of small angle reflections (i.e., smectic layer reflections) was analyzed by Hosemann's theory of lattice distortion of the second kind, and an average smectic domain size or correlation length of 482 Å (very long range order) with a relative lattice dimensional fluctuation of 1.14% (very small) were obtained from the analysis. Generally, the smectic layer normal was oriented in the sample thickness direction (i.e., perpendicular to the substrate surface).

Finally, B5A could be polymerized at room temperature to high conversion and showed little density change upon polymerization. The low polymerization shrinkage with added advantage of room temperature polymerization has important implications if the monomer is used in composites or potting compounds. The interfacial stresses will probably be very low due to low polymerization shrinkage and low polymerization temperature (i.e., no thermal shrinkage).

ACKNOWLEDGMENT

The authors would like to thank EPIC (Edison Polymer Innovation Center) for its continuous financial support for this study.

REFERENCES

1.   Y. B. Amerik and B. A. Krentsel, *J. Polym. Sci.*, C-16, 1383 (1967).
2.   C. M. Paleos and M. M. Labes, *Mol. Cryst. Liq. Cryst.*, 11, 385 (1970).
3.   A. C. DeVisser, K. DeGroot, J. Feyen, and A. Bantjes, *J. Polym. Sci.*, A-1, 9, 1893 (1971).
4.   L. Strzelecki and L. Liebert, *Bull. Soc. Chim. France.*, 2, 597 (1973); 2, 603 (1973).
5.   A. Blumstein, R. B. Blumstein, G. J. Murphy, C. Wilson, and J. Billard, in *Liquid Crystals and Ordered Fluids*, Edited by J. F. Johnson and R. S. Porter, Plenum Press, p. 277, 1974.
6.   Gy. Hardy, K. Nyitrai, F. Cser, Gy. Cselik, and I. Nagy, *Europ. Polymer J.*, 5, 133 (1969).
7.   D. J. Broer, J. Boven, G. N. Mol, and G. Challa, *Makromol. Chem.*, 190, 2255 (1989).
8.   D. Demus, L. Richter, C.-E. Rurup, H. Sackmann, and H. Schubert, *J. Physique Colloq.* 36, C1-349 (1975).
9.   J. C. Dubois and A. Zann, *J. Physique Colloq.* 37, C3-35 (1976).
10.   J. J. Mallon and S. Kantor, *Macromolecules*, 22, 2070 (1989).
11.   X. Qian and M. Litt, in preparation.
12.   R. Hosemann and S. N. Bagchi, in *"Direct Analysis of Diffraction by Matter"*, North-Holland Publ., Amsterdam, 1962.
13.   R. Bonart, R. Hosemann, and R. L. McCullough, *Polymer*, 4, 199 (1963).
14.   R. Hosemann, and W. Wilke, *Faserforsch. Textiltech.*, 15, 521 (1964).
15.   P. Scherrer, *Göttinger Nachrichten*, 2, 98 (1918).

INDEX